UNDERSTANDING WIND POWER TECHNOLOGY

UNDERSTANDING WIND POWER TECHNOLOGY

THEORY, DEPLOYMENT AND OPTIMISATION

Edited by

Alois Schaffarczyk
University of Applied Sciences, Kiel, Germany

Translated by

Gunther Roth
Adliswil, Switzerland

First published under the title Einführung in die Windenergietechnik by Carl Hanser Verlag
© 2012 Carl Hanser Verlag, Munich/FRG. All rights reserved. Authorized translation from the original
German language edition published by Carl Hanser Verlag, Munich/FRG

This edition first published 2014
© 2014 John Wiley & Sons, Ltd

Registered Office
John Wiley & Sons, Ltd, The Atrium, Southern Gate, Chichester, West Sussex, PO19 8SQ, United Kingdom

For details of our global editorial offices, for customer services and for information about how to apply for
permission to reuse the copyright material in this book please see our website at www.wiley.com.

Library of Congress Cataloging-in-Publication Data

Schaffarczyk, Alois.
    Understanding wind power technology : theory, deployment and optimisation / by Alois Schaffarczyk.
        p.   cm.
    Includes bibliographical references and index.
    ISBN 978-1-118-64751-6 (cloth)
1. Wind power.   2. Wind energy conversion systems–Design and construction.   I. Title.
    TJ820.S33 2013
    621.31′2136–dc23
                                                2013022982

A catalogue record for this book is available from the British Library.

Set in 10/12pt Times by SPi Publisher Services, Pondicherry, India
Printed and bound in Singapore by Markono Print Media Pte Ltd

1   2014

Contents

6   The Drive Train   **202**
Sönke Siegfriedsen

9 Control of Wind Energy Systems   340

Reiner Johannes Schütt

Preface

Although nearly 20,000 windmills dotted Germany's landscape by the end of the eighteenth century, the era of modern wind energy began in 1983 when the aptly-named GROWIAN prototype (a German abbreviation for Grosswindanlage, or 'large wind turbine') started operation in the German state of Schleswig-Holstein. By the end of 2011, almost 23,000 modern wind turbines had been erected in Germany and supplied nearly 10% of the country's electricity demand. It took only 30 years for the modern wind industry to develop to the extent that turbines the size and power of the once-colossal GROWIAN had become standard and mass-produced.

At the request of the Carl Hanser Verlag publishing company and under the umbrella of CEwind eG – the consortium for wind energy research between Schleswig-Holstein's universities – authors from the wind community in Schleswig-Holstein and the Netherlands have collaborated to compile this introductory text on wind energy. Over 11 chapters the interested reader will become familiar with the modern state of this technology.

This text begins with a brief history and then supplements this with an explanation of the importance of wind energy in the international energy policy debate. Following chapters then introduce the aerodynamic and structural aspects of blade design. Then the focus shifts to the flow of energy and loads through the wind turbine, through the powertrain and also the tower-foundation system, respectively. Next, the electrical components such as the generator and power electronics are discussed, including control systems and automation. Following is an explanation of how wind turbines are integrated into the electricity grid, despite the highly fluctuating nature of both this energy source and the grid load; this particular topic is especially relevant for Germany's transition to renewable energy. The final topic covers one of the youngest and most promising aspects of wind energy: offshore technology.

Kiel, February 2012

For CEwind eG: A.P. Schaffarczyk
English translation: Gunther Roth

About the Authors

Dr H.C. Jos Beurskens previously led the Department for Renewable Energy and Wind Energy at the Dutch research institute for energy (ECN) for over 15 years. He was awarded the Poul-la-Cour Prize for lifetime achievement at the European Wind Energy Association (EWEA) 2008 conference. He is now an independent consultant for technology development and research strategies.

Prof. Dipl.-Ing. Lothar Dannenberg has over 10 years experience with rotor blades and offshore foundations. He has taught classes at the Kiel University of Applied Sciences in these areas, as well as the topics of ship construction and design, fibre composites, and underwater vehicles.

Frank Ehlers has been involved with the development of the German grid connection codes since the passing of the German Renewable Energy Act (in German: Erneuerbare-Energien-Gesetz, EEG), for which he was a member of the federal approval committee. Today he is responsible for the planning and expansion of grid and distribution networks at energy supply company EON Hanse.

Prof. Dr.-Ing. Torsten Faber has served since November 2010 as the Director of the Wind Energy Technology Institute (WETI) at the Flensburg University of Applied Sciences in Germany. He has 10 years of experience in the certification of wind turbines.

Prof. Dr.-Ing. Friedrich W. Fuchs leads the faculty of Power Electronics and Electronic Drives at the Christian-Albrechts-University in Kiel, Germany. One of his group's main research goals is supporting the transition to renewable energy. He also has 14 years of industrial experience, most recently as Research Director for CONVERTEAM (later renamed General Electrical Power Conversion). He is a founding member and board member of CEwind eG.

Dr Hermann van Radecke has specialised in liaison managing technology transfer between the wind industry and his employer, the Flensburg University of Applied Sciences, for over 20 years. He is also a founding member of CEwind and a lecturer for courses in physics and wind energy. Additionally, he was active in the shaping of the CEwind MSc Wind Engineering programme, previously serving as the program director for Flensburg.

Dr Klaus Rave leads the department of Energy Economics in Schleswig-Holstein and served as an officer in the region's investment bank for many years. He has been active in many organisations in the international wind community.

Prof. Dr A.P. Schaffarczyk has been involved with wind turbine aerodynamics since 1997. He is a founding member and previous manager of CEwind eG and teaches courses in the CEwind MSc Wind Engineering programme.

Prof. Dr Reiner Johannes Schütt has previously served as the Head of Development and Technical Director of the turbine manufacturer ENERCON NORD Electronic GmbH in Aurich, Germany. Today he teaches and researches in the field of controllers, electrical drives, and wind energy technology at the West Coast University of Applied Sciences in Heide, Germany.

Dipl.-Ing. Sönke Siegfriedsen founded the Aerodyn company in 1983 and still serves as its general manager. His company has developed more than 25 complete wind turbine designs, from which approximately 27,000 examples have been produced (amounting to 31,000 MW of capacity) to date.

Dr Sven Wanser leads the Grid Operation business unit at electricity provider Schleswig-Holstein-Netz AG and teaches the subject of electrical energy technology at the West Coast University of Applied Sciences in Heide, Germany.

1

The History of Wind Energy

Jos Beurskens

1.1 Introduction

Wind has been used as a source of energy for more than 1500 years. In times when other sources of energy were unknown or scarce, wind energy represented a successful means for industrial and economic development. Wind energy became a marginal source once cheaper, easier to exploit and easily obtainable sources of energy became available. From the point of view of the contribution of wind energy to economic development, one can divide the history of wind energy into four overlapping time periods (see Figure 1.1). Except in the first period, the emphasis here is the generation of power by wind:

- **600–1890: Classical period**. Classic windmills for mechanical drives; more than 100,000 windmills in northwestern Europe. The period ended after the discovery of the steam engine and because of the ready availability of wood and coal.
- **1890–1930: Development of electricity-generating wind turbines**. The development of electricity as a source of energy available to everyone leads to the use of windmills as an additional possibility for generating electricity. Basic developments in the field of aerodynamics. The period ended due to cheaper fossil oil.
- **1930–1960: First phase of innovation**. The necessity of electrifying rural areas and the shortage of energy during the Second World War stimulated new developments. Advances in the field of aerodynamics. The period ended because of cheaper gas and fossil oil.
- **From 1973: Second innovation phase and mass production**. The energy crisis and environmental problems in combination with technological advances ensure a commercial breakthrough.

Understanding Wind Power Technology: Theory, Deployment and Optimisation, First Edition.
Edited by Alois Schaffarczyk. Translated by Gunther Roth.
© 2014 John Wiley & Sons, Ltd. Published 2014 by John Wiley & Sons, Ltd.

Figure 1.1 Historical development of the use of wind as a source of energy. The first and last periods have had the greatest effects on society

During the classical period, the 'wind devices' (windmills) converted the kinetic energy of the wind into mechanical energy. After direct current and alternating current generators were invented and came to be used for public power supply, windmills were used for electrical power generation. This development began effectively in the late nineteenth century and, after the energy crisis in 1973, became a great economic success.

In order to differentiate clearly between the different plants, they are called windmills or wind turbines in this book.

1.2   The First Windmills: 600–1890

Water mills were very probably the precursors of windmills. Water mills, again, were developed from devices that were operated by people or animals. The devices that are known to us from historical sources possessed a vertical main shaft to which cross bars were attached in order to drive the main shaft. The cross bars were operated by farm animals such as horses, donkeys or cows. It seems only logical that the vertical windmills developed from these devices. However, there are few historical sources to provide proof of this. More sources can be found on the 'Nordic' or 'Greek' water mills that evolved from the animal-operated devices (see Figure 1.2). These types of water mills had their origin about 1000 BC in the hills of the Eastern Mediterranean area, and were also used in Sweden and Norway [1].

The first windmills with vertical main shafts were found in Persia and China (See Figures 1.4). In the middle of the seventh century AD, the building of windmills was a highly prized trade in Persia [3]. In China, vertical windmills were introduced by traders. The first European to report on the windmills in China was Jan Nieuhoff, who travelled there in 1656 with one of the Netherlands ambassadors. Figure 1.3 shows an illustration by Jan Nieuhoff [4]. Similar windmills were in use in China until quite recently (see Figure 1.4).

Other types of devices were treadmills that were operated by the bodily strength of people or animals. Spades were arranged radially to the main shaft. The horizontal water mill developed from the treadmill by the replacement of people or animals by flowing water. A further development in the first century AD was the so-called Vitruvian water mills, which were introduced by the Roman Vitruvius. This water mill can be seen as the prototype for the under-shot water mill that can be found throughout Europe in rivers and streams with low water-level differences. Also, it is assumed that the Vitruvian wheel is the forerunner of the horizontal windmill [1].

Figure 1.2 Water wheel with vertical axis of rotation near Göteborg, Sweden. From Ernst, *The Mills of Tjorn* (1965) published by Mardiska Museet, Stockholm [2]. Reproduced with permission of Mardiska Museet, Stockholm

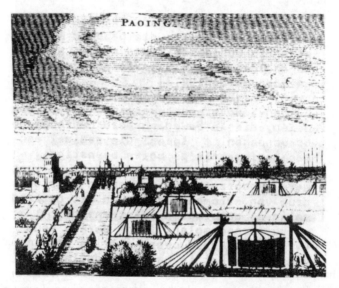

Figure 1.3 Drawing of Chinese windmills in Paoying (Chiangsu) by Jan Nieuhoff, 1656 [4]. Reproduced with permission of Cambridge University Press

The first horizontal windmills were found during the crusades in the Near East and later in northwest Europe. These windmills possessed a fixed rotor construction that could not be rotated in the wind (yawing). The rotor blades of these windmills were similar to those that can be seen today, for instance, on the Greek Island of Rhodes. By about 1100 AD there were reports of fixed post mills that were positioned on the city walls of Paris. It is unclear whether

Figure 1.4   Left: Chinese wind wheels at Taku that pump brine solutions for the extraction of salt (Hopei [3]). Reproduced with permission of E & F N Spons; right: schematic depiction of the function of a Chinese windmill. Solid lines represent blades and dash-dotted lines represent sails [5].

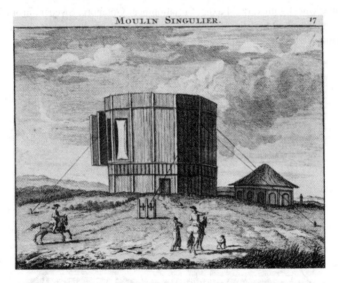

Figure 1.5   A vertical axle windmill from the year 1718 [2]. Reproduced with permission of Hugh Evelyn

the windmills that were widely distributed came from the Near East to Europe or were reinvented in Western Europe. Some authors even doubt the existence of horizontal windmills in the Near East during the Crusades [3,6]. Others, again, speak only of vertical windmills at that time [4,7].

The assumption that the windmills of Western Europe were invented independently of those of the Near East is supported by documents that have been found in the archives of the Netherlands Province of Drenthe. In these documents that originate from the year 1040, at the time of the Crusades, there is mention of two windmills (Deurzer Diep and Uffelte). During the Renaissance some vertical windmills were also built in Europe (see Figure 1.5). Especially well known was the windmill built by Captain Hooper in Margate, England [2].

1.2.1   Technical Development of the First Horizontal Windmills

The first windmills possessed no yaw mechanisms and the blades consisted of a frame of longitudinal and lateral bars through which sailcloth was tied (see Figure 1.6, left). The power output was controlled in that the sail was wholly or partly furled up by hand (see Figure 1.6, right).

For reasons of statics, the main shaft had an angle of inclination (dimension of the milling building, the axle load on the axial sliding bearing, the possibility of erecting a load-bearing building or a conical tower for stabilisation).

The development of the classical windmill in Western Europe will be described before investigating the global development of windmills into wind turbines with which electrical power is generated today.

Although the wind comes mainly from a particular direction in the windy regions of Europe, the wind direction varies so strongly that a yaw mechanism makes sense in order not to lose too much energy with side flows of the wind. This requirement led to the first post windmills (see Figure 1.7), which could be yawed into the wind. These windmills were used for milling corn. By means of a strong beam attached to the mill building, the whole house, which stood on a fixed substructure, could be turned until the rotor was vertical to the wind.

Often the support beams of the substructure were covered with wooden planking so that a storeroom was created. The millstone and the gear wheels were situated in the rotating mill building. One of the first depictions of this type of windmill, dating from the year 1299, comes from a convent in Oedenrode, in the Noord Brabant region in the Netherlands.

Another attempt to turn the rotor into the wind was attempted by building a windmill on a floating platform. The platform was fixed by means of a joint to a pile that was sunk into the ground of a lake in the north of Amsterdam in 1594. Probably because of the lack of stability, such a windmill was never built again, but the concept can be taken as the first attempted offshore wind turbine.

The so-called 'Wip' (Dutch) or 'Koker' (German) windmill was developed from the post windmill (see Figure 1.8). After 1400, windmills were used in the flatter regions of the Netherlands not only for grinding corn, but also for pumping seas and marshlands dry. The pump arrangement, usually a bucket wheel, was attached to a fixed position of the mill building. Only the transfer elements of the windmill were positioned inside, which made the rotating part of the windmill markedly smaller. By the start of the sixteenth century, there was a requirement for more pumping capacity and so the 'Wip' windmill was replaced by a mill with a rotating cover. Only the bevel gear drive was situated inside the cover, with the result that this part weighed relatively little. As the demand for power output increased, windmills were built whose only rotating part was the cover. The drive machinery was able to be positioned in the mill building and no longer needed to be placed in the movable part (e.g. as with the post windmill) or in the open (as with the 'Koker' windmills. The sketches in Figure 1.9 show the development of the main features of the classical windmill.

With the increasing number of windmills, the pressure to use them more efficiently increased. Improvements resulting from this motivation were integrated into the mills. One improvement was the automatic yawing of the windmill rotor into the wind with the aid of a windrose: a rotor whose shaft was attached vertically to the main shaft of the windmill. In England in 1745, Edmund Lee fastened a windrose to a windmill. The windrose was a wooden

Figure 1.6 Left: a post windmill of the early fourteenth century (British Museum) [2]. Reproduced with permission of Hugh Evelyn; right: 'power control' of a classical windmill [8]

Figure 1.7 Post windmill, Baexem, Netherlands [8]

Figure 1.8 'Koker' windmill from South Holland. Photo (right) from [8]

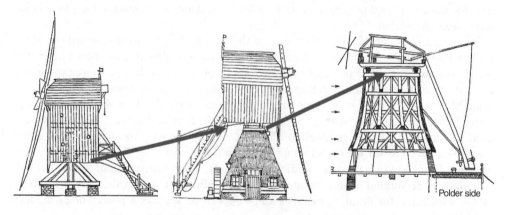

Figure 1.9 The development of the classical 'Holland mill'

construction that was mounted to the rotating part of the windmill in order to turn the rotor into the direction of the wind. John Smeaton, also English, invented a windrose that was attached to the rotating cover of the windmill.

This innovation was so successful because it was used on a large number of windmills, especially in England, Scandinavia, north Germany and in the eastern part of the Netherlands.

Figure 1.10 Wind direction follower with sensor on an early Lagerwey wind turbine [8]

This concept was retained up to the era of the wind turbines of the later nineteenth century and even into the late twentieth century. At the start the transmission was fully mechanical, and later the windrose acted only as a sensor in order to send a control signal to the yaw mechanism (see Figure 1.10).

In the first phase of the classical period of the windmills, they were primarily used for milling corn and for dewatering. More and more wind was also used as source of energy for all possible industrial processes. The wind played a great role as a source of energy for industria and economic development, especially in regions where no other easily available energy carriers such as wood and coal were available. This was the case especially in 'de Zaanstreek', north of Amsterdam, and in Kent in England. Windmills were used for sawmills, for the production of paper, oil and colours, for dehusking rice and crushing, as well as for the manufacture of mustard and chocolate [6]. Besides this, they were used for the ventilation of buildings (England). The construction of windmills was especially concentrated in suitable areas. The multiplicity of windmills, as in the gallery of the windmills, for draining marshes and seas can be seen as a precursor of modern wind farms.

Further innovations in the area of performance and control of the rotor were continuously implemented. The sail that was wound through the blade bars was replaced by textile strips that were fixed to the front of the blades. The vacuum on the lee side kept the cloth in place, whereby it obtained an aerodynamic profile. Power was controlled in that the wooden frame of the blade was partly covered. In order to reduce maintenance, the wooden bars and frames were eventually replaced by iron and steel structures.

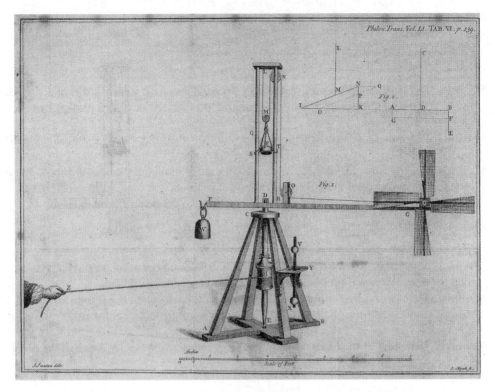

Figure 1.11   Smeaton's test rig for determining the performance of windmill rotors

The path to a substantial increase in the aerodynamic efficiency was supported by scientific research from the middle of the eighteenth century. The most fascinating work was carried out by John Smeaton (1724–92) and can be seen as a precursor of modern research. His work was based on experiments with the apparatus that can be seen in Figure 1.11. With the pull on the sail the vertical shaft begins to rotate exactly as the arm at whose end the model of a windmill rotor is fixed. The rotor is acted upon by wind velocity flow that is equal to the blade tip speed of the arm. During the rotation, the rotor lifts a weight. By changing the rotor properties the optimum 'efficiency' (in modern terminology known as the 'performance') can be determined.

Smeaton presented the results of his experiments, *On the Design and Efficiency of Windmill Blades*, in a classic treatise that was presented to the Royal Society in 1759. The 'efficiency' was equal to the product of the weight and the number of revolutions that were carried out by the rotor in a particular period of time, whereby friction losses of the apparatus were to be equalised.

Smeaton determined the best form and 'weather' of the blades. In classical windmill technology, 'weather' designates the angle between the blade section and the rotation level. Today 'weather' is designated as the twist of the rotor blades. Later, Maclaurin investigated the local prevailing angle of attack with the aid of a distance function that describes the angle between the cross-section of the plant and the axis of the rotor. It is interesting to compare the work of Smeaton with that of current research, and for this reason the following paragraph will give the results of his experiments or 'Maxima' verbatim [3].

Maxim 1: The speed of mill blades for the same form and position is almost that of the wind. In this it is insignificant whether they are unloaded or loaded in such a manner that they produce a maximum.

Maxim 2: The maximum load is almost but a little less than the square of the wind speed insofar as the form and position of the blade are the same.

Maxim 3: The performance of the same blade with maximum power output is almost, but a little less, than the speed of the wind to the power of three.

His results from his theoretical contemplations:

Maxim 6: With blades with similar form and position, the number of rotations in a particular time period remains anti-proportional to the radius or length of the blade.

Maxim 7: The maximum loading that blades with similar form and position can take at a particular distance from the point of rotation has a value of the radius to the power of three.

Maxim 8: The effect of the blade with similar form and position has the value of the radius squared.

Besides the automatic wind-tracking yaw and the improved configuration of the blades, the efficiency of the windmills was improved by means of further innovations. For example, in 1772 Andrew Meikle obtained a patent for lamellae in the rotor blades in order to control the power output automatically. In 1787 Thomas Mead introduced the automatic control of these rotor blades by means of a centrifugal regulator.

With the invention of the steam machine (Watt), it was possible to generate power at will. The supply of energy could be perfectly adapted to the demand. Besides this, fuels such as coal and wood were relatively cheap. This had devastating effects on the use of windmills. During the nineteenth century, the overall number of windmills in northeastern Europe was reduced from an original 100,000 to 2000. Thanks to the active maintenance policy of the Verening de Hollandsche Molen (Dutch Mill Association), 1000 of the almost 10,000 windmills were able to be retained. These classical Holländer windmills are still capable of operation.

1.3   Generation of Electricity using Wind Farms: Wind Turbines 1890–1930

When the first electrical dynamos and alternating current generators were put into operation (see Box 1.1), use was made of all possible sources of power in order to drive the generators. The generators were operated by treadmills, wood- or coal-fired steam machines, water wheels, water turbines and wind rotors. In this, the wind was seen as only one of

---

Box 1.1   Dynamos

The original name for the direct current generator was the 'dynamo'. In contrast to this was the alternating current generator that generated alternating current via a slip ring or a rotor magnet. The first operational electric power station was built in New York in 1880. It consisted mainly of dynamos and operated arc lamps in a 2-mile-long circuit. There was strong competition between the proponents of direct current systems under the leadership of the American inventor Thomas Alva Edison, and the proponents of the alternating current systems under the leadership of the American industrialist George Westinghouse. Direct current had the advantage that the power could be stored in electrochemical batteries. The great advantage of the alternating current was that the voltage could easily be converted to a higher voltage level in order to reduce transmission losses and then could be converted back to a lower voltage level at the user end. Eventually the alternating current systems won the battle.

---

many possibilities for generating energy. In 1876, for instance, the improved direct current generator by Charles Brush was driven by a treadmill that was operated by horses.

With the discovery of the dynamo it became possible to supply business users and individual households with energy by means of electricity from afar. Electricity could simply be transmitted from a central generator to the users. After the introduction of the first central electric power station, the demand for primary energy grew very quickly.

The development of the power-generating windmills (in the following called wind turbines) was not independent but overlapped with the availability of the first electric power stations and the first local power grids.

The first person to use a windmill for the generation of power was James Blyth, Professor at the Anderson College in Glasgow. His 1887, 10 m high wind turbine, whose blades were covered with sail cloth, was used to charge the batteries for lighting his holiday home.

In 1888 Charles Brush, the owner of a machine tool company, constructed a 12 kW wind turbine with a diameter of 17 m at his house in Cleveland, Ohio (US). In comparison with its rated power, the plant had a very large diameter. The rotor area was fully covered by 144 smaller rotor blades, which meant the speed of revolution was slow. This resulted in a very high transfer relationship from the rotor shaft to the generator. The power output was automatically controlled by a so-called 'ecliptic controller'. The rotor was turned from the increasing wind out of the wind by a wind flag that was positioned vertically to the main blade wheel, whilst the main blade wheel was fixed to a slanted joint. A picture from *Scientific American* of 20 December 1890 (see Figure 1.12) shows the features of the plant.

Wind turbines were also used on board ships to generate power. The plants were erected on the deck and operated a dynamo by means of belt transmissions. The power was then used to load the batteries on board. The rotors possessed blades covered with sail cloth. Two examples of this are the *Fram*, the ship with which Fridtjof Nansen sailed to the Antarctic in 1888, and the *Chance* out of New Zealand (see Figure 1.13).

A WEEKLY JOURNAL OF PRACTICAL INFORMATION, ART, SCIENCE, MECHANICS, CHEMISTRY, AND MANUFACTURES.

Vol. LXIII.—No. 25.]
Established 1845.]
NEW YORK, DECEMBER 20, 1890.
[$3.00 A YEAR.
Weekly.

1. Windmill in the park.   2. Vertical section of the tower.   3. Dynamo.   4. Storage batteries.   5. Regulating apparatus.

THE WINDMILL DYNAMO AND ELECTRIC LIGHT PLANT OF MR. CHARLES F. BRUSH, CLEVELAND, O.—[See page 389.]

Figure 1.12   Page from *Scientific American* magazine, 20 December 1890. Reproduced with permission of Scientific American/Wikimedia Commons

Figure 1.13 Electrical generators on ships at the turn of the last century. Sailing ship *Chance*, New Zealand 1902. Reproduced with permission of National Library NZ/Wikimedia Commons

In 1891 Professor Poul la Cour constructed his first wind turbine in Askov, Denmark, in order to generate power which he used for various purposes. He connected his wind turbine, with four remote-controlled louvre blades, to two 9 kW dynamos. The power generated was used to load the batteries of the Askov Folk High School, and hydrogen was generated by the electrolysis of water with which gas lamps were operated. La Cour's developments were based on wind tunnel measurements at his school (see Figure 1.14).

The louvre blades used by Poul la Cour (see Figure 1.15) were invented and used in 1772 by Andrew Meikle in Great Britain. Meikle replaced the sail cloth with rectangular lamellae. With gusting wind, the lamellae automatically opened against the force of steel springs. This was the first possibility of automatic control and made the work of the miller much more comfortable. However, the stress of the springs still had to be set by hand. For this purpose the windmill had to be completely stopped.

Later, the lamellae were either controlled automatically or manually by means of a rod that ran through the hollow main shaft of the mill. In this way the mill could be controlled without constantly having to be stopped. The system was patented in 1807 by William Cubitt. The lamellae were controlled by means of a spider-like construction that can still be found today in classical windmills, among others in northern Germany, the UK and Scandinavia (see Figure 1.16).

Although the blades of La Cour's wind turbine possessed some innovations, the aerodynamic design was based on that of the classical windmill. It took about two decades for efficient aerodynamic profiles, developed from aviation, to be used for wind turbines.

Resulting from the experiments carried out by La Cour in Askov (see Figure 1.15), he made suggestions for their practical implementation from which, among others, the Danish manufacturers Lykkegaard and Ferritslev (Fyn) developed commercial wind turbines. By 1908, Lykkegaard had erected 72 wind turbines and by 1928 the number had risen to about

Figure 1.14 Poul la Cour's wind tunnel. Reproduced with permission of Poul la Cour Fonden

Figure 1.15 Poul la Cour's wind energy test rig in Askov, Denmark; right: wind turbine from the year 1891; left: larger plant from 1897 [9]. Reproduced with permission of Poul la Cour Fonden

120. The maximum diameter of the La-Cour-Lykkegaard wind turbines was 20 m. They were equipped with 10–35 kW generators. The plants generated direct current that was fed to small direct current grids and batteries. As fuel prices had increased greatly, the development of wind technology in Denmark continued during the First World War.

Figure 1.16   Louvre blades in the mill at the Wall in the centre of Bremen [8]

Between the two world wars, attempts were made in the Netherlands to improve the efficiency of the classical windmills. At the TU in Delft, the helicopter pioneer Professor A.G. von Baumhauer and A. Havinga carried out measurements on four-blade rotors of classical windmills [10]. The stonework of the classical windmill, which was meant to support the improved rotors, would not have stood up to the axial forces as these also increased with the higher efficiencies of the rotors. In the 1950s and 1960s further experiments were carried out, all of which failed due to lack of structural integrity or for reasons of economy (Prinsenmolen; de Traanroeier, Oudeschild, Texel).

In 1920 in Germany, the leader of the Aerodynamic Experimental Institute in Gottingen, Albert Betz, published a mathematical analysis of the theoretical maximum value of the performance coefficient of a wind turbine with Zhukowsky, but after Lanchester (1915) (this is usually called the Lanchester-Betz coefficient and is 16/27 = 59.3%). It is based on the axial flow model. In addition, Betz also described wind turbines with improved aerodynamic blades [11]. Figure 1.17 shows a fast-rotating four-blade wind turbine from the Aerodynamo Company in Berlin.

The plant had brake flaps on the low-pressure sides of the blades. Immediately after the First World War it was Kurt Bilau who wanted to improve the efficiency of his four-blade Ventimotor by giving the aeroprofile of the blades a streamlined form. He even asserted that he had reached a higher efficiency than Betz stated later as the maximum value for the performance coefficients. Besides this, Bilau erected test plants in East Prussia and in southern England.

After the First World War the availability of fossil fuels rose substantially, which meant that interest in wind energy declined. In the Western industrial world, the further development of wind energy was carried out in a very low-key manner until the Second World War. However, this was not the case in the Soviet Union, because there the Stalin Regime instituted a large programme for electrification of remote areas. The little information from those times show that the Soviet engineers made use of the latest developments of aerodynamics for their concepts. An example of this is shown in Figure 1.18.

Figure 1.17 Fast-rotating wind wheel of the Aerodynamo A.G. Berlin, Kutfürstendam. The figure shows the brake flaps used by this company on the suction side of the blades [12]. Photograph originally taken by Betz, A. and used with thanks

The rotor blades, developed by the Central Aerohydrodynamic Institute (ZAGI), can be adjusted by means of a small auxiliary blade at the rear edge of the main blade. In 1931 an experimental wind turbine was built near Sevastopol on the Crimea that was operated in parallel with a peat-fired 20 MW electric power station. The WIME D-30 plant had a rotor diameter of 30 m and a nominal output of 100 kW. It remained in operation until 1942. The wind turbine (Figure 1.19) possessed similar aerodynamic features as the smaller model in Figure 1.18.

1.4   The First Phase of Innovation: 1930–1960

Various countries resumed the development of wind turbines during, and immediately after, the Second World War. The reason for this was that strategic resources such as fossil fuels were becoming scarce. In this period many innovations were introduced, which probably permitted the widespread introduction of wind turbines for power generation parallel to the power grid. These innovations, primarily on the structure of the rotor, were based mainly on the innovations of the previous era.

Figure 1.18 ZAGI turbine with Sabinin's auxiliary blades. Rotor diameter 3.6 m, with complete control of the blade adjustment angle [8]

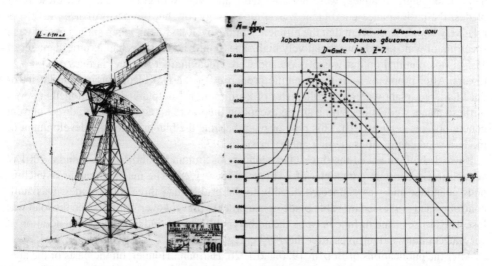

Figure 1.19 Left: concept of the wind turbine near Sevastopol on the Crimea: in operation from 1931 to 1942, rotor diameter 30 m. Right: measured performance factors and torque coefficient as a function of the revolution speed [8]

Figure 1.20 An out-of-service Smidth-Aeromotor in Denmark. Rotor diameter 24 m, nominal output approximately 70 kW. Photo by Paul Smulders in 1972

The most important developments occurred in Denmark, the US and Germany. During the Second World War the F.L. Smidth Company in Copenhagen developed wind turbines for generating electricity. As Denmark did not possess its own fossil fuels, wind energy was one of the few ways of generating power. The Smidth plants possessed two-blade rotors; the blades had a fixed angle of attack, were not adjustable and were stall-controlled. With these rotor blades, the performance coefficient was comparatively low but the performance curve was relatively broad. This meant that the efficiency of the overall system, when spread over a wide spectrum of wind speeds, was relatively high. The Smidth aeromotors had a rotor diameter of 17.5 m (the nominal output was 50 kW) and were erected either on steel lattices or concrete towers. After problems with the dynamic properties of the two-blade rotors, Smidth introduced a larger plant with a rotor diameter of 24 m (nominal output 70 kW). Altogether seven of these plants were built. With a single exception they were all equipped with direct current generators (see Figure 1.20). This type of plant became the blueprint for the development of modern wind energy after the first energy crisis in 1973.

It was J. Juul who used the three-blade concept of Smidth around 1957 to build a 200 kW version in Gedser with a diameter of 24 m (see Figure 1.28). The machine had an asynchronous generator and was connected directly to the grid. It had three rotor blades, was stall-controlled and possessed movable blade tips in order to prevent overturning when the load was lost. The Gedser wind turbine became the archetypal 'Danish wind turbine' of a generation of very successful wind turbines after the 1973 energy crisis.

After the publications of Betz in 1920 and 1926, Hermann Honnef, on the basis of the analytical results of Betz and others, conceived a very large structure with several rotors. This was possibly the first concept that was based completely on scientific knowledge. His concept had

Figure 1.21   Vision of a 5 × 20 MW, 5 × 160 m wind turbine by Hermann Honnef, 1933

five rotors, each with a diameter of 160 m, and six blades each. Each rotor was meant to drive a 20 MW generator. The rotors consisted of two counter-rotating wheels. On 80% of the rotor surface they each carried a ring (see Figure 1.21). The rings were part of a giant 'ring genera-tor'. The concept was far in advance of what at that time was technically feasible, since it is only now (2012) that plants of similar size are being designed and built (see also Figure 1.35).

With the support of the Board of Trustees for Wind and Water Power, which was founded in 1941 in order to support inventors in the search for sources of energy, Honnef built a model of the Multirotor with two blades on a test field on the Mathiasberg, northwest of Berlin. After the war the test field was destroyed by the Soviets and the plant was melted down in the furnaces of Hennigsdorf.

With the collapse of the Third Reich, Honnef had to end his work in the field of wind energy in March 1945. Previously the search for independence in the supply of energy led in 1939 to the founding of the Reichsarbeitsgemeinschaft Windkraft (RAW: Reich Working Group for Wind Energy), in which scientists, inventors and industry collaborated. A project supported by the RAW was the three- or four-bladed wind turbine planned by Franz Kleinhenz, equipped with a rotor diameter of 130 m and a nominal output of 10 MW and was designed in coopera-tion with the MAN company. However, the war hindered the planned construction in 1942.

From the end of the war to the resumption of research and development for wind energy during the oil crisis in 1973 and beyond, Professor Ulrich Hütter was a leading figure in Germany, and led a test rig of the Venti-motor GmbH in Webicht, Weimar. There he obtained

Figure 1.22   Allgaier WE 10 wind turbine

much practical experience in the design of smaller wind turbines. Hütter received his degree in December 1942 at the University of Vienna with his dissertation *Beitrag zur Schaffung von Gestaltungsgrundlagen für Windkraft-werke* [Contribution to the creation of design basics for wind turbines]. During his career he worked alternately in aviation technology and wind energy technology. The first wind turbine to be built after the end of the war was by Hütter in 1947.

In 1948 Erwin Allgaier wanted to build his wind turbine in series (three rotor blades, 8 m rotor diameter, 13 kW nominal output). Slightly larger plants (11.28 m rotor diameter, 7.2 kW nominal output) were exported to South Africa, Ethiopia and Argentina. Because of a relatively high tip speed ratio of 8, the wind turbines were very light. Also the equipped capacity per unit of covered rotor surface was very small so that the wind turbine was suitable for low wind regimes and, at the same time, delivered a relatively high equivalent full load hours (capacity factor) (see Figure 1.22).

In order to provide power in remote areas, the brothers Marcellus and Joseph Jacobs began to develop wind turbines for loading batteries in the US in the early 1920s. After experiments with two-blade plants they introduced a three-blade wind turbine with a rotor diameter of 4 m and a directly-driven direct current generator. Several thousand of these plants were sold between the early 1920s and the first years after the 1973 oil crisis (see Figure 1.23).

With the extension of the rural power grid, the supply of power to rural areas presented no great problem anymore and the development of wind energy turned towards large plants for the operation of the grid. During the Second World War, wind turbines seemed to be a potential strategic technology for the utilisation of sources of energy that could be used in times of crisis.

Figure 1.23　Wind farm with Jacobs wind turbines, Big Island, Hawaii, 1988 [8]

Figure 1.24　Smith-Putnam wind turbine on Grandpa's Knob near Rutland, Vermont, US [13]. Reproduced with permission of John Wiley & Sons

The first megawatt plant ever built was the Smith-Putnam wind turbine designed by Palmer C. Putnam and built by the S. Morgan Smith Company (York, Pennsylvania) that was erected on Grandpa's Knob, a 610 m high hill near Rutland, Vermont (see Figure 1.24). This plant consisted of an idler with a rotor diameter of 53.3 m. It was equipped with individually adjustable rotor blades and the nominal output of the synchronous generator was 1.25 MW.

The power output was controlled by means of hydraulically adjusted blade angles. The rotor had no blade twist and a constant blade width. The plant was in operation from 1941 to 1945, and during its 1000 hours running time fed electricity into the grid of the Central Vermont Public Service Company. After the plant lost a rotor blade on 26 March 1945, it was taken out of service as the financial means for repairing the rotor were not available. It took until the oil crisis for Putnam's experience of realising a whole series of large wind turbines to be used in the US.

Among the reasons for the development of wind energy after the Second World War were [14]:

- The rapidly rising demand for electricity, whereas most communities had no local sources of energy;
- Energy sources near the users were already depleted; and
- Poverty after the war and the political conditions led to the search for local sources of energy, instead of relying on imported fuels.

The knowledge of aerodynamics and materials that was made available by engineers occupied during the war in the military industry, and now active in civilian industry, eased the conditions for continuing with the development of wind turbines. Because of the new technologies, the possibility of building even more successful plants than that on Grandpa's Knob was opened up.

Later, in the 1950s, critical scientists and politicians recognised that coal and oil should not be burnt for the purpose of power generation but were more suited as materials. A further fact was cause for consideration: the thought of being dependent on only one source of energy (oil) that had to be imported from politically unstable regions. This consideration was the first indication of a political debate about democracy, limits of growth and utilisation, diversity and environmental loading for industrial development that began in the 1960s, and peaked with the publication of the study *The Limits to Growth* (Meadows *et al.*) of the Club of Rome [11].

From the 1950s until the onset of the first energy crisis in 1973, it was not only Denmark, the US and Germany that contributed to further development of wind energy, but also countries such as France and the UK took part. Surprisingly, the Netherlands, the country known as the Land of the Windmill, did not take part in the development for modern utilisation of wind energy, but attempted to use classical windmills for power generation.

In 1950 the John Brown Company built a three-rotor plant with a nominal output of 100 kW and a rotor diameter of 15 m that was operated in parallel with a diesel aggregate on the Scottish Orkney Islands for the North of Scotland Hydroelectric Board [3]. The rotor blades were fixed to a hub by means of blade flapping hinges. However, the complex rotor failed after some months.

At the same time the French engineer Andreau designed a two-rotor plant with a very extraordinary transmission technology. The rotor blades were hollow and had an opening at the ends. The rotor thus acted as a centrifugal pump that pulled air in through the opening at the base of the tower. The air passed through an air turbine that was positioned at the foot of the tower and drove a generator. This caused a soft transmission that was an alternative to the rigid drive trains that were based on directly-connected synchronous and induction generators. In 1951, De Havilland Propellers built a prototype plant for Enfield Cables Ltd. in St Albans (Hertfordshire) (see Figure 1.25). As it was impossible to operate the plant economically due

Figure 1.25 Left: Sketch of the operation of the Andreau-Enfield wind turbine; right: Andreau-Enfield wind turbine in Algeria

to the low wind speed at the site and a low transmission efficiency of 20%, it was taken down in 1957 and re-erected in Grand Vent in Algeria, but taken out of service again after a short period of activity. Although the concept was extraordinary, it was not unique. In 1946 Ulrich Hütter had previously presented a hollow one-blade plant that was meant to act as a centrifugal pump. As a counterweight, the air turbine and the generator were fastened on the opposite side of the rotor blade (see Figure 1.26). R. Bauer designed a one-blade plant with a rotor diameter of 3 m that was operated from 1952 by the Winkelstraeter GmbH.

Besides Andreau, various other French engineers were involved in the design of wind turbines. In 1958 L. Romani built an 800 kW, three-blade pilot plant with a rotor diameter of 10.1 m for the EDF (Electricité de France) energy supplier in Nogent-le-Roi near Paris (see Figure 1.27). The so-called Best-Romani plant was equipped with a synchronous generator and was taken out of service in 1963 when a rotor blade dropped off.

At the same time Louis Vadot designed two wind turbines with similar equipment as the Best-Romani plant and erected them for the Neyrpic in Saint-Rémy-de-Landes (Manche). One of the plants had a rotor diameter of 21.1 m (132 kW), the other had a diameter of 35 m (1 MW), and both possessed induction generators. As EDF lost interest in wind power, the plants were taken out of service in 1964 and 1966.

Simultaneously with the technical development, the researching industrial associations of various countries were requested by international organisations to present their results at

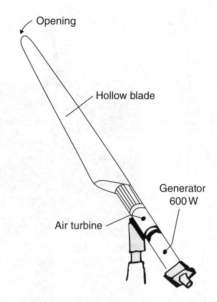

Figure 1.26 Hütter concept of the single-blade rotor with aerodynamic force transmission. The blade length is 6 m.

Figure 1.27 Wind turbine by Best-Romani in Nogent-le-Roi (Eure et Loir), France

international conferences. Among these institutes was UNESCO, the Organization for European Economic Cooperation (OEEC) and the World Meteorological Organization (WMO). The proceedings of these organisations [see 15–17] provided an excellent overview on the progress of wind turbine technology from the end of the war to the start of the oil crisis.

Interestingly, the use of wind power expanded again from industrial regions to remote arid regions and developing countries. Besides the advances in theory and technology, the further potential for wind power was being increasingly discussed. These international conferences were the first modest steps towards comprehensive international grids in the field of wind power that exist today.

From today's point of view, it is surprising how many modern discoveries were made and tested in the period from 1930 to 1960. The developments were based less on analytical methods than on experiments. Not all experiments were successful and many technical errors were made. There were no financial means available for repairs or the continuation of wind power technology. Most of the pilot projects that were carried out in France, the UK, Germany, the US and Denmark were stopped and the equipment was destroyed, with the exception of the wind turbine in Gedser, Denmark. Why this plant was so successful will be explained in the next section. The development of wind power was completely stopped due to the lack of financial means. One reason was the fact that fossil fuels became particularly cheap and nuclear power more popular. This optimistic scenario came to a sudden stop with the publication of the study *The Limits to Growth* [11] and the outbreak of the oil crisis in 1973.

Table 1.1 provides a short overview of the modern concepts that were developed in this time. The overview also includes an indication of the success of the invention.

1.5   The Second Phase of Innovation and Mass Production: 1960 to Today

Almost all important technological developments in wind power were ended by the middle of the 1960s. Fossil fuels were freely available and very cheap, and nuclear power was seen as the solution to all future energy problems. Although in the circles of the decision-makers there was little discussion on supply security or about environmental and safety matters, society had doubts regarding the limitless growth of the economy and its influences on permanently available resources. The publication of the study *The Limits to Growth* in 1971/1972 commissioned by the Club of Rome [11], the resulting discussions, and the outbreak of the oil crisis in 1973 as the result of a further Near-East conflict, turned the supposed future problems into current and present problems.

The political reactions to the crisis led to a new energy policy that was based on the following key problems:

- The dependency on energy monopolies (oil) was to be limited by diversifying energy supply options, among others by the use of local sources of energy with a simultaneous increase in energy efficiency.
- Fossil sources of energy were to be reserved for the manufacture of materials and should not be burnt for the generation of energy faster than they could regenerate themselves.

About a decade later, environmental concerns (fossil fuels, nuclear waste) and security considerations (nuclear energy at Three Mile Island, Chernobyl) became loud in the political debate.

Table 1.1 Innovation and development from 1930 to approximately 1960 (selection)

No.	Description	Inventor/developer	Country	Application today
	Rotor			
1	High-speed machine: three rotor blades	Smidth Aeromotor	(1942)	Yes
		Jacobs (1932)	US	Yes
2	High-speed machine: two rotor blades	Smidth (1942)	Denmark	Yes
3	Stall control: three rotor blades	Smidth (Gedser)	Denmark	Yes, medium-sized turbines
4	Blade tip brakes for stall control	Juul (Gedser)		
5	Stall control: two rotor blades	Smidth (1942)	Denmark	Yes, limited size
6	Single blade rotor	Bauer (1945)	Germany	Yes, limited
7	Complete blade angle setting, active	John Brown	UK	Yes
		Neyrpic-Vadot (1962–64)	France	Yes
8	Complete blade angle setting control supported by auxiliary blade	WIME	USSR (1932)	No
		ZAGI	USSR (1930)	No
9	Flettner rotor on ships	Flettner (1925)	Germany	No (with the exception of the Enercon experiments)
	Flettner rotor on rails [18]	Madaras (1932)	US	No
10	Counter-rotating rotor	Honnef (1940)	Germany	No
11	Introduction of GFK for rotor blade materials	Hütter	Germany	Yes
	Capacity			
12	>1 MW nominal power, 4 rotor blades	Draft by MAN-Kleinhenz (1942)	Germany	No
13	>1 MW nominal power, 2 rotor blades	Smith-Putnam (1.25 MW, 1945)	US	Yes
14	>1 MW nominal power, 3 rotor blades	Neypric-Vadot (1 MW, ca. 1960)	France	Yes
	Pilot			
15	Multirotor	Honnef (1932)	Denmark	Yes, strongly limited size in NL
16	Vertical axle rotor	Darrieus (patent 1930)	France	Yes, limited size
17	Idler	Kleinhenz (1942)	Germany	Yes, limited size
18	Concrete tower	Smidth Aeromotor	Denmark	Yes
19	Rotor works as centrifugal pump; generates flow for the air turbine drive generator; two rotor blades	Andreau Enfield	France, UK	No
20	Rotor works as centrifugal pump; generates flow for the air turbine drive generator.	Hütter	Germany	No

Within the framework of the new energy policy, many countries turned immediately to renewable sources of energy. This included solar energy and other sources of energy such as wind power, biomass, and the generation of energy from the warmth of the ocean. Investigations were also carried out in the area of further sources such as geothermal and tidal energy. The first research programmes on a national level were introduced by 1973, and wind power played an important role in many of them. There were many similarities in the programmes: resource availability, selection of sites, technological options, requirements for research and development, potential influences on the national energy balances, (macro-)economic and social influences and implementation strategies. However, the specific approaches and projects differed substantially from country to country. Looking back, it can be said that the most successful projects were realised where time and money provided a balance between technological developments, market development (on the demand as well as on the supply side) and political support (research funds, laws, and infrastructure). However, this was not foreseeable on the middle of the 1970s. Some countries started to develop wind turbines from basics and carried out all kinds of analysis without taking the market or the infrastructure into account. Examples of this took place in the UK, the Netherlands, Germany, Sweden, the US and Canada. Without exception, they all preferred large wind turbines as a basis for long-term energy scenarios.

Because of experiences in the past, however, this was not particularly surprising. Although not all of them were successful, the experiments and analyses of Hütter, Kleinhenz, Palmer Cosslett Putnam, Juul, Vadot, Honnef, Golding and others all tended in the same direction: the introduction of wind power on a large scale would only be economically realisable when very large wind turbines with a capacity of several megawatts were used.

The construction of such large plants was seen not only as technically feasible, but also economic. This becomes clear from the minutes of the American Congress from the year 1971 that refer to a corresponding study of 1964 [19]. A research group commissioned by the government under the direction of Ali B. Cambel came to the following conclusions:

'Sufficient knowledge is available to build a prototype with a capacity of 5,000 to 10,000 kW which permits a realistic estimate of wind power usage. A design study and a meteorological investigation of the possible sites would have to precede the actual construction. Such a program would provide important information on the economics of wind turbines and their integration into the power grid [...]. Wind energy provides a reliable source of energy also in the long term [...]. It is inexhaustible and has no negative effect on the environment as it produces no damaging and undesired side products.'

Because of this, the multimegawatt wind turbines became the technical basis of all state-supported developments. The only country that did not follow this general trend of focusing on large wind turbines was Denmark. There, risky technical experiments were avoided from the start, market introduction was stimulated, and politics supported the introduction of institutional framework conditions.

Independently of the parallel state-supported programmes, pioneering companies began with the development and sales of small wind turbines for water supply, for loading batteries and for connection to the grid. Many of these companies were inspired by E.F. Schuhmacher's 'Small is Beautiful' vision of 1973. In the early 1980s about 30 companies were active in the market in Denmark and about 20 in the Netherlands. A special case is the Tvind wind turbine in Ulfborg, Denmark. Between 1975 and 1978 teachers and students of the Tvind international school centre built a 2 MW, 54 m rotor diameter wind turbine with three manually (!) pitch controlled blades. The construction of the blades and the materials used was trend setting for modern manufacturing technology.

Also non-governmental organisations (NGOs) participated with the aim of using wind power for water supply for households, for irrigating fields and for watering farm animals in developing countries. The utilisation of their own sources of energy for covering the main demands were important for further development, and required hardly any outside capital. These organisations often had connection to universities. Examples of this are the SWD/CWD in the Netherlands, the ITDG in the UK, the BRACE institute in Canada, the Folkcenter in Denmark, the IPAT in Berlin, as well as several associations in various countries as well as the US, that were networked with universities.

In the following section, the various technical developments are discussed in greater detail. As the developments that took place after the oil crisis were so far-reaching and varied, it is not possible to describe them in as great a detail as the development in the historical epochs. The following description limits itself to the general tendencies and presents the most important cases.

First described are the state-supported developments of wind turbines. The best sources for historical details are the reports issued annually since 1985 by the IEA (International Energy Agency) wind energy programme.

National projects often made contributions to the IEA programme. Almost simultaneously with the state-supported development of large wind turbines, small pioneering companies developed small wind turbines. The consequent extension and enlargement of these wind turbines formed the foundation of the present market. In the following, current developments such as wind farms, offshore farms and the connection to the power grid are discussed.

1.5.1   The State-Supported Development of Large Wind Turbines

The first experiments were carried out in Denmark and formed a collaborative project between Denmark and the US. As already mentioned, although the wind turbine in Gedser was taken out of operation in 1966, it was not pulled down. The first step to a resuscitation of wind power development was the restarting of the wind turbine in Gedser in 1977 (see Figure 1.28).

The results of the mutual Danish-American measurements and tests served as a starting point both for the development of the wind power research programme of NASA, as well as for Danish research and economic activities. The design principles of Ulrich Hütter were also an important component of the American research and development programme. Besides Denmark and the US, important research and development programmes were being kicked into life in the late 1970s in the Netherlands, Germany, Sweden, the UK, Canada, and later in Italy as well as Spain. Smaller programmes, or rather projects, were also in existence in Austria, Ireland, Japan, New Zealand and Norway. The first group of countries adopted the development of large wind turbines. The two main questions in the conception of the first large pilot plants were:

- What potential was offered by wind turbines with vertical axles (Darrieus rotor) in comparison with wind turbines with horizontal axles?
- What strategy was to be followed in order to develop cheap wind turbines in the medium term?

Should one first adopt the more risky but also potentially more advantageous direction to light, fast- running turbines with corresponding rotor concepts? Or should one go first for reliability and step-by-step improvement of the already-tested designs from before 1960?

Figure 1.28 Gedser wind turbine in the early 1990s [8]

The US, the Netherlands, the UK and Germany carried out a systematic analysis of the potential of horizontal axle plants compared with plants with a vertical axle. From the start Canada focused on plants with vertical axles and Denmark was the only country that followed the first strategy. The first large wind turbines were put into operation in 1979 (Nibe 1 and 2 in Denmark) and the last purely experimental, non-commercially operated wind turbine was completed in 1973. In total, in the various countries about 30 pilot-only plants that had strong state support were built.

Tables 1.2 and 1.3 provide an overview of a selection of these wind turbines. The commercial prototypes are also shown. Figures 1.29 to 1.33 show some of these wind turbines, whose technical designs were very diverse. Many innovations of the past (Table 1.1) were newly designed and implemented. Substantial improvements were achieved by incorporating glass-fibre reinforced plastics into the rotor structure, and new electrical conversion systems were utilised.

The finite element method (FEM), even if not as developed as today, was used for improving the design of the sensitive components of the wind turbine, especially the structure of the hub. The basics of comprehensive design methods were very incomplete. In aerodynamics there were at best only imprecise simulations of flow separations, three-dimensional effects, aeroelastic modelling, and so on. The same applied to wind description at the rotor level, the effects of turbulence on performance, and the mechanical loading as well as the modelling of the flow tails.

Table 1.2   Selection of state-supported pilot plants

Wind turbines (land)	Rotor diameter, m	Rated capacity, MW	Year of start-up	Commercial successor
Nibe 1 (Nibe, Denmark)	40	0.63	1979	No
Nibe 2 (Nibe, Denmark)	40	0.63	1979	No
25-m-HAT (Petten, Netherlands)	25	0.4	1981	No
5 × MOD-0 (Sandusky, Ohio; Clayton, New Mexico; Culebra, Puerto Rico; Block Island, Rhode Island; Kuhuku Point, Oahu-Hawaii)	38.1	0.1–0.2	Since 1975	No
WTS-75 (Näsudden, Sweden)	38.1	0.1–0.2	Since 1975	No
WTS-3 (Marglarp, Sweden)	75	2	1983	No
WTS-4 (Medicine Bow, Wyoming, US)	78	3	1982	No
MOD-1 (Boone, North Carolina, US)	61	1	1979	No
5 × MOD-2 (3 in Goodnoe Hills, Washington State; Medicine Bow, Wyoming; Solano, California, US)	91	2.5	1980	No
MOD-5B (Kahuku Point, Oahu-Hawaii, US)	97.5	3.2	1987	No
ÉOLE (Cap Quebec, Canada)	100	4	1980(?)	No
GROWIAN (Kaiser-Wilhelm-Koog, Germany)	100.4	3	1982	No

Table 1.3   First large European wind turbines – development and pilot series

Wind turbines (land)	Rotor diameter, m	Rated capacity, MW	Year of start-up	Commercial successor
European programme WEGA I				
Tjæreborg (Esbjerg, Denmark)	61	2	1989	No
Richborough (UK)	55	2	1989	No
AWEC-60 (Cabo Villano, Spain)	60	1.2	1989	No
European programme WEGA II				
Bonus (Esbjerg, Denmark)	54	1	1996	Yes
ENERCON E-66 (Germany)	66	1.5	1996	Yes
Nordic (Sweden)	53	1	1996	No
Vestas V63	63	1.5	1996	Yes
WEG MS4	41	0.6	1996	No
THERMIE European presentation programme				
Aeolus II (Germany and Sweden)	80	3	1993	No
Monoptoros	56	0.64	1990	No
NEWECS 45 (Stork, Netherlands)	45	1	1991	No
WKA-60 (MAN, Helgoland, Germany)	60	1	1989	No
NEG-MICON (Denmark)	60	1.5	1995	Yes
NedWind (Netherlands)	53	1	1994	Yes

Figure 1.29 Left: Heidelberg wind turbine with vertical axle in Kaiser-Wilhelm-Koog [8]; right: GROWIAN wind turbine in Kaiser-Wilhelm-Koog [8]

Figure 1.30 Left: 25-M HAT wind turbine in operation in Petten, the Netherlands; middle: Canadian EOLE wind turbine with Darrieus rotor; right: WTS-75 wind turbine in Näsudden, Gotland, Sweden [8]

Included in the wind turbine concepts were, among others:

- Rotors with one to three rotor blades for wind turbines with horizontal axes, and two or three rotor blades for plants with vertical axles.
- Rigid hubs, teetering hubs and movable hubs.

Figure 1.31   Left: Nibe wind turbine in Jutland, Denmark; right: WEST wind turbine with single-blade rotor in the foreground and two-blade plant in the background in the Alta-Nurra pilot field, Sardinia, Italy [8]

Figure 1.32   MOD-2-wind turbine in Solano, California, US [8]

Figure 1.33　Disassembled AWCS-60 in Kaiser-Wilhelm-Koog, Germany [8]

- Rigid rotor blades, stall control and complete or partial control of the blade adjustment angle.
- (Almost) constant and variable revolution transfer systems.

In addition, a spectacular series of installation techniques were used. The methods ranged from the conventional installation with the aid of a crane, to the use of the tower of the plant as a lifting device for platforms that were used to position the motor housing and the rotor blades.

After a modest start with regard to financial means, by 1988 the European programme started to strengthen support for the development of large wind turbines. These changes in politics were preceded by exhaustive discussions with scientists and representatives of the industry on the optimum industrial strategies and market potentials.

The results of these talks were that the manufacturers, who were already active earlier in the production of smaller plants and who had a serious interest in commercialising large wind turbines, reacted to the initiative of the European Commission and closed contracts for the development and construction of the first commercial prototypes of megawatt wind turbines. The participation of commercial companies in this programme changed the industry permanently. The programme for testing and evaluation, financed by individual governments and implemented by large construction and aviation concerns, slowly came to an end. The physical end for some wind turbines was quite spectacular: MOD 2, GROWIAN and Aelous II were dynamited.

The design philosophy of the commercial prototypes was based on the gradual enlargement of smaller plants, and was developed and commercialised by some of the pioneer companies that survived a serious crisis in the 1980s.

The most successful wind turbines at the start were not the most advanced, fast-running models, but rather those that possessed many features of the well-known 'Danish concept'. This concept was based on the blueprint of the Gedser wind turbine. Figure 1.34 offers an overview of the WEGA I and II European programmes.

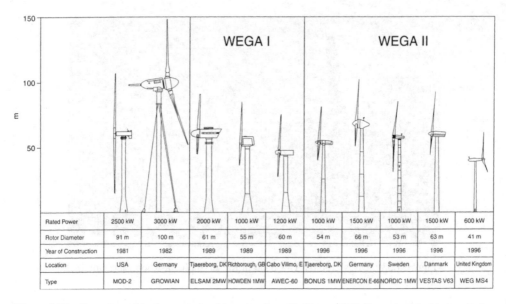

Rated Power	2500 kW	3000 kW	2000 kW	1000 kW	1200 kW	1000 kW	1500 kW	1000 kW	1500 kW	600 kW
Rotor Diameter	91 m	100 m	61 m	55 m	60 m	54 m	66 m	53 m	63 m	41 m
Year of Construction	1981	1982	1989	1989	1989	1996	1996	1996	1996	1996
Location	USA	Germany	Tjaereborg, DK	Richborough, GB	Cabo Villmo, E	Tjaereborg, DK	Germany	Sweden	Danmark	United Kingdom
Type	MOD-2	GROWIAN	ELSAM 2MW	HOWDEN 1MW	AWEC-60	BONUS 1MW	ENERCON E-66	NORDIC 1MW	VESTAS V63	WEG MS4

Figure 1.34 Overview of the large wind turbines developed in Europe [21]. Reproduced with permission of Keesing Media Group

The WEGA and THERMIE programmes ushered in the steady expansion of wind turbines (see Figure 1.35). Typical for this development phase was the consistent expansion of the smaller, commercially successful wind turbines. Many of the advanced technical designs such as teetering hubs, idlers, and fast-moving rotors with one or two rotor blades were given up by the industry. Their first prototypes were rather conservative as the customers were primarily concerned with reliability and not so much with advanced systems with potential for future cost savings. The innovations that were later developed differentiated themselves from the first generation of plants. The greatest further development was the performance electronic converter, as it allowed much improved control of the turbines. With these conversion systems, together with control of the blade adjustment angle and advanced multiparameter control strategies, the modern plants correspond to the demands of the power grid. Critical development and design procedures and the use of new materials led to a reduction in weight and thus to a reduction in the costs of power generation.

With the enlargement of the wind turbines, there was also an immense increase in the market volume of the wind turbine (see Figure 1.36). The product life of a particular type of wind turbine, when based on the turbine capacity, is usually six years and has extended with the growth of the turbine size since 2002 [20].

Technical knowledge has increased especially in the fields of aerodynamics, the modelling of flow loops in wind farms, aeroelasticity, finite element modelling, construction dynamics, measuring techniques, system modelling and control techniques.

The purely analytical results needed to be certified, whereby, besides laboratory installations, experimental wind turbines were also erected on the open air. The laboratory installations consisted mainly of test rigs for rotor blades, experimental stands for materials, as well as test rigs for drive trains and wind tunnels. Most of these research arrangements originated from

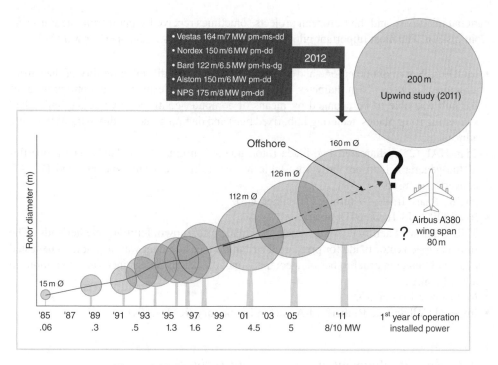

Figure 1.35   Enlargement trend of modern wind turbines

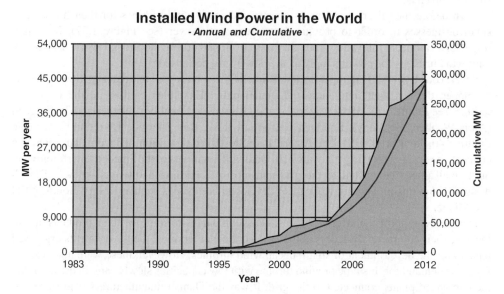

Figure 1.36   Growth of the world market for wind turbines

national initiatives, and the European projects sometimes received support from the European Commission. The most important pilot plants for wind turbines in the open air were:

- MOD-0 (38 m in diameter, one and two-blade rotors, US): adjustable rigidity of the carrier structure, Uniwecs (16 m diameter, two-blade rotor, idler, Germany): the configuration of the hub was able to be changed by means of computer-controlled hydraulics (individual swivelling rotor blades, teetering hub, fixed hub) and damping and rigidity parameters were adjustable.
- 25 m HAT (25 m diameter, two-blade rotor, upwind turbine. Netherlands. The properties of the generator load are fully adjustable in that a direct current generator and DC–AC converter are used.
- NREL, Phase II, III, IV turbines (Boulder, US).
- Risø, TELLUS Turbine (Denmark).
- TUD Open Air Facility (10 m diameter, two-blade rotor, upwind turbine, Netherlands). The fully equipped rotor blade for pressure distribution measurement was also able to be tested in a wind tunnel in order to permit a comparison with exactly determined flow conditions in a wind tunnel.
- Mie University (Japan).
- Imperial College and Rutherford-Appleton Laboratory (RAL) (UK).

1.5.2   The Development of Smaller Wind Turbines

In order to fully understand modern wind energy and its marketing success, one must also consider the role of the wind energy pioneers. This refers to individual persons as well as small companies.

Even before the oil crisis, pioneers constructed small wind turbines for their houses and small businesses in order to provide themselves with power (see Figure 1.37). In certain respects the pioneers followed a trade that was developed during the Second World War. Many built wind turbines for power supply as the power grid collapsed.

Several do-it-yourself attempts from this time of the war are still in existence. Some of these developed designs went into industrial manufacture. Examples of this are the Lagerwey turbine (see Figure 1.38 [Henk Lagerweij and Gijs van de Loenhorst, Netherlands]), Enercon (Alois Wobben, Germany), Enertech (see Figure 1.39, Bob Sherwin, US), and Carter (see Figure 1.40, Jay Carter, US). The Americans followed the tradition of Jacobs, but seen as a concept, the design was fairly innovative. In Denmark, members of the Smedemesterforeningen built small plants connected to the grid and mostly according to the well-known 'Danish design'. From this pioneering work there developed companies such as Vestas, Bonus, NEG and Micon.

In order to support the efforts of the smaller companies to improve their products and to increase their commercial success in the US and the Netherlands, the governments organised tenders for the development of small, cheap wind turbines. However, these were not successful as the demand for this type of wind turbine with low capacities slowly sank in favour of the medium-sized plants connected to the grid. It was the Danish manufacturers who started to sell their plants on a large scale in their local markets and in various European countries. At the same time they exported very successfully to the US.

Figure 1.37   Dutch do-it-yourself pioneer in the field of wind energy, Fons de Beer, with his passively controlled wind turbine [8]

Figure 1.38   The early passively controlled wind turbine by Lagerwey van de Loenhorst with variable speed. Diameter: 10 m [8]

With increasing demand for large wind turbines, the manufacturers increased the capacities of their well-known concepts step by step. At the start of the 1990s they achieved the same size as the smaller plants supported by the state in the 1980s. With the support of the European Commission (WEGA and THERMIE programmes), it was possible for them to improve their designs, to test their plants and to write success stories. This was a necessary condition for successful market introduction. At this point the two, originally separated, lines of development

Figure 1.39   Enertech wind turbine in a Californian wind farm [8]

Figure 1.40   J. Carter's passively controlled idler [8]

merged into one. Because of the demand for very large wind turbines, the enlargement trend continued as these seemed to be potentially more competitive in the offshore region.

Almost none of the pioneer companies from the 1970s and 1980s emerged from this phase unchanged. Some went bankrupt and others were successful. They either remained completely independent (ENERCON) or, as independent companies, attracted the attention of foreign investors (Vestas). Others merged or were taken over by larger companies. Examples of well-known mergers are NEG and Micon.

1.5.3   Wind Farms, Offshore and Grid Connection

Together with the increasing capacity of wind turbines, wind energy projects (wind farms) (see Figure 1.42) increased to such an extent that their total capacity rose from one megawatt to several hundred. In order to plan wind farms efficiently, it was first necessary to have knowledge of handling wake flows and flow conditions within the wind farm. Research in these areas began at the start of the 1980s with the physical analysis of wake flows and experimental investigations in wind tunnels (see Figure 1.41). This research field became of increasing

Figure 1.41 Early wind farm measurements carried out by David Milborrow, Electrical Research Association, UK. Photo: ERA

importance from the first investigations on flow conditions in wind farms. External influences such as wind shear, intensity of turbulence and the stability of the atmosphere play a large role in the distribution of flows. As these influences are very different on land and offshore, the economic planning of offshore wind farm depends to a large extent on the investigation results on the flow conditions in wind farms.

The importance of offshore wind energy became greater as the best sites for wind energy generation on land in northwest Europe were already developed. Offshore is the only possibility for coastal countries to increase the contribution of wind energy for energy supply (>20%). A further reason for the growing importance of offshore wind energy is public resistance to the erection of wind turbines in protected and old landscapes, especially in the UK and Sweden. The potential of offshore wind energy was recognised right at the start of the modern wind energy era. Offshore wind energy has been investigated since 1978 within the framework of the IEA wind energy programme (see Figure 1.43).

In 1991 the first large commercial offshore wind farm with a total capacity of 4.95 MW was erected 2.5 km from the coast of Vindeby in Denmark. This comprised 11 Bonus wind turbines of 450 kW each. The second Danish offshore wind farm was erected at Tuno Knob in 1995.

Figure 1.42   Multimegawatt plant of ENERCON, North Germany [5]

This wind farm has a total capacity of 5 MW and is made up of 10 Vestas wind turbines of 500 kW each. Both wind farms were erected in protected and flat waters. The first offshore wind farm near Copenhagen, Middelgrunden, was erected with the help of landscape architects and has a total capacity of 40 MW. It comprises 20 Bonus wind turbines that produce 2 MW each, are arranged in a curve and at a water depth of between 5 and 10 metres.

The first step in the direction of wind energy utilisation in the rough environment (see Figure 1.44) of the North Sea was taken in 2002 with the erection of the Horns Rev Offshore Wind Farm. The installed capacity was 160 MW, making it the first wind farm whose total capacity exceeded 100 MW. The farm comprised 80 wind turbines of 2 MW each, lying between 14 and 17 km from the coast at a water depth of 6–14 m. One at a time, more countries have joined the offshore community. In the middle of 2012 there were wind turbines with a total capacity of 4100 MW. These were spread over the following countries: Denmark, Sweden, the Netherlands, the UK, Ireland, Belgium, Germany and China.

The rapid growth in wind farm capacity is shown by the following figures:

Average capacity: 43 MW/farm
Average capacity of the 10 smallest, older wind farms   8 MW/farm
Average capacity of the 10 largest, newer wind farms   198 MW/farm

Taking into account the capacities of offshore wind energy, the European states are planning further offshore wind farms, and this in an area without an existing electrical infrastructure. In this way the necessity of a new concept for grids becomes clear. This necessity becomes

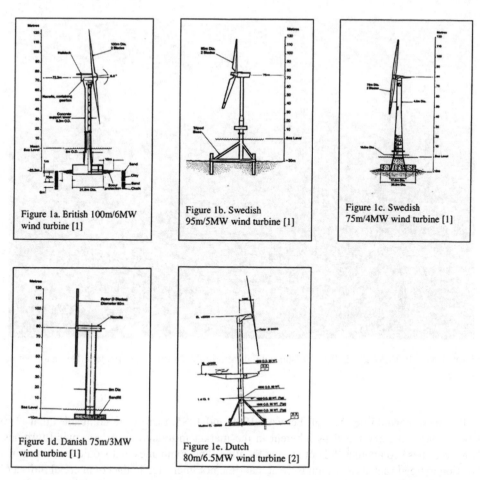

Figure 1a. British 100m/6MW wind turbine [1]

Figure 1b. Swedish 95m/5MW wind turbine [1]

Figure 1c. Swedish 75m/4MW wind turbine [1]

Figure 1d. Danish 75m/3MW wind turbine [1]

Figure 1e. Dutch 80m/6.5MW wind turbine [2]

Figure 1.43 Offshore technology has been investigated since 1978 within the framework of the IEA-LS-WECS programme. The participating countries were Denmark, the Netherlands, Sweden, the UK and the US

more urgent with a view to future renewable sources of energy of variable power. This affects mainly wind turbines, concentrated solar energy plants and hydroelectric plants. As the lead time for extension of grids is very long, usually ten years or more, the expansion of the grid should have already started ten years ago in order to meet today's demands. The expansion of the grids could and should have been a part of the implementation, but this was not the case. It is clear that the bottleneck for further expansion of offshore wind energy is the grid.

1.5.4   International Grids

Looking back, there arises the interesting question of whether the programmes instituted by the governments of 1979 were useless, as there were no directly resulting commercial applications.

Figure 1.44   160 MW Horns REV Offshore wind farm in the North Sea in Esbjerg, Denmark. Photo: ELSAM

The answer would be 'Yes' if, between 1979 and 1985, there were companies that were large enough to carry the risks inherent in the market introduction of large wind turbines. However, these companies did not exist at the time. The companies that would have been in a position to tread such a risky path were at that time not interested in the commercial introduction of large wind turbines.

The answer would be 'No' of one takes into account the enormous quantity of collected data on design and operation that was gained from accidents and events. The manner in which this knowledge contributed to the new generation of wind turbines cannot be directly proven as this development did not take place in sequential steps. The learning process was carried out by experts who changed to the private sector and brought their knowledge with them. Added to this there were also conferences (e.g. organised by the EWEA, the European Commission and the AWEA) and seminars (e.g. organised by IEA).

The national research associations constantly created establishments for the collection of expert knowledge on wind turbines. These establishments (at the start, the national research laboratories Risø, ECN, SERI/NREL), universities and newer university networks such as ForWind and CEWind in Germany were supported in the long term by national governments. Without these institutions, the access to knowledge on wind turbines would have been locked up or even lost. The interconnection of the various aspects of this knowledge, such as resources management, structural dynamics, electrical and mechanical energy transmission and others, would also have been lost.

The fact that this knowledge is still coherent, up-to-date and accessible is due in large part to the degree of organisation of the scientific world and the intensive contacts between science and industry. In this, governments played only a minor role. Although they and the European Commission promoted the cooperation, it was the wind energy sector that started or coordinated many initiatives, distributed contracts and developed strategic scientific programmes. Examples of these initiatives from Europe are the EWEA (European Wind Energy Association), EUREC (European Renewable Energy Agency), EAWE (European Academy for Wind Energy) and MEASNET (Measurement Quality Assurance Network). In comparison to today's national associations, which have been in existence for a long time, these international networks present a new dimension.

1.5.5   To Summarise

When one follows the long path of history, from the start of the use of windmills by the Persians up to the large-area usage of wind energy, one can differentiate two very successful periods: within the first period (1700–1890) the use of wind energy allowed the industrialisation of parts of Northern Europe, especially in the Netherlands. In the second period (after the oil crisis of 1973), wind energy was of increasing importance worldwide. The close meshing of theory and practice in the development of the technology, the introduction of reliable market incentives and political support were shown to be key to the success of today's developments in the wind energy sector.

References

[1] Greaves, W.F. and Carpenter, J.H. 1969. *A Short History of Mechanical Engineering*. Longmans Green and Co., London.
[2] Reynolds, J. 1970. *Windmills and Watermills*. Hugh Evelyn, London.
[3] Golding, E.W. 1955. *The Generation of Electricity by Wind Power*. E. and F.N. Spon, London.
[4] Needham, J. 1965. *Science and Civilization in China*, Vol. IV, Section 27. Cambridge University Press, Cambridge.
[5] Beurskens, J., Houët, M., van der Varst, P. and Smulders, P. 1974. *Windenergie*. Technical College Eindhoven, Department of Nature Studies, Working group Transport Physics, Eindhoven.
[6] Stokhuyzen, F. 1972. *Molens*. Unieboek, Bussum.
[7] Notebaart, J. 1972 *Windmühlen*. Mouton Verlag, Den Haag, Paris.
[8] Beurskens, H.J. Private collection of photographs.
[9] Poul la Cour Foundation, Mollevej 21 (There should be a / through the o), Askov, DK 6600 Vejen.
[10] Rijks-Studiedienst voor de Luchtvaart. Rapport A 258, A.G. von Baumhauer. *Onderzoek van molen-modellen (Messungen an Windmühlenmodellen)*. Rapport A 269, *Proefne-mingen met modellen van windmolens (Untersuchung von Windrädern)*.
[11] Meadows, D.H. Meadows, D.L., Randers, J. and Behrens III, W.W. 1972. *The Limits to Growth. A report for the Club of Rome's project on the Predicament of Mankind*. Universe Books New York.
[12] Betz, A. 1926. *Windenergie und ihre Ausnützung durch Windmühlen*. Vandenhoed & Ruprecht, Göttingen.
[13] Palmer Cosslett, P. 1948. *Power from the Wind*. Van Nostrand Reinhold Company, New York.
[14] Hau, E. 2000. *Wind Turbines. Fundamentals, Technologies, Application, Economics*. Springer Verlag, Berlin.
[15] Organisation for European Economic Co-operation. Technical papers presented to the Wind Power Working Party.

[16] *Procès Verbal des Séances du Congrès du Vent*. Carcasonne, September 1946.
[17] UNESCO. 1954. *Wind and Solar Energy*. Proceedings of the New Delhi Symposium. Volume I, Wind Energy. (English, Spanish, French)
[18] *Electrical World*. 1932. Will towers like these dot the land? *Electrical World*, May 28.
[19] Congressional Record. 1971. Proceedings and Debates of the 92nd Congress, Vol. 117, No. 190. Washington, DC.
[20] Ender, C. 2011. Windenergienutzung in Deutschland; Stand 31-12-2010. *DEWI Magazine*, February 2011.
[21] Husslage, G. 1968. *Windmolens*. Keesing, Amsterdam.

2

The International Development of Wind Energy

Klaus Rave

Wind energy is a universal resource. It can satisfy a multitude of global energy problems not only theoretically but also in actuality. It can be used to generate electricity, the leading source of energy of the twenty-first century. From the finite nature of fossil fuels, their geographically uneven distribution, and the resulting climate change due to its burning, to the dangers of the nuclear sector, recently experienced so dramatically in Fukushima, these problems are increasingly being answered through the generation of electricity by means of regenerative sources, especially with the use of wind energy.

2.1 The Modern Energy Debate

The international energy debate reached a new dimension with the publication of Meadows' *Limits to Growth*, the report of the Club of Rome [1]. The shock of the first oil crisis in 1973 hit the industrialised world hard: fuel shortages and distributions, and in Germany there was even a prohibition on driving on Sundays. These were the scenarios that demanded a realignment of resources.

The nuclear accidents of Harrisburg and especially Chernobyl (1986) marked a further dimension in the hazards caused by energy supply. The so-called 'peaceful use of nuclear energy' was increasingly questioned with regard to its risks [see 2–4]. Citizens' movements formed, and the 'Green' political movement came into existence in many countries as a result of these protests.

The third great challenge of the international energy supply was the climate debate that began in the 1980s. Awareness grew that anthropogenous effects has an ever-increasing

Understanding Wind Power Technology: Theory, Deployment and Optimisation, First Edition.
Edited by Alois Schaffarczyk. Translated by Gunther Roth.
© 2014 John Wiley & Sons, Ltd. Published 2014 by John Wiley & Sons, Ltd.

influence on the Earth's climate (see early popular science publications: [5]; also [6], currently [7]). Questions of security – geostrategic military challenges as well as security of supply – dominated the current debate [8].

Even though these three debates took place at different times, it was attempted, for instance, to use the climate debate to drive more rapid expansion of nuclear energy as a supposedly CO_2-free form of power generation. Yet it was the continual development of renewable energy carriers, especially wind energy, that was an important and growing part of the argument (see Table 2.1 for global growth). This connection was also made by the UN General Secretary Ban Ki-moon when he made the request for 'Sustainable Energy for All' with the target year 2030 (see *New York Times* of 11 January 2012 in the preliminary report to the Future World Energy Forum and the General Meeting of the IRENA – see below – in Abu Dhabi).

Otherwise, for the hazards from climate change as well as for the nuclear risks, the formula 'avoid the unmanageable and manage the unavoidable' applies.

Two issues supported and strengthened this trend. One was the historical experience with the use of wind energy as part of the development of human civilization since the eighth century. The other is an already decades-long tradition of research and development in this field, whether in the US, in Germany, Denmark, the Netherlands or the UK (see the detailed discussion in Chapter 1).

In 1993 the breakthrough of a limit was celebrated at the European Wind Energy Conference in Schleswig-Holstein: 1000 MW had been installed. The year 2011 saw a new dimension: throughout the world 240,000 MW had been erected, and of this more than 1000 MW in 21 individual countries (see Table 2.2).

Not only have the size and number of turbines continually grown, but also the number of countries in which wind energy is used. The USA, Denmark, Germany and Spain are pioneers in the early development of wind energy. For a long time it had been feared that this group of

Table 2.1 Total of the worldwide installed capacity (in MW)

Year	Capacity	Growth
1996	6100	1280
1997	7600	1520
1998	10,200	2520
1999	13,600	3440
2000	17,400	3760
2001	23,900	6500
2002	31,100	7270
2003	39,431	8133
2004	47,620	8207
2005	59,091	11,531
2006	74,052	15,245
2007	93,820	19,866
2008	120,291	26,560
2009	158,908	38,793
2010	197,039	38,265
2011	237,669	40,564
2012	282,469	44,799

Table 2.2   The wind countries of the world

		Wind energy generation capacity		
		To 2010	New 2011	2011 total
Africa and the Near East				
	Cape Verde	2	23	24
	Morocco	286	5	291
	Iran	90	3	91
	Egypt	550	–	550
	Others[1]	137	–	137
	Total	1065	–	1093
Asia				
	China	44,733	17,631	62,364
	India	13,065	3019	16,084
	Japan	2334	168	2501
	Taiwan	519	45	564
	South Korea	379	28	407
	Vietnam	8	29	30
	Others[2]	69	9	79
	Total	61,106	20,929	82,029
Europe				
	Germany	27,191	2086	29,060
	Spain	20,623	1050	21,674
	France	5970	830	6800
	Italy	5797	950	6737
	Great Britain	5248	1293	6540
	Portugal	3706	377	4083
	Denmark	3749	178	3871
	Sweden	2163	763	2970
	Netherlands	2269	68	2328
	Turkey	1329	470	1799
	Ireland	1392	239	1631
	Greece	1323	311	1629
	Poland	1180	436	1616
	Austria	1014	73	1084
	Belgium	886	192	1978
	Others[3]	2807	966	3708
	Total	86,647	10,281	96,606
	Of this EU 27	84,650	9616	93,947
Latin America and Caribbean				
	Brazil	927	583	1509
	Chile	172	33	205
	Argentina	50	70	130
	Costa Rica	119	13	132
	Honduras	–	102	102
	Dominican Republic	–	33	33
	Caribbean[4]	91	–	91
	Others[5]	118	10	128
	Total	1478	852	2330

(*continued overleaf*)

Table 2.2 (*continued*)

		Wind energy generation capacity		
		To 2010	New 2011	2011 total
North America				
	USA	40,298	6810	46,919
	Canada	4008	1267	5265
	Mexico	519	50	569
	Total	44,825	8127	52,753
Pacific	Australia	1990	234	2224
	New Zealand	514	109	623
	Pacific Islands	12	–	12
	Total	2516	343	2859
Worldwide		**197,637**	**40,564**	**237,669**

*Preliminary data: (1) South Africa, Israel, Nigeria, Jordan, Libya, Tunisia; (2) Bangladesh, Indonesia, Philippines, Sri Lanka, Thailand; (3) Romania, Norway, Bulgaria, Hungary, Czech Republic, Finland, Lithuania, Croatia, Ukraine, Cyprus, Luxembourg, Switzerland, Latvia, Russia, Faroe Islands, Slovakia, Slovenian Republic, Macedonia, Iceland, Liechtenstein, Malta; (4) Jamaica, Cuba, Dominica, Guadalupe, Curacao, Aruba, Martinique, Bonaire; (5) Colombia, Ecuador, Nicaragua, Peru, Uruguay.

four would not expand to include further countries. The danger grew that political change in only one country would trigger negative consequences for overall development. Today the use of wind energy has spread to more than 75 countries. This growth is associated with technological and geographical diversity. In 2010, for the first time, the non-OECD states, led by China, became the drivers of growth. The quantitative element has been supplemented and lifted to a new level.

2.2   The Reinvention of the Energy Market

The very uneven regional and thus political distribution of conventional fuels was, and remains, a formative cause of the crisis. Whether oil, gas, coal or uranium, the presence of fuels on national territory and its import/export potential or dependence determines a region's welfare, development and economic growth. The OPEC price cartel, 'cheap' and highly subsidised coal mining, the highly dangerous nuclear fuel cycle of uranium, and the specific dependencies of the capacity-linked gas supply, have all led in different ways to external political stresses, even wars, and political distribution disparities accompanied by serious internal conflicts [8, p. 227].

In the modern history of humanity – the history of industrialisation – the energy markets have always been politically designed or influenced. Government energy suppliers, monopolitical and oligopolitical raw material supporters shape the global energy economy. The ten largest (measured by reserves) oil and gas companies in the world are in government hands (*The Economist*, 21 January 2012).

The intensity of regulation of the energy market is shaped in an extremely differentiated manner; a scale of 0 to 100 would be fully used up. The correlation with the Global Corruption Index of Transparency International is eye-opening. The spectrum of political influence ranges from 'Atoms for Peace' through the 'Kohlepfennig' (coal penny) to the oligarchical Gazprom, from research programmes to the EU rules and regulations. The regulatory configuration of the EU domestic market is one of the most important and complex political processes of the present time.

However, this is far overshadowed by the interminable negotiations on an international climate protection agreement. The 2° target of Copenhagen has – so far – no binding international effect. The intention to come to an agreement on a target by 2015 must be seen as some sort of success in connection with the conference in Durban, in order not to feel more hopelessness (see [9], p. 258 for development and an overview of agreed-to instruments).

Some global problems – the security, environmental and social difficulties in energy demand coverage – remain without an international framework. The analogy to the situation of the financial markets is forced upon one. The remark 'If the climate were a bank, they'd have saved it by now' was occasionally heard at the peak of the post-Lehmann Brothers crisis. In both fields, binding agreements and regulatory standards are urgently needed and equally far away. In both fields the potential danger for the state, the economy and society grows constantly. For both areas it is the case that only enforceable, international binding agreements can have a permanent effect. The protection of the climate and the regulation of the financial markets present the international community with new dimensions of cooperation, but are necessary for survival of the global ecology [10] as well as for the economy.

The change in the energy markets is currently taking place on different levels:

- At the level of the actors, new players are appearing on the field and traditional providers are changing. On the one hand, old monopolies are being broken up for the creation of real markets with the separation of network operation and power generation, as in the EU, and on the other hand, the number and importance of independent investors in the generation of

---

Box 2.1   The New Design of Ocean Use

The use of the offshore region is linked with great opportunities but also large risks. At present the hazards of oil spills, as recently seen in the Gulf of Mexico and near the coast of Brazil, are being discussed. To an even greater extent these threats apply to all developments in the Arctic and Antarctic regions which are still protected by 'frozen claims' as well as by a framework of international agreements. Here there is a responsibility for the international community. This also applies to the world's oceans outside the 200 mile exclusive economic zone established by the United Nations Convention on the Law of the Sea. Therefore it seems encouraging that the US and the EU Commission have presented a green paper – better a blue paper – from which an independent ocean policy is to be developed. The approval process for offshore wind farms in the North and Baltic seas, as is implemented in a most demanding manner by the BSH, has set legal planning standards regarding environmental consequence estimates, among others, and has led to 'staking claims' to the ocean [11]. In this way the use of the ocean can be carried out in a responsible manner. As regards the unique ecological system of our blue (!) planet, mankind would then leave the condition of the hunter and gatherer and, in a quantum jump, would then have arrived in the maritime industrial age [12, 13].

energy is growing continuously. In addition, new services are developing such as the trading of electricity on the stock exchange or the supply of storage capacity of different types (pump and compressed air stores, electro-mobility, 'Power to gas'; attention should be paid to the special role of Norway with the Statoil and Statkradt undertakings and their strategy of the role of a 'Battery of Europe').

- At the level of raw material acquisition of conventional energy sources, the offshore region is of increasing importance, and so-called unconventional sources such as shale gas are accessed and recovered using new, environmentally questionable methods.
- At the level of the end-energy supply, electricity is gaining systematically in market share and will become the key energy of the twenty-first century (see [8] p. 714), not least because of its importance in the sector of information technology as well as energy carrier, and the technological interaction of information technology and energy supply ('smartgrids', 'smart metering', 'smart home').
- At the level of the transport of energy, especially the transport of electrical current, new HVDC technology is being exploited to reduce resulting losses over longer transport distances.
- At the level of price formation it can be assumed that medium- to long-term price rises on the raw material side as well as for the end energy will apply.

The growth of importance of precious electrical energy will be underlined by the fact that the end energy can be generated from diversified sources.

2.3 The Importance of the Power Grid

Because the provision of power in secure grids is indispensable for the growth of the international communication paths, its special strategic importance is becoming increasingly clear [14]. The Internet and the electrical grid represent different technologies. Through the resulting innovative symbiosis of both strategic infrastructure investments, they are key to modernising and consistently developing international economies from the local, via the regional and national, up to the global level [15].

From 'Supergrid' (see Figure 2.1), as Eddy O'Connor, founder of Airtricity, now Mainstream Renewable Power, saw it, up to the 'smartgrid' and 'smart metering' as it is conceived at present in Europe (Italy, Sweden and Germany), innovative technology and cross-over applications lead to new dimensions [9]. The pure supply side of the generator, compared with the user, is replaced with demand-oriented control by the optimisation of the resources used. In this way the renewable energy carriers can achieve their full development. Of lesser importance is on how many levels the current is transported, and by whom it is distributed and delivered to the end customer. This is finally a question of costs, since a profit must be made at every level: in Germany, for instance, on three levels, but in France or Italy on only one.

A 'global link' is no longer a utopia. Quite rightly the US National Academy of Engineering has celebrated large-scale electrification by means of power grid operation as the greatest engineering achievement of the twentieth century. The potential of information technology in connection with the electric grid and power generation from renewable sources will require engineering performance that can set standards for the twenty-first century. In his article in the New York Times, the UN General Secretary Ban Ki-moon pointed out that just 20 years ago the worldwide spread of mobile phones was unimaginable (11 January 2012). According to the principle of emergence, new things will come into existence with the linking of energy

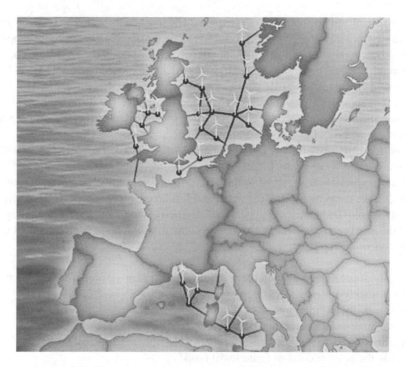

Figure 2.1 European wind energy current linkage network (Supergrid)

technology with information technology. My thesis: analogous to the development of cloud computing, a type of 'cloud generation', carried by a worldwide linkage, the 'global link'.

Thus the three megatrends are linked to each other: globalisation, decentralisation and decarbonisation. The second triad of scarcity, security and quality will be considered, as well as the third already mentioned: finite resources, risks of nuclear utilisation, and hazards related to climate change. With all the asynchronies of international development, especially the widely diverging development conditions, per-head income and CO_2 emissions, wind and solar energy in addition to hydro-power, and the potential of biomass form the overarching solution within the framework of an intelligent grid.

Access to the grid, and the attachment of unstable island grids by means of linkages, is a great challenge for the developing international economies. At present about a quarter of the 7 billion people on Earth do not have this access. Yet in China, India, the whole southeast Asian region, the now-independent former Asian Soviet Republics, the African continent and in Latin America, the use of necessary technology is the key to economic and individual development. A secure grid is therefore indispensable.

Warnings of the 'digital divide' have been given often enough, but it is often disregarded that in the access to electricity, the security of a grid is an important step to development [16].

Whereas in Europe, new 'top-down' forms of cooperation need to be tested, and in North America the strengthening of the north–south axis must face the challenge of an east–west linkage, it is valid in the developing national economies to arrive at system integration in a 'bottom-up' manner. Common for both developments is that only a technically demanding network operation provides supply security. In this, Europe, and especially Germany, must bring their special and positive experience to bear and must contribute. The German grid, with its background as a transit and industrial country, is by far the best configured and operated in the world. Downtimes are around 20 minutes per year, whereas the next best country is 4 hours, and the national economy in the US suffers damage of approximately US$150 billion per year caused by blackouts [14].

The power supply will also undergo a change in the future. The electricity supplied to the socket must always be of the same quality and always be available. Therein lies the reason for the modernisation task of grid configuration and operation.

The age of burning fossil fuels to provide energy has passed. The vision of an everlasting nuclear fuel cycle is no more. As before, investors are interested in the prevention of 'stranded investments', especially with regard to the high capital costs of nuclear power stations. There is still no storage capacity for nuclear waste and fast breeder technology. If the conversion to new energy sources and new ways to generate electricity is to be successful, then the modernisation of the grid is an important factor. Hence to my formula 'no transition without transmission', the following applies:

* to link natural potential with suitable technologies in a manner proper for demand;
* to achieve supply security across borders;
* to devise socially friendly prices;
* to achieve price security and thus economic stability by means of:
* calculating work on the upfront costs as is possible with wind energy by:
* preventing the volatile costs of finite fossil fuels.

In this, the extension of the grids is comparable in its strategic importance for the twenty-first century with the expansion of the rail and road networks and telephone grids in the late nineteenth and twentieth centuries. With regard to the latter, for a long time a transatlantic cable was seemingly impossible.

Without the vision of linkages and networks the history of mankind would have developed differently. But why should the trade in electricity not be globalised and be physically transported by international means? The Silk Road or the Trans-Siberian Railway can serve as comparable references.

There are questions of financing just as of organisation – governmental, private, or mixed economy. By means of the long-term, regulated (see below), ensured 'return on investment', there is an attractive investment object here that, especially in times of political instability, can be of great importance, for instance for pension funds. There is also space for private investment: security on a long-term basis, with better interest than a savings account.

There are also indications of the opportunities for not only political, but material and financial, participatory energy policy. Citizen wind farms were the precursors, and now there are multiple forms of participation. Citizen networks are being erected, and the opening of this form of participation can be an active answer to the widespread St Florian's principle or 'NIMBY' effect ('not in my backyard'). Whereas in Germany it is repeatedly pointed out that it takes more than ten years from planning to realisation, the wind developers on the Baltic Sea island of Fehmarn have shown the world otherwise: in 11 months the island, in the course of repowering, was completely cabled and reconnected with the mainland, and all privately financed. The Arge (Coop) Netz in Nordfriesland developed and operated not only the grid for over 1000 MW wind farm capacity, but also offers on its land area access to fast Internet via broadband supply [14]. Why not issue citizens with shares in network operators, and thus make available acceptance for investments for the common good?

In view of the plans for extension of wind energy as the main source of energy, the Danish government and Parliament have combined the grid operation of the whole country, integrating both grids of Nordel and UCTE, and transferred them to a state company. (This company now consults worldwide on behalf of the government when there are questions of the integration of wind power into the grid, although also with the aim of the promotion of Danish turbine manufacturers such as VESTAS, as seen in 2011 in China with the presentation of a report to its Government on the occasion of the China Wind Power Conference and Exhibition.)

The operation of grids is becoming more challenging, and wind energy is a decisive driver in this respect. Initiated by the Fördergesellschaft Windenergie [Association of Wind Energy] and scientifically supported by the Kassel ISET, the first conference on the subject of large-scale integration was held in 2000 and marked the start of a qualified and intensive discussion in the industry on this subject. The organiser was the European Wind Energy Association (EWEA), the oldest and largest trade association, which then continued to work on this subject in close cooperation with the Vereinigung der Stromnetzbetreiber (ENTSO-E) [Association of Power Operators] on a European level, and provided important momentum to an international congress in 2010 in Berlin (see also [9] p. 173; [17]). This cooperation is urgently needed. At present there are no legal instruments on the basis of which, or with its help, it is possible to advance the physical planning and the energy-economic agreements for an expansion of the grid throughout the EU, irrespective of the international dimension.

The electricity grid is a natural monopoly and therefore needs more stringent regulation so that, on the one hand, the correct drivers for investment are provided, by satisfactory yields with the highest technical standards, and, on the other hand, independent power producers are not discriminated against. As is well known, the breakthrough for wind energy in Germany was achieved by the Feed-in Law of 1989 that for the first time provided access to the grid and also secured a defined remuneration.

2.4   The New Value-added Chain

As the leading energy source of the twenty-first century, it is my theory that electricity will change current market characteristics significantly. A new global value-added chain will be created:

- production from diversified regenerative sources;
- transport as an independent service with an original business model;
- storage as a strategic component of the supply chain;
- trading to achieve the optimum of offer and demand;
- suitable pricing.

In this, this new constellation of market actors will act on all three defined marketplaces, although not at the same time and to the same extent, but in an overarching manner. The mature onshore market as well as the developing onshore markets such as China, India and Brazil, and also the new offshore markets, are already seeing, or will see in the future, the appearance of new actors. The traditional view as expressed by BTM Consult that some markets are environmentally driven and others economically driven can no longer be maintained. This, too, is a symptom of global development and market exploitation. Scarce raw material resources, climate protection and nuclear risks are not evaluated internationally in the same way in their hazard potential, but together are the drivers of the worldwide growth in wind energy utilisation. The replacement of inefficient power stations or the first contraction of a power supply are at the same time a reason for a test of the meaningfulness of investment in wind energy.

A careful risk estimate is essential for any investment. This applies especially in the energy sector where the long-term investment and amortisation attract a high political risk. Great hopes are placed on solar energy as well as on wind energy. There is a worldwide consensus that the use of renewable energy should be encouraged and increased. This acceptance, as well as the decentralisation of power generation, therefore allows one to speak of the contribution of renewables to the democratic legitimisation of energy policy (see [9] p. 399).

A multitude of legal determinations and support vehicles are already available. Often, however, they only have a short-term effect or are bound to the annularity principle of public budgets (see, for instance, for the EU: [9], p. 231). In the context of this discussion it is therefore not possible to provide an overview with the promise of completeness.

Emphasis should be placed on the two formative models: the fixed price system as well as the system of quantity regulation (see in greater detail [18] pp. 12 and 288). This system is supplemented, for instance, by the trade in 'Green Certificates' or emission rights, or by tax incentives and regulations with regards to CO_2 content of a portfolio of an energy supplier. Especially as regards investment safety, the fixed price systems influenced by the current German Energy Law have been shown to be decisive. In addition, as determined by the courts, complaints by German energy suppliers in the EUGH failed. Rather, it is acknowledged that a higher than normal remuneration is appropriate as the emission of CO_2 is prevented. (On the part of the EU Commission a surcharge of 5 ct. is recognised; Hohmeyer's study on internalising the external costs of energy generation has achieved enormous practical relevance. For more detail see [9], p. 370; [19] p. 111; [20] p. 213.)

The methods of financing are as diverse as the sources. However, the project financing is predominant:

'Project financing is the financing of an intent, in which the loan provider focuses on the credit-worthiness testing on the cash flow of the project as the only source of financial means through which the credits are served.' (Nevitt, Fabozzi according to [18], p. 14)

The further process is described in the variants of BOT, BOOT, BLT (build, own, transfer; build, own, operate, transfer; build, lease, transfer). The relationship of external to own capital is risk-dependent. In developed onshore markets it is 80:20, and in the offshore area at least 30% own capital is required.

There is a large degree of plurality on the part of equity, as well as for outside capital providers. Practising farmers and closed funds, private and cooperative banks and savings banks, development institutions and development banks with long-running cheap interest rates move in this market with different country-specific shareholdings. Due to the ongoing crisis in the international financial markets, the proportion of public lenders has clearly increased in recent years. It ranges from the US Federal Financing Bank through the China Development Bank, the IFC or the IDB and ADB up to the EIB, the ERBD and the KfW. The UK has recently created a state-owned 'green investment bank' (for the role of development banks, see the details in the *Global Wind Report* p. 4).

Box 2.2   Plea for a New Agreement

The political attempts for a follow-up agreement to the Kyoto Protocol are well known. As much as expert opinion ever more increasingly points to the hazards of climate change, with representatives of the IEA such as Fatih Birol pointing to the small time-window that is available for changing direction, little is accelerating the international process for understanding the global climate change targets. The importance of energy generation and use is too comprehensive, the economic situation or the national per-head consumption as well as the CO_2 emissions released by them, are too dissimilar for a quick agreement to be expected. Lack of courage and perspective have set in, not only with environmental activists or progressive government representatives, but especially by the peoples such as the Pacific island states or Bangladesh who are hit particularly hard by accelerating global warming and the associated rise in sea-level.

As a positive reference, the global problem of the ozone hole should be remembered, caused by the emission of FCKW. This led in a relatively short time to an internationally binding prohibition. The geostrategic environmental hazard was stopped. Replacement materials were at hand. Damage to the growth of the developing as well as developed national economies was prevented.

Replacements for coal, oil, uranium and gas are also available: they are the renewables, the force of the wind, the energy of the sun, as well as biomass and hydropower. Solutions for further causes of the climate problem, such as deforestation of rain forests or mass animal husbandry, must naturally also be developed.

My proposal: Instead of setting agreements based on prohibition or reduction, an international agreement should provide and assure global perspectives for investment in renewables. This would be a new type of agreement: a 'renewable generation contract'. There is also a wider trend to which the international community needs to find answers: demographic change is not only a European problem but also a challenge for Japan and for China with its 'one-child' policy. Investment in wind and solar energy today ensures that the higher up-front costs can be shouldered by today's well-earning generation, thus avoiding the volatility risk of rising energy prices and allowing for higher social costs. In this way a positive social balance by investment in wind energy is created.

2.5   International Perspectives

The potential for wind energy has been described in many scenarios: for Germany, the European Union and in a global perspective. The year 2050 is seen as the target for 100% renewable energy; intermediate targets are to be reached in the years 2020 and 2030.

At the start, two key words that play an important role in the international debate should be mentioned. One is the argument about 'local content' and the other is the question of 'good governance'. Local content, as will be shown, is always a subject for local acceptance and support for political decisions. However, there is no agreement on what is to be understood by that. In the political arena, the focus is very much of the creation of jobs in industrial production (towers, generators, drives, blades). Economically, the entire value chain has to be taken

Table 2.3   Top ten wind energy companies according to BTM Consult

Company	Market share
Vestas	14.8%
Sinovel	11.1%
GE Wind	9.6%
Goldwind	9.5%
Enercon	7.2%
Suzlon Group	6.9%
Donfang	6.7%
Gamesa	6.6%
Siemens	5.9%
United Power	4.2%

into account. The building of the plant is always local, as are the follow-up services that are indispensable for operation.

Incentive systems such as in the Canadian province of Ontario, which demand a certain 'local content', or the credit conditions of the Brazilian BNSDE that link their loans to 'local content', lead to complaints from manufacturers to the World Trade Organization (WTO), and have already led to overcapacities with negative economic consequences for manufacturing companies in the industry. There is a need here for international agreement in order to prevent stranded investment, to the disadvantage of the industry and thus the whole development.

The internationally active companies are organised in very different ways. With wind companies (see Table 2.3 for global market share), some are departments of multinational companies (Siemens, GE), some have a long stand-alone tradition (Vestas and Enercon) or have a high assembled content (Repower, now belonging 100% to the Suzlon company), whereas others are state-owned and have a high local market share (such as the Chinese companies Sinovel, Goldwind); the supplier industry is also differently structured. The rapid growth of the industry continually makes new demands upon them: new countries and diverse climates like desert sand and ice storms, onshore and offshore technologies, and large, medium-sized and small turbines must be handled at the same time. Sufficient equity and good credit for raising debt capital are prerequisites for being able to master this growth. The consolidation process has not yet ended.

'Good governance' (i.e. the reliable implementation of regulatory framework conditions in good administration practice) is the political foundation for continual expansion in other countries. This is not only because the energy markets – as mentioned above – have always been politically driven. A reliable management culture is also key to the implementation of investments in long-term economic goods and their unhindered operation. The attractiveness of the investment depends directly upon this. The guarantee of suitable public services is a relevant part of the risk assessment in this regard.

Reports present the wind regimes and thus the potential profits of a project, while 'due diligence' evaluates technical risks, but 'good governance' is difficult to evaluate. Only a comparison between the states that have a legal instrument similar to the German Energy Supply Law, and that have the installed plants, show the importance of good administrative practice (see also [18], p. 12). Although there are rating instruments for the creditworthiness of states, these do not go into the required depth and width in order to deal with the problem of 'good governance'. Here too there is scope for action at the international level.

Box 2.3   On the Special Importance of Denmark and Schleswig-Holstein for International Development

The first political objectives for the expansion of wind energy were made in Denmark and Schleswig-Holstein almost at the same time, at the start of the 1990s, and not simply by chance (see [21–23], [24], p. 246, as well as the yearbook *Ökologie* 1992, p. 352, 'Windenergie in Schleswig-Holstein'). Linked with a clear strategy based on the 'three columns' – saving energy, energy efficiency and expansion of renewables – they were formulated by a cabinet decision as an objective for the time period to 2010 of 20–25% electricity demand coverage from wind energy. Climate protection, as well as the phasing out of nuclear energy, were decisive mutual elements of this energy policy. Mutual agreements on knowledge transfer and cooperation were signed by energy ministers. The Danish slogan 'Atomkraft – Nej Tak' ['Nuclear power – no thanks'] was seen around the world. From Copenhagen, the Swedish nuclear power plant Barsebaek was clearly visible [25]. In Schleswig-Holstein the nuclear power plant at Brokdorf became the symbol of the anti-nuclear power plant movement [3]. The targets and policies were seen as extremely ambitious, but they were very successful in their implementation. In both countries the aims were achieved long before the advised times, namely seven to eight years earlier. Besides the clear formulation of political will, it was crucial that there was a clear regulatory framework. Both countries followed fixed price systems. Key for investors, as well as for local acceptance, was a long-term planning strategy.

Thus, in Schleswig-Holstein new procedures were established to define exclusive areas for the installation of wind farms. The limiting to 1% of the state's surface area had the effect of cementing acceptance. Consistent financing conditions by linking regional credit institutes as well as special credit lines, and the development of local ownership in the design of citizen wind farms, had the effect of steady growth. The regional and local value-added effects due to income, jobs and taxes became clearly visible. A wind economy based on medium-sized companies also partly developed due to cooperation between the maritime and agriculture sectors (for mutual development see [22]).

A syndicate provided the Schleswig-Holstein Chamber of Agriculture with data (generated power, average wind speeds per site, farm or turbine) from 1990. In this way a wealth of knowledge was produced over a uniquely long period of time. Nowhere else in the world are operating results of plants and sites compiled, processed and published in such a detailed and transparent manner. For investors, this is a source of information of particular quality (see also [22] p. 41). The finance of repowering thus has been made much easier. As the assessment of the wind potential was key to setting the very first target of its kind in Germany, it is no surprise that the first overall wind maps were made available there (see history of the wind test Kaiser-Wilhelm-Koog GmbH).

The continuity and bipartisanship of the Danish energy policy is remarkable. Presently this is personified by the EU Commissioner for climate protection Connie Hedegard. Almost in the same cycle the governments in Copenhagen and Kiel – with broad parliamentary support – have expanded their targets in 2011. For Denmark the aim is 50% and for Schleswig-Holstein 100%, and both for the year 2020. Both countries, Germany and Denmark, are of great international importance for knowledge transfer. Academic education and research institutes of vital importance are those from Risoe and BZEE (leaders in primary, further and advanced training; see among others the Windskill project). Husum also includes the most important wind fair anchored there, and its long background of experience is a worldwide attraction.

2.6   Expansion into Selected Countries

This special development should be placed in the context of the general and global expansion of wind energy. This phase has been analysed by Beurskens from a technological point of view and celebrated as the age of modern wind power utilisation. The US, and especially California, were pioneers of this modern trend. Still today – unfortunately – the plants of the first boom years can be seen in the desert near San Francisco, more standing than turning, written off in every sense of the word. Because the 'bust' followed the boom, tax incentives were cancelled and thus further development was abruptly stopped. Wind companies such as Vestas went into insolvency from which, nevertheless, they restructured and emerged strengthened by the progressive Danish insolvency laws. US Windpower, a dominant player, finally had to retire from the market. Thus it is all the more remarkable that, despite this negative experience with tax-driven systems, the US is still a leader.

There were also very early experiments in the use of wind energy in China. In the 1980s Germanische Lloyd – today GL GarradHassan – with the support of the BMFT, operated a test site in Inner Mongolia (see also [22] p. 130). German and Danish wind pioneers made their way to India in order to make a contribution to rural electrification. By means of programmes by German Development Aid, the 50 kW two-blade turbine of MAN, the Aeroman, was supplied to many developing countries (see the depiction of these beginnings in [22]). Yet these attempts were always selective or were seen as niches. They brought about no substantial changes. The breakthrough in China (see below) was only achieved through self-manufacturing and targets in the 5-year plan.

The UK was also a pioneer in the field of research and development. There was a test site near Glasgow until the 1990s. Smaller companies undertook the manufacture and operation of plants. Special mention here should be made of the development and spread of mobile small plants, which, for instance, were used by nomadic people in the Asian steppes. The concentration on the development of the financial centre of London and the associated trend to deindustrialisation stopped these attempts. Various policies for stimulating the local use of onshore wind failed. The best prices could be obtained near the coast on the highest points and these caused intensive resistance especially from influential land-owners. Worthy of special mention is the contribution of British scientists and entrepreneurs in the founding and development of international associations such as the EWEA.

A very special path in the development of wind energy was also taken by Germany. At the start the research and development (R & D) programmes stood in the foreground. The GROWIAN was meant to overtake the Danish development rather than catching up with them. This two-bladed turbine with 100 m rotor diameter was erected in Kaiser-Wilhelm-Koog as the world's biggest plant. It was designed by the engineers of the DLR in Stuttgart; aircraft manufacture and aviation played a dominating role in the development. This very expensive experiment failed technically but provided the industry with much important knowledge for further development [22]. This was followed by the 100 MW programme. Eligibility was based on the three criteria of diversity of turbines, diversity of sites, and diversity of operators. A bonus of the 8 Pfennigs (about 4ct.) for each generated kWh was granted in addition to what the utility would pay to the independent power producer. The initiative was declared to be a R & D measure in the sense of a wide test (in order to bypass a notification at the EU Commission). Actually its purpose was to allow German industry to catch up with the Danish, world-market leaders of the time who were successful because of a very pragmatic approach of step-by-step upscaling of the turbines (making good use of numerous components made in Germany; experts estimated a value addition of 50%, see [22]). The result of this programme was the installation of numerous wind turbines of different technologies: from single-bladers to vertical axis rotors, from 'bush and tree' concepts (one large turbine surrounded by various smaller ones) up to citizen wind farms. Big industry players participated – but just for the fun of it with no commercial interest. But it kick-started a medium-sized industry as many engineers left the big companies when they realised that their expertise was no longer in demand after the programme came to an end. However, the breakthrough for the whole industry came with the Electricity Feed-in law in which access to the grid was guaranteed and a fixed remuneration was assured.

This law, which was initiated by a small group of members of the Bundestag across the political spectrum, introduced the system change upon which the present renewable energy law is based. It has taken up a commanding position in the international energy-political debate (see also [8] p. 536). Also external programmes linked to foreign aid such as El Dorado and Terna should be mentioned. However, these had rather negative effects, because although installations of turbines for developing countries were financed (almost 100%), there was hardly any accompanying programme for education or attempts to link local structures. As such, due to lack of maintenance, these wind turbines were often abandoned a short time later.

The third European activist for wind energy is Spain. Here it was clear industrial policy that stimulated the industry. This policy dealt with the creation of new jobs in the country's regions. Manufacturing companies sprang up. In contrast to Germany, there were hardly any 'independent power producers' as investors, but utilities, especially Iberdrola. Today this company is one of the world's leading investors, with excellent experience of wind farm monitoring and grid integration.

Independent development took place in Brazil, shaped by strong growth in the economy, political stability and excellent energy resources (from hydropower via oil and gas reserves and biomass) as well as wind regimes. Whilst in a first introductory phase there was a law similar to the Energy Feed-in law in Germany, with a legal and support regime (Proinfa), it was replaced by a strictly competitive tendering process in 2010. Meanwhile two tenders have been issued for a total capacity extension of approximately 7 GW by 2015. The prices, which have been made public, are very low in comparison, caused by the high capacity factors: a challenging development.

2.7   The Role of the EU

The development of policies related to energy and renewable energy in the EU has a long track record. The creation of a common market for coal and steel, as well as agriculture, was in the foreground. Some contracts for nuclear usage were closed. From the six founding states, France, Germany, Italy and the Benelux countries, the EU quickly became 27 countries with the fall of the Iron Curtain. The Eurozone has 17 member countries. Because of the crises in finance and the economy, this community of states is undergoing a historic testing period. It is to be hoped that with the mobilisation of mutual strengths, these weaknesses will be overcome.

Table 2.4 National Renewable Energy Action Plans

	NREAP on	NREAP off	NREAP total
Belgium	2320	2000	4320
Bulgaria	1256	–	1256
Denmark	2621	1339	3960
Germany	35,750	10,000	45,750
Estonia	400	250	650
Finland	2500	–	2500
France	19,000	6000	25,000
Greece	7200	300	7500
Great Britain	14,890	12,990	27,880
Ireland	4094	555	4649
Italy	12,000	680	12,680
Latvia	236	180	416
Lithuania	500	–	500
Luxembourg	131	–	131
Malta	145	95	109
Netherlands	6000	5178	11,178
Austria	2578	–	2578
Poland	6150	500	6650
Portugal	6800	75	6875
Romania	4000	–	4000
Sweden	4365	182	4547
Slovakia	350	–	350
Slovenia	106	–	106
Spain	35,000	3000	38,000
Czech Republic	4365	182	4547
Hungary	750	–	750
Cyprus	300	–	300
EU	170,054	43,324	213,378

These strengths are:

- great diversity;
- good experience in dealing with decentralisation;
- the integration of different cultures;
- diverse experience in change management, whether by means of industrial change…
- or by means of system change after the failure of state communism.

In the 1980s the EU commission began to support wind energy within the framework of its research programmes as well as in European conferences (from which, among others, the EWEA was founded; see below for the history of its founding). This decade-long tradition was transformed in 2005 into the Wind Technology Platform (TP Wind, see [9]), within which all research activities were coordinated with the aid of the EU via EWEA. In a second phase, the first Green Book came into existence in the 1990s, mostly organised by Arthouros Zervos who came as a specialist from Athens University and worked in the General Directorate of Research. He has been the President of EWEA for many years as well as the EREC (European Renewable Energy Council) and founded the GWEC (see below for the story of its founding).

A further development step was achieved by the 20-20-20 target. This aim was first formulated at the EREC conference in Berlin in 2005, then taken up by the EU Commission and the European Parliament and finally decided on by the Council of Ministers: 20% energy supply from renewable sources in 2020, connected with a 20% reduction in CO_2 emissions. There were long arguments on the obligations to the target between Commission and Parliament as well as the Council. The result of this dispute was a new instrument. Each member state was required to present a National Renewable Energy Action Plan by the autumn of 2010 (see Table 2.4) with information on how, at a national level, this common aim was to be achieved. (Almost) all member states have presented their plans on time, so that for the first time in European history a detailed and verifiable framework for the expansion of renewable energy has become available. In the sector of power generation, the wind is the most important driver for the change to a CO_2-free demand coverage both onshore and offshore (see below).

The next development step was presented to the public in December 2011 by the Energy Commissioner Oettinger: a 2050 roadmap. In this document various scenarios were presented: both further expansion of nuclear energy as well as the extensive use of the now only laboratory-proven CCS technology is considered. Besides this, it includes a scenario of increased expansion of renewable electricity generation. For the year 2050 this is seen as feasible. Of this, a dominant portion comes from the addition of wind turbines. This new map is now up for public criticism and in the trilateral agreement process between Commission, Parliament and Council, one or more changes are expected.

2.8   International Institutions and Organisations

Besides the numerous lobby and interest groups, international communication is placing greater importance upon questions of energy in general, the role and potential of renewables and especially upon two institutions:

* the International Energy Agency (IEA) and;
* the International Renewable Energy Agency (IRENA).

In Vienna there is also the seat of the International Atomic Energy Organization (IAEO) as a specialised global institution.

The founding of the IEA in 1973 was an immediate and prompt reaction to the oil crisis of that year. Sixteen industrial nations took the initiative. This institution was then situated in Paris in 1974 as an autonomous unit of the OECD. Since 2011 the Executive Director has been the previous Dutch Foreign minister Maria Van der Hoeven, and the already-mentioned Fatih Birol is the Chief Economist. Despite regular and sometimes heavy criticism of estimates and forecasts, the annually published *World Energy Outlook* (last issue in November 2012) is the most influential global publication for the industry and thus also for the renewables sector (see below for the scenarios). Despite the emphasis on the conventional energy economy by the industrial countries, the estimate in past years of the potential of wind energy has become more important (for example see [26]).

One reason for this was the founding of the IRENA: a positive competition has taken place. The story of the development of the IRENA is lengthier than that of the IEA. The first proposals were found in the 'Brandt report' (named after the former German Chancellor Willy Brandt who led the UN-initiated North–South Conference). At the following UN conference in Nairobi in 1981, this proposal entered the closing document. A tireless fighter and proponent for this idea was Hermann Scheer, founder and long-time chairman of Eurosolar who, in his role as SPD Member of Parliament, was also successful in having the proposal taken up after the formation of the first Red-Green coalition agreement. In Bonn in 2004 the proposal for the first International Conference for Renewable Energies was emphasised in the closing resolution. This was followed by a preparatory conference in Berlin in April 2008, and by the founding conference in January 2009 in Bonn. The statutes came into force in July 2010: from the original 75 it increased to 86 signatory states as well as the EU; in total, 148 states have now signed as well as the EU. The seat of the institution is Abu Dhabi, and the General Secretary is currently the Kenyan Adnan Amin, an experienced UN diplomat. Three subjects are defined as tasks for the working programme:

- knowledge management and technological cooperation;
- political consulting and capacity expansion;
- innovation and technology.

At the general meeting traditionally held in January, many state and government chiefs including the UN General Secretary gave positive reports on the expansion potential of renewables, as did the Chinese head of government who provided remarkable expansion figures and targets for his country in wind energy.

REN 21:   The official title of this network based in Paris is 'Renewable Energy Policy Network for the 21st Century'. It is carried by the German GIZ and by the UNEP programme and supported by the IEA. Governments, NGOs and international institutions are participants. The chairman is Mohamed El-Ashry; Christine Linz is the General Secretary. Inspired by the above-mentioned Bonn conference of 2005, this network among others is mandated to service international voluntary agreements. The *Global Status Report of Renewable Energies* issued annually in June serves as the standard work of the industry.

WindMade: Windmade, the first global user label for wind power, was started on the initiative of the Danish wind turbine manufacturer VESTAS. The independent foundation with its seat in Brussels sees itself as a communication platform. Companies and organisations as well as products can apply to be taken up under this seal of quality. The WWF and the GWEC support this initiative, among whose founder members are Bloomberg New Energy Finance, LEGO and the Deutsche Bank.

Global Wind Day: Since 2007, 15 June has been Global Wind Day. The EWEA took this initiative, which is organised at present with the GWEC. In 2012 more than 40 countries on five continents took part in this event. The aim is to bring the importance of wind power to public attention. The forms of this public relations event are numerous: photographic competitions, viewings, climbs, kite flying and drawing competitions. The number of countries participating has grown continuously. In this way the international reach is becoming visible.

EWEA, AWEA, GWEC and Co.: EWEA is the oldest and largest association in the wind energy industry. Founded in 1982, this association has more than 700 members in more than 60 countries. It is a characteristic of the association that the membership includes all the actors of the industry: manufacturers as well as suppliers, investors, project developers, almost all European associations and the most important research institutes. This 'broad church' differentiates it from other lobby groups and interest associations and, due to its membership structure, possesses specific knowledge. Besides *Wind Directions*, it regularly publishes background reports such as *Wind in Our Sails*, *EU Energy Policy to 2050* or *Pure Power – Wind Energy Targets for 2020 and 2030*. The most important Europe-wide events are the conferences, whether it is the EWEA Conference or the Offshore Conference. The EWEA has a special responsibility for the EU Commission in taking over the management of the 'technology platform', within which the future technological development, and aid of the research framework programmes of the EU, are discussed and defined. The focus of the political work is oriented mostly towards four subject areas:

- the creation of a real internal market for energy;
- the establishment of a stable regulatory framework;
- the achievement of a 'level playing field' for wind power use;
- the continuous increase in wind power capacity and use.

AWEA, with its seat in Washington, is the 2500-member-strong association of the interests of the US wind industry, and as such is an important driver for centralised and decentralised policies. Publications and events enjoy great demand and interest.

CREIA stands for the Chinese Renewable Industry Association, IWTA for the Indian Wind Turbine Manufacturers Association, BWE for the German Association of Wind Energy, and DWIA for the Danish Wind Industry Association: all four associations are very influential in their countries. Both the German association and its Danish equivalent have performed pioneering work in their own countries and beyond. There are further associations in both countries, some with a long history as regards technical standards, the FGW (Society for the Promotion of Wind Energy), or some purely industrial associations such as the VDMA (Association of German Manufacturers) or the Danish Association of Plant Operators. Thus, once again the special importance of the EWEA becomes clear: only this association includes all the actors of the industry. Also increasing in influence and importance are the associations in Brazil, Mexico, Canada, Japan and Korea.

These organisations are all members of the Global Wind Energy Council (GWEC). This was founded in 2006 as an international umbrella organisation and is situated in the Brussels Wind Power House with representatives in Washington, DC, and Peking. The General Secretary since its founding is Steve Sawyer, formerly of Greenpeace. The members of the GWEC are listed on www.gwec.net. Besides national and continental associations, all important industrial companies and further actors in the value-added chain are members.

2.8.1 Scenarios

Because of the long-term nature of the investment cycles and because of the strategic importance of energy – especially electricity – for national economies, scenarios traditionally are of
great importance. Besides the prognosis of price developments, the availability of raw materials (see, among others, the debate on 'peak oil') and the strategic allocation of resources play
an important role. Mention has already been made of the publications of the EU Commission.
The vehement controversy on the utilisation of nuclear energy in Germany, with the decision
to withdraw from it by the government of Schröder/Fischer, the revision of this decision by the
Merkel government and the revision of this revision by the present Federal Government after
the disaster of Fukushima resulting from the underwater earthquake and tsunami, have resulted
in a particularly intensive preoccupation with scenarios. Mention should be made here of the
SRU (Advisory Council on the Environment) which considered an energy supply based solely
on renewable energy by 2050 as possible and feasible. Also the IPCC has presented a
comprehensive 1000-page study in which 164 scenarios were evaluated. Representatives of
194 countries came to a consensus; 80% is to be the portion of renewables in the energy supply by 2050.

In order to bring the voice of the wind energy sector into concert with the scenario propos-
ers, the EWEA went on the offensive in 2000 and published, together with Greenpeace, the
first *Windforce 10* report that was continued in the following years as *Windforce 12*, and was
replaced after its founding by the *Global Wind Energy Outlook* of the GWEC (Global Wind
Energy Council).

In its scenarios, *Global Wind Energy Outlook* investigates the wind energy potential for the
periods 2020, 2030 and 2050. It is published in cooperation with Greenpeace and the DLR
(Deutsche Anstalt für Luft- und Raumfahrt [German Aerospace Centre], which has been
active since the pioneering days of wind energy [22]). Three scenarios are discussed:

- The assumption of the *World Energy Outlook* of the IEA (extrapolated by the DLR to the
  year 2050); this is seen as the conservative scenario.
- In a second scenario – with reference to the already developed support instruments – it is
  assumed that they will be implemented and that national expansion targets have been
  achieved; this is the moderate scenario.
- In the third it is assumed that the most advanced expansion targets have been realised; this
  is the progressive scenario (see Table 2.5).

Table 2.5 Prospective market development: global (information in MW)

Year	Conservative	Moderate	Progressive
2007	93,864	93,864	93,864
2008	120,297	120,297	120,297
2009	158,505	158,505	158,505
2010	185,258	198,717	201,657
2015	295,783	460,364	533,233
2020	415,433	832,251	1,071,415
2030	572,733	1,777,550	2,341,984

2.9 *Global Wind Energy Outlook 2012* – The Global View into the Future

Looking back it can be established that national targets as a rule have been exceeded or reached before their due date. This applied to the initial targets as they were set for Denmark and Schleswig-Holstein (see above). It also applied to all the forecasts of the EWEA for the EU member states, but only for the development of the onshore market. This was, and is, decisive but is also to be viewed differently today. A uniform development as was seen in the 1990s and the first half of the 2000s no longer applies.

As mentioned above, three major trends can be differentiated:

1. The growth of the mature or maturing onshore markets, In this, exceeding the 1000 MW mark is to be defined as an important step that moves the wind power market out of the niche and into the 'mainstream'.
2. The specific framework conditions for the development of offshore markets. Not only are new paths trodden here, but planning this requires adapted technology. The financial factors are decisive: the costs per erected plant are a factor of at least 2.5 higher than an onshore wind turbine, and the respective finance required is of the order of three-figure millions, so that the consequences of the financial market crisis are clearly noticeable here.
3. To meet the demand of the undersupplied national economies and peoples, that is, the fifth of humanity – 1.4 billion – without access to power because of their poverty levels, who can make no financial contribution, there is a need for instruments such as micro-credits for small (wind) plants and publicly financed projects (after careful evaluation of the programmes already carried out for rural electrification).

The global estimates of the growth in capacity in *Global Wind Energy Outlook* are based on historical data. Current trends are taken into account as important market developments. The IEA scenario for 2010, for instance, assumed a reduction of the installed capacity to only 26.8 GW, but in fact more than 38 GW were installed. This dimension verifies the moderate scenario. It assumes an annual added construction of 40 GW that could be increased to 63 GW from 2015.

2.9.1 *Development of the Market in Selected Countries*

2.9.1.1 China

The development in China is especially remarkable. After three years in succession of having the leading position in the annual market by a wide margin, in 2012 the People's Republic ceded the No. 1 position to the US. 'Only' 12.96 GW were erected in 2012, a decrease of 26%. Nonetheless, China still maintains its position as a global wind leader in cumulative terms with a total of 75.32 GW. Meanwhile, four Chinese manufacturers are among the worldwide top ten (see also Table 2.6). The strong local market serves as a base for global market exploitation, with the focus on Africa and Latin America. Considerable effort has been used to develop relations between China and Brazil. A great advantage in this is that there is no shortage of liquidity even for large projects. The China Development Bank accompanies the manu-

Table 2.6   Prospective market development in China (information in MW)

Year	Conservative	Moderate	Ambitious
2009	44,733	44,733	44,733
2010	62,364	62,364	62,364
2015	114,001	125,835	134,687
2020	179,498	214,445	230,912
2030	279,017	400,130	499,614

Table 2.7   Prospective market development in the US (information in MW)

Year	Conservative	Moderate	Ambitious
2009	45,085	49,329	49,648
2015	75,585	119,190	140,440
2020	106,085	220,041	303,328
2030	141,085	410,971	693,958

facturer on its path towards internationalisation. It is also an advantage that clear expansion planning has been allowed for in the 12th five-year plan. By 2015 the installed capacity is meant to be 100 GW. There is also a statement on the 'magic' year 2050. By then, 17% of electrical power in China should come from wind (see *Windlog*, January 2012).

2.9.1.2   USA

Further development of the second strongest market is marked by great uncertainty. The US is constantly involved in a pre-election battle. The future of PTC is very controversial, and shale gas is changing the American energy market dramatically. Despite the cheap gas price, and the power of the conventional energy lobby, the interests of the wind sector are significant (Table 2.7), but states in the Mid-West are not linked into the grid, and nobody can make an accurate prognosis. Present expansion is determined by pull-forward effects. With newly installed capacities of 13 GW, 2012 was a record year, and the US passed the 50 and 60 GW milestones in one year. Despite this, climate protection is an election-losing subject.

2.9.1.3   Canada

The North American neighbour to the US, Canada, has made progress too. In 2012 more than 6200 MW were installed. Although expansion has taken place in all provinces, Ontario (at more than 2000 MW) is far ahead. Quebec wind farms made up about 46% of the total new installations in 2012.

Table 2.8 Annual newly installed capacity in the EU (in MW)

Year	Onshore	Offshore
2001	4377	4
2002	5743	51
2003	5203	170
2004	5749	259
2005	6114	90
2006	7528	93
2007	8201	318
2008	7935	373
2009	9929	582
2010	8764	883
2011	8750	966
2012	10,729	1166

Table 2.9 Proportions for the EU countries (at end 2012)

Country	Capacity (GW)	Share
Germany	31.3	30%
Spain	22.8	21%
UK	8.4	8%
Italy	8.1	8%
France	7.6	7%
Portugal	4.5	4%
Denmark	4.2	4%
Sweden	3.7	4%
Poland	2.5	2%
Netherlands	2.4	2%
Romania	1.9	2%
Others	8.5	8%

2.9.1.4   The EU

There is uncertainty also in the European Union, especially with respect to the ambitious off-shore targets. The debt crisis has left its mark; the Spanish market has come to a stop. Grid expansion plans are announced but await implementation. However, the climate protection targets are clear, and the NREAP exists; small dips in development must not overshadow the general trend. More than 11 GW of new capacity was added in the EU in 2012. This was an increase of 23% compared with the previous year. Around 10% of these new installations were located offshore (Tables 2.8 and 2.9).

Table 2.10 Prospective market development in Central and South America (information in MW)

Year	Conservative	Moderate	Ambitious
2010	1489	1489	1489
2011	2330	2330	2330
2015	2330	9296	15,303
2020	6241	21,903	47,970
2030	13,868	56,075	134,411

Table 2.11 Prospective market development: India (information in MW)

Year	Conservative	Moderate	Ambitious
2010	13,065	13,065	13,065
2011	16,084	16,084	16,084
2015	23,784	31,499	37,436
2020	32,933	59,351	89,299
2030	66,400	124,826	191,711

2.9.1.5 Central and South America

One can also view the Latin America market as positive. In 2012 Brazil installed slightly more than 1 GW of new capacity, and has almost 7 GW in the pipeline to be completed by 2016. A total of 3.5 GW has been installed on this continent at the end of 2012 (Table 2.10). About 16.5 GW of installed capacity is expected by the end of 2017. Chile has an enormous need for electricity, but as yet there is no regulatory framework. Argentina has excellent potential but the south of the country is not linked to the grid. These are the two main reasons why this part of the globe remains far below its capabilities.

2.9.1.6 India

This applies in a comparable sense also to India (Table 2.11). Lack of power supply has long been understood as an important hindrance to growth. It was not least the industrialists who invested in the first wind farms in order to avoid power outages by means of energy generation. The targets defined up to now by the government have remained far behind the exploitable potential (see GWEC India Report). But the expansion of coal power is advancing massively: US$130 billion have flown into this industry in the past five years (*The Economist*, 21 January 2012). Although in 2011 the Indian wind industry experienced its strongest annual growth to date, 2012 was a slower year, with just over 2 GW of new installations. This is mainly due to a lapse in policy. It can be assumed that the government in Delhi will correct its targets upward (however, there are fears of changes in the tax system with negative effects for the future). The situation and potential are described comprehensively in *Indian Wind Energy Outlook 2011* (Global Wind Energy Council, with World Institute of Sustainable Energy and Indian Wind Turbine Manufacturers Association).

2.9.1.7 South Korea

There were 76 MW of newly installed capacity in South Korea in 2012, which summed to a total of 483 MW installed capacity. A Renewable Portfolio Standard was introduced in 2012. It is expected that this will expedite the development of new wind projects in the future. From their experience in steel and maritime construction, the national economy is beginning to see the potential for wind energy. This applies especially to the offshore industry. Here the emphasis will be in the country itself. Korean companies are starting to involve themselves internationally in this industry. The government and the economy are presenting scenarios that are similar to the procedures of the GWEC (moderate to ambitious). Target years have been defined. At least 1.9 GW are expected for 2015, and at most 7.4 GW; for 2020 the curve extends from 3 to 7.5 GW; for 2030 from 7.4 to 23 GW. The latter is the target of the industry. In this way 10% of the electricity demand would be achieved (see Korean Wind Energy Industry Association).

2.9.1.8 Africa

For a long time Africa has been a barely reachable and exploitable market. Although it is especially on this continent that a lack of power supply is evident, besides North Africa, specifically Morocco, Tunisia and Egypt, there have been hardly any attempts to use the abundant supply of wind energy. Not only is there a lack of a power grid or a legal base, but also the trade and industrial services are not available locally. For this reason it is an especially important signal that now South Africa, in two rounds of tenders, has given the green light for the construction of several wind farms. This country was successful in politically damming up the strong coal and uranium lobby in order to give renewables a chance. Because of the industrial strength of the country, positive spill-over into neighbouring states is to be expected. It has just been announced in Kenya that a 300 MW project is to be implemented there. Morocco has excellent resources not only because of its long Atlantic coast but also because of the necessity of expansion of electricity generation for its important tourist industry. Almost 300 MW has already been installed. A further expansion is the subject of a government programme. Egypt started very early with a pilot field and wind farm, and by 2011 achieved more than 550 MW. The political change that has become known as the Arab Spring has interrupted this development.

2.9.1.9 Turkey

Turkey is by far the fastest-growing national economy in the region. Thus there is a growing demand for energy, especially electricity. Also this state is a transit country for pipelines. The current energy mix is coal-based, but the country is beginning to see and realise the prospects

Table 2.12   Countries with more than 1 GW wind energy capacity

		Capacity in GW
Asia		
	China	75
	India	18.4
	Japan	2.6
Europe		
	Germany	31.3
	Spain	22.8
	UK	8.5
	Italy	8.1
	France	7.5
	Portugal	4.5
	Denmark	4.2
	Sweden	3.8
	Poland	2.5
	Netherlands	2.4
	Turkey	2.3
	Romania	1.9
	Greece	1.8
	Ireland	1.7
	Austria	1.4
	Belgium	1.1
North America		
	US	60
	Canada	6.2
Latin America		
	Brazil	2.5
	Mexico	1.4
Pacific		
	Australia	2.6

for using wind power. In 2012 Turkey added slightly more than 500 MW of newly installed capacity, for a total of 2312 MW. The installations are expected to reach 1000 MW per year from 2013 onwards.

2.9.1.10   Australia and New Zealand

Australia has driven wind capacity from approximately 71 MW in 2001 to 2584 MW at the end of 2012. Despite enormous coal reserves and the corresponding influence of this industry on the national energy policy of the country, the Australian government has decided upon ambitious climate protection targets. Thus, one can count on continued expansion. The states of South Australia and Western Australia are in the forefront in this. New Zealand is notching up an approximately 25% annual growth and has crossed the 600 MW divide. A continuing expansion is forecast.

2.10 Conclusion

*   The number of countries in which wind energy is used for the generation of electricity has grown steadily to 79. One can expect a further spread of this technology.
*   At the same time the proportion of countries in which the installed capacity has exceeded the critical mark of 1 GW has risen. In all, 24 countries have reached this milestone, 16 European and 8 others (see Table 2.12).
*   The formulation of targets is decisive for a politically driven energy market. There have never been so many clearly defined targets as in 2011 (EU, China, Korea, India).
*   With the start of the activities of IRENA, it can be assumed that the learning curve will become even steeper, and the process will accelerate. For the first time an international institution is concentrating on spreading the knowledge of renewables.
*   The number and the economic and political weight of the actors on the worldwide market of wind energy have reached an independent dynamic.
*   The cross-over technologies of energy generation and information processing open up new perspectives and accelerate the change.
*   The modern energy age has begun.

References

[1] Meadows, D.H. Meadows, D.L., Randers, J. and Behrens III, W.W. 1972. *The Limits to Growth. A report for the Club of Rome's project on the Predicament of Mankind*. Universe Books. New York.

[2] Altenburg, C. 2010. *Kernenergie und Politikberatung. Die Vermessung einer Kontroverse*. VS Verlag für Sozialwissenschaften. Wiesbaden.

[3] Aust, S. 1981. Brokdorf. Symbol einer politischen Wende. Hoffmann und Campe. Hamburg.

[4] Traube, K. *et al.* 1986. Nach dem Super-GAU. Tschernobyl und die Konsequenzen. Rowohlt. Reinbek bei Hamburg.

[5] Lyman, F. 1990. *The Greenhouse Trap: What We're Doing to the Atmosphere and How We Can Slow Global Warming*. World Resources Institute. Beacon Press. Boston.

[6] *Erster Bericht der Enquete-Kommission "Schutz der Erdat-mosphäre"*, 12. Wahlperiode. Deutscher Bundestag. 1991. Economica Verlag. Bonn.

[7] Stern, N. 2007. *The Economics of Climate Change*. Stern Review. Cambridge University Press. Cambridge, UK; New York.

[8] Yerkin, D. 2011. *The Quest. Energy, Security and the Remaking of the Modern World*. Penguin Books. New York.

[9] EWEA. 2009. *Wind Energy: The Facts. A Guide to the Technology, Economics and Future of Wind Power*. EWEA. Taylor & Francis Ltd. London.

[10] Odum, E.P. 1991. *Prinzipien der Ökologie – Lebensräume, Stoffkreisläufe, Wachstums-grenzen*. Spektrum der Wissenschaft. Heidelberg.

[11] Mann-Borghese, E. 1975. *Das Drama der Meere*.

[12] Runge, K. (ed.) 2002. *Coastal Energy Management. Integration Erneuerbarer Energieer-zeugung an der Küste*.

[13] Zierul, S. 2010. *Der Kampf um die Tiefsee. Wettlauf um die Rohstoffe der Erde*.

[14] Rave, K. and Erdgas, S. Breitband. 2010. Netzinfrastrukturen in Schleswig-Holstein im Wandel, in *Arbeitspapier 92* (eds Rave, K., Schlie, K. and Schliesky, U.). Lorenz-von-Stein-Institut. Kiel.

[15] Rave, K. 2001. Information technology plus energy technology. A new approach, in *World Market Series, Business Briefing*. World Bank. p.127–130. World Markets Research Centre. London.

[16] Rave, K. 2003. Grüner Strom statt Blackout. *Solarzeitalter*, Volume **15**, No. 4/2003.

[17] EWEA. 2010. *Powering Europe: Wind Energy and the Electricity Grid*. EWEA.

[18] Böttcher, J. (ed.) 2012. *Handbuch der Windenergie. Onshore-Projekte: Realisierung, Fi-nanzierung, Recht und Technik*. Oldenbourg Wissenschaftsverlag. München.

[19] Cairncross, F. 1991. *Costing the Earth: The Challenges for Governments, the Opportunities for Business.* Harvard Business School Press: Boston MA.

[20] Hohmeyer, O. and Ottinger, R. (eds) 1991. *External Environmental Costs of Electric Power.* Springer-Verlag: Berlin, New York.

[21] Rave, K. 1988. Programmarbeit – und sie bewegt doch!, in *Demokratische Geschichte, Jahrbuch zur Arbeiterbewegung und Demokratie in Schleswig-Holstein III*, p. 611. Neuer Malik Verlag Kiel, Kiel.

[22] Rave, K. and Richter, B. 2008. *Im Aufwind – Schleswig-Holsteins Beitrag zur Entwick-lung der Windenergie.* Wachholtz, Neumünster.

[23] Schleswig-Holsteinischer Landtag (ed.) 1993. Energieversorgung in Schleswig-Holstein bis zum Jahr 2010.

[24] Rave, K. 2010. The emergence of the wind economy in Germany, in *Sparking a World Wide Energy Revolution*, p. 264. AK Press, Oakland, CA.

[25] Rieder, S. 1998. *Regieren und Reagieren in der Energiepolitik. Die Strategien Dänemarks, Schleswig-Holsteins und der Schweiz im Vergleich.* Haupt Verlag: Bern.

[26] IEA. 2004. *Renewable Energy. Market and Policy Trends in IEA Countries.*

[27] GWEA. 2011. Global Wind Report – Annual Market Report 2011. Brussels.

3

Wind Resources, Site Assessment and Ecology

Hermann van Radecke

3.1 Introduction

This chapter deals with production and environmental aspects of wind energy. How the wind resource is determined using current methods and reanalysis data will be discussed first. This is followed by a simple calculation of the annual energy production of a wind turbine. It is shown how, in a wind farm, the opposite-side weakening is calculated. Then the emissions of the wind turbines are described and noise and shadow emissions will be discussed in detail. Two programs with which wind farms are planned are mentioned. The technical guidelines of the FGW and the IEC 61400 that prescribe the measuring procedures and other processes in connection with wind turbines are mentioned. The environmental influences are regulated by means of the Federal Emission Law (BImSchG), and the approval procedure based on it is sketched out. Exercises and their solutions are given.

3.2 Wind Resources

Wind resources, their calculations and the resulting energy production are presented in the following.

3.2.1 Global Wind Systems and Ground Roughness

The main direction of the wind in Europe is southwest by west. This is a result of the global wind system. The sun and the slant of the Earth's surface relative to the sun, described with

Understanding Wind Power Technology: Theory, Deployment and Optimisation, First Edition.
Edited by Alois Schaffarczyk. Translated by Gunther Roth.
© 2014 John Wiley & Sons, Ltd. Published 2014 by John Wiley & Sons, Ltd.

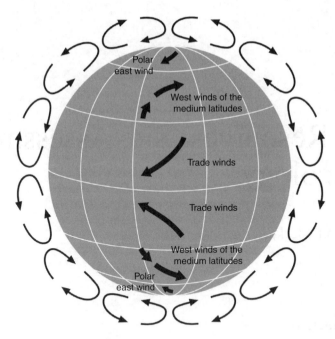

Figure 3.1 Global wind system

the cosine of the geographical width plus the declination of the sun, drive the system by means of differential warming. At the Equator it is warm, the air density is low, the air rises, and a low-pressure area develops. At the poles it is cool, the air sinks, and a high-pressure area exists. Differences in air pressure generate flows of air at great heights from the equator to the North Pole and the same to the South Pole. On the ground the air flows back in the opposite direction. The air circuit is closed. The rotation of the Earth generates the Coriolis force. In the northern hemisphere this generates a deviation to the right. At the latitude of the Azores the deviation is so great that the air, high up, no longer continues on its way to the North but comes to an aerodynamic short circuit. The air sinks and the high-pressure belt of the Azores is created. Because there is a high at the pole, there must be a low-pressure belt between these two high pressure regions; these are the polar low pressure regions. The result of this global wind system is that at the ground north and south of the equator the trade winds flow from the south, and Europe is situated in the region of the west wind drift. Thus, the main direction of the wind in Europe is southwest to west (see Figure 3.1). The global wind system is overlaid by local and regional high- and low-pressure regions that move through, or are substantially stable in time and space. In Schleswig-Holstein this leads to a secondary prevaling of wind direction from the east.

The ground roughness is the unevenness of the Earth's surface (see also Section 3.2.2). It retards the wind velocity. Despite waves, the roughness over the ocean is very small, which leads to the wind velocity over the ocean being high. Therefore, all coastal regions of the North and Baltic seas are subject to high wind velocities. Further inland, the wind velocity decreases with increasing distance from the coast. At greater heights in the Alps, the wind velocity is greater, which is the result of the height profile. Between mountains there are also greater wind velocities in connection with land-sea winds. In this the mountains limit the air flow horizontally, as in a nozzle. An example of this is the Mistral in the south of France between the Alps and the Pyrenees. Valley winds from heights can generate local winds driven by vertical temperature differences (at the top it is colder than down below). An example of this is the Bora fall wind in the coastal strip of Croatia.

Graphic depictions of the description can be found in the *European Wind Atlas* [1], in the German translation [2], and comprehensively on the home-page of the *European Wind Atlas* [3].

3.2.2 Topography and Roughness Length

The wind has a height profile. Because of ground adhesion, the wind velocity directly on the ground must be zero, whilst at great heights the velocity is fully formed. The wind at great height, at which ground friction has no effect, is called the geostrophic wind.

In between this, a boundary layer is formed in which the horizontal velocity v is dependent on the height h, $v = v(h)$. This function is called the height profile. Winds are turbulent, as can be seen immediately with the Reynolds number, which again is linked to the low kinetic viscosity of the air at $15 \times 10^{-6} \, \text{m}^2/\text{s}$. Two equations are used for the description of the profile: the exponential wind profile according to Hellmann, and the logarithmic wind profile.

According to Hellman two velocities at two heights have the following connection:

$$\frac{v_2}{v_1} = \left(\frac{h_2}{h_1}\right)^{\alpha}$$

(3.1)

where v_2 is the wind velocity (m/s) at a height of h_2 (m) above the ground, v_1 is the wind velocity (m/s) at a height of h_1 (m) above the ground, and α is the Hellmann height exponent. v_1 is often the measured reference velocity v_{ref} and the measured height h_1 is the associated reference height h_{ref}.

The universal logarithmic wall law by Prandtl [4] applies to turbulent boundary layers on a small and large scale, thus in turbine spades, in flow channels and in the atmosphere. It is described by the following equation:

$$v(h) = \frac{u_\tau}{\kappa} \ln\left(\frac{h}{z_0}\right)$$

(3.2)

The term u_τ is the nominal shear stress velocity and is hard to define. In the FGW guidelines [5] and in the IEC 61400 [6] the term u_τ/κ is replaced by the wind velocity v_{ref} measured at a height of h_{ref}. The logarithmic height profile thus takes on the easily used form:

$$v(h) = \frac{v_{ref}}{\ln\left(\dfrac{h_{ref}}{z_0}\right)} \ln\left(\frac{h}{z_0}\right)$$

(3.3)

$v(h)$ is the wind velocity (m/s) at height h, H is the height above ground-level (m), z_0 is the ground roughness (m), v_{ref} is the wind velocity measured at height h_{ref} (m), and ln is the natural logarithm. In some depictions of Equation (3.3) one can find the logarithm to base ten, or log, in place of the natural logarithm ln, which is identical as the logarithm is in both the numerator and the denominator and shortens the conversion factor.

Both equations (3.1) and (3.3) describe a very similar profile. There is an interconnection between the Hellmann height exponent α and the ground roughness z_0. This is not uniform and is dependent upon the height of the reference measurement. As an approximation $\alpha = \alpha(z_0)$, can be given, for instance, according to Equation (3.4) [7] or Equation (3.5) [8]:

$$\alpha(z_0) = \frac{1}{\ln\left(\dfrac{15.25\,\text{m}}{z_0}\right)} \tag{3.4}$$

$$\alpha(z_0) = 0.096 \, \log_{10}\left(\frac{z_0}{1\text{m}}\right) + 0.016\left(\log_{10}\left(\frac{z_0}{1\text{m}}\right)\right) + 0.24\,\text{K} \tag{3.5}$$

It is important to note that the unit of ground roughness length is z_0 metres. Equation (3.4) better describes the interconnection of α and z_0 shown in [9]. In Equation (3.5), an equation between units (see below), only the numerical value of z_0 (in metres) need be given. This equation better describes the connection with standard conditions $z_0 = 0.1$. The reason for the unclear connection is that the height exponent is weakly dependent on the height. The best method of determining the height exponent is in the definition of the height exponent in Equation (3.1) and two measured velocities at two heights α.

For practical evaluation, roughnesses are divided into classes so that there are classes for landscape forms. The classes range from open sea through flat open country to urban community areas. The connection presented here is clearly shown in [9]. The classes will be discussed further below.

The standard condition is $\alpha = 0.159$, which is comparable to $z_0 = 0.1$ m. These are the parameters for flat open country used for agricultural purposes with some low wind obstacles. It corresponds in many places to the landscape form of Schleswig-Holstein with its hedgerows. This is where wind turbines are erected so that the standard depicts the typical site of a wind turbine. As a first approximation it is possible using the named parameters α or z_0 that a wind velocity measured at a desired height can be converted to another desired height (e.g. the hub height of the wind turbine).

In Figure 3.2 the height profile is depicted for standard conditions. Standard conditions are the abovementioned roughness length or their exponent and the wind velocity 5.5 m/s measured at a height of 30 m. The logarithmic profile(solid line) and the profile according to Hellmann with $v_2 = v_{ref}$ and $h_2 = h_{ref}$ are shown. It can be seen that the profiles are not completely identical but very similar.

This standardised logarithmic profile in the FGW Guidelines [5] serves for the calculation of the reference energy production, with which the duration of the increased allowance is calculated (see Section 3.2.10).

3.2.3   Roughness Classes

The roughness classes with their roughness length are presented based on the described roughness length z_0, according to the European Wind Atlas [2]. Five classes are defined from 0 to 4, whereby broken classes are also possible; see Figure 3.3 and Table 3.1. Relevant

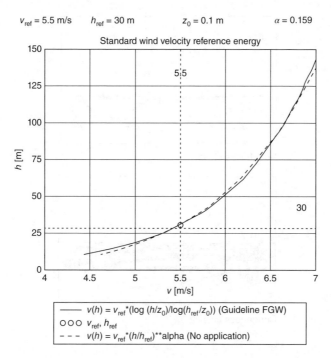

$v_{ref} = 5.5$ m/s $h_{ref} = 30$ m $z_0 = 0.1$ m $\alpha = 0.159$

Standard wind velocity reference energy

— $v(h) = v_{ref}*(\log (h/z_0)/\log(h_{ref}/z_0))$ (Guideline FGW)

○○○ v_{ref}, h_{ref}

--- $v(h) = v_{ref}*(h/h_{ref})**$alpha (No application)

Figure 3.2 Height profile, height above ground as vertical axis, horizontal velocity as horizontal axis, logarithmic profile with standard value for roughness length $z_0 = 0.1$ m, profile according to Hellmann with corresponding standard value for height exponent $\alpha = 0.159$, both for the wind velocity $v = 5.5$ m/s at $h = 30$ m (reference value). These are the standard conditions for a profile

as a site for wind energy plants are Class 1, an open agricultural landscape with very few wind obstacles and Class 2, which is an agricultural landscape with many buildings and bushes. Class 1 is seldom seen in Schleswig-Holstein, Class 2 is often found as a site. Class 3 represents villages or woods, among others. Villages are not erection sites, and installation in woods are not allowed in some regions even when the wind energy plants are high enough. However, this class is often found in the extended wind input region and must therefore be taken into account in some circumstances. Also Class 4, described as large towns with dense development, is relevant for the wind input region. Class 0 is water surface and is also relevant for onshore plants that are erected on the wind input region of the ocean or large bodies of water. The formation of waves that increase the roughness length is ignored. It is expected that the rise of turbulence over offshore water surfaces and over woods is only relevant above 12 m/s [11]. The wind input region for roughness is accounted for up to a radius of 20 km.

The calculation of the roughness of a landscape with houses is given in the *European Wind Atlas* [2], according to Equation (3.6):

$$z_0 = \frac{\frac{1}{2}h^2\,bn}{A} + z_{0T} \tag{3.6}$$

Roughness class 1

Roughness class 2

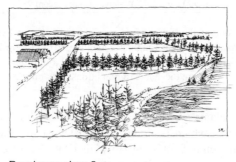

Roughness class 3

Figure 3.3 Roughness classes according to *European Wind Atlas* [2]. Reproduced by permission of DTU Wind Energy

where h is the height of the houses (m), b is the width of the houses lateral to the wind (m), n is the number of houses, A is the horizontal area being viewed (m^2), and z_{0T} is the roughness length between the houses (m). Agriculturally used areas are given the value $z_{0T} = 0.03$. The roughness z_{0T} between the houses that describes the roughness of the ground is also called the background roughness. In some user programs (see also Section 3.6) a background roughness must be given.

One can assume a coarse estimate of the roughness length z_0 that the real heights of the obstacles are 30 times as large as z_0. This can be seen in the picture of roughness class 1 with corn fields with a height of 1 m whereby the roughness length is 0.03. It is important with this very coarse estimate to understand that the ground with its obstacle height determines the profile, but that the roughness length is a mathematical dimension in the logarithmic equation. Originally both are linked, but because of their different meanings are on different scales.

Table 3.1 Roughness classes according to *European Wind Atlas* [2,10]

Roughness class	Roughness length	Type of landscape
0	0.0002 m	Water surfaces
0.5	0.0024 m	Open terrain with smooth surface, e.g. concrete, landing strips at airports, mown lawn
1	0.03 m	Open agricultural land without fences and hedgerows, possibly with well spaced houses well apart, very gentle hills
1.5	0.055 m	Agricultural land with some houses and 8 m high hedges with spacing of approximately 1250 m
2	0.1 m	Agricultural land with some houses and 8 m high hedges with spacing of approximately 500 m
2.5	0.2 m	Agricultural land with many houses, bushes, plants or 8 m high hedges with spacing of approximately 250 m
3	0.4 m	Villages, small towns, agricultural land with many or high hedges, woods and very rough and uneven terrain
3.5	0.8 m	Large towns with high buildings
4	1.6 m	Large towns with high buildings, skyscrapers

Many wind consultants prefer the information on the roughness not as a roughness class z_0 but as Class Kl. For the conversion $z_0 = z_0(Kl)$, there applies for classes greater or equal to 1 [2]:

$$\ln\left(\frac{z_0}{1m}\right) = 1.333 \times Kl - 4.9 \text{ for } Kl \geq 1 \tag{3.7}$$

3.2.4 Contour Lines and Obstacles

In the following the influence of contour lines and obstacles with correction factors resulting from rises or dips is described.

3.2.4.1 Contour Lines

A difference in height along the flow of the wind, such as a gentle hill (Figure 3.4), changes the profile. It acts like the lower half of a nozzle and represents a cross-sectional narrowing, which, according to the continuity equation $vA = $ constant, with v as velocity and A as the cross-sectional area, leads to an increase in the velocity. The increase in velocity is described by a correction factor Cor or S (speed-up).

The profile in front of the hill is described as a logarithmic profile $v_0(h)$ according to Equation (3.8), whereby index 0 points to the profile of the hill that is undisturbed in front of the hill. The hill has a height H and at half the height, the width is $2L$ (i.e. the half-value width at half the height is L). For the profile of the crest of the hill $v(h)$, the height coordinate h is counted from the crest.

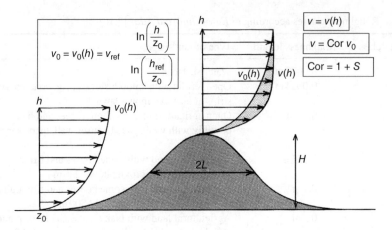

Figure 3.4 Height differences change the profile, from *European Wind Atlas* [2] with the author's supplements. Reproduced by permission of DTU Wind Energy

The new profile is made up of the undisturbed profile $v_0(h)$ with unchanged height coordinates h and the correction factor Cor or S according to the following equations [12]:

$$v_0(h) = v_{ref} \frac{\ln\left(\dfrac{h}{z_0}\right)}{\ln\left(\dfrac{h_{ref}}{z_0}\right)} \tag{3.8}$$

$$v(h) = \text{Cor}(h) \times v_0(h) \quad \text{Cor}(h) = 1 + S(h)$$

$$S(h) = \frac{\frac{1}{2}H}{L} \cdot 2 \frac{\ln\left(\dfrac{L}{z_0}\right)}{\ln\left(\dfrac{h}{z_0}\right)} \frac{1}{\left(1+\dfrac{h}{L}\right)^2} \tag{3.9}$$

The increase in S is proportional to the rise $\frac{1}{2}H/L$ of the hill. In this method the tangent must not be greater than 0.3, which corresponds to an angle of rise of 16°. In programs such as WAsP and WindPRO, one can check whether this meaning is fulfilled on the basis of the RIX index (ruggedness) (see also, among others, [13]). The middle factor, $\ln(L/z_0)/\ln(h/z_0)$, describes the influence of the height of the hill on the profile indirectly over the length and the rise. The factor $1/(1+h/L)^2$ is derived from the potential theory. This is the reason that there must be no turbulence for the applicability of this equation, and is again the reason for the required limitation on the steepness of the hill. The speed-up factor $S(h)$ is applicable for heights h that are greater than a tenth of the half-value width. $h > L/10$ must apply. Wind energy plants normally fulfil this condition. The contour lines of an area are also designated as orography (see Sections 3.2.5 and 3.2.6).

3.2.4.2   Obstacles

Obstacles weaken the average wind velocity behind the obstacle. According to the *Wind Atlas* [2], the height H of the obstacle is important. The depth of the obstacle has a secondary influence and is not taken into account here; for the sake of simplicity the depth is taken as zero.

In order to take wind-permeable obstacles into account, the porosity P is introduced. Trees have a porosity of 0.5, houses a porosity of 0. The weakening at height h above the ground and at a distance x behind the obstacle is dealt with in a similar manner as above, with a correction factor Cor, or here with a weakening factor a, Equation (3.10). The relative weakening of the wind velocity is calculated according to [14] with Equation (3.11), where $\alpha=0.14$ as height exponent is a fixed value. The roughness length is given in the example calculation from [14] with $z_0=0.1$.

$$v(h, x) = \text{Cor}(h, x) \times v_0(h) \quad \text{Cor}(h, x) = 1 - a(h, x) \tag{3.10}$$

$$a(h, x) = \frac{\Delta u}{u} = 9.75 \left(\frac{H}{h} \right)^{\alpha} \frac{H}{x} (1 - P)\, \eta \, \exp\left(-0.67 \eta^{1.5} \right) \tag{3.11}$$

with
$$\eta = \frac{h}{H} \left(\frac{0.32}{\ln\left(\dfrac{H}{z_0} \right)} \frac{x}{H} \right)^{-\frac{1}{2+\alpha}} \qquad \text{and} \quad \alpha = 0.14$$

where a is the weakening behind the obstacle, x is the horizontal distance downstream behind the obstacle (m), h is the height above ground (m), H is the height of obstacle (m), P is the porosity of the obstacle, α is the fixed height exponent ($\alpha=0.14$), z_0 is a fixed roughness ($z_0=0.1$ m), and η is an intermediate result.

The range of the weakening is surprisingly large. The height H of the obstacle is the deciding dimension both vertically and horizontally. The weakening is verifiable up to three times the height and horizontally up to 40 times, thus it is a very significant effect (Figure 3.5).

Not all obstacles need to be taken into account. According to the above calculation for Figure 3.5, it can be seen that weakening can be verified only up to three times the height of the obstacle. This means that downdrafts of obstacles that are only a third as high as the height of the tip of the rotor blade in its lower position pull through. This means again that these obstacles need not be taken into account. With the hub height NH and the rotor diameter D, all obstacles H for which $H < ⅓ (NH - ½D)$ applies can be ignored. Because of the large spread of the weakening, obstacles of up to approximately 25 m height are investigated in distances up to 1000 m [13]. Obstacles at greater distances are not seen as obstacles but as roughness elements (see Section 3.2.3).

3.2.5   Wind Resources with WAsP, WindPRO, WindFarmer

With the above-presented interlinkages between height profile, roughness length, contour lines and obstacles, WAsP is used to calculate the wind potential for the site of a wind turbine or a wind farm [1–3]. The abbreviation stands for Wind Atlas Analysis and Application Program.

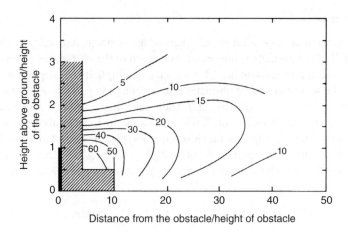

Figure 3.5 Wind weakening behind an obstacle. Black: the obstacle has a height of 1 and a depth of 0. Grey: no statement possible. Numbered contours show the weakening in %, according to *European Wind Atlas* [2]. Reproduced by permission of DTU Wind Energy

Starting with the data of a meteorological measuring station (reference station) (Figure 3.6) that has measured wind data for at least ten years, and whose data for 12 direction sectors are known as frequency distributions (see Section 3.2.8), calculations are carried out, taking into account the local obstacles (lower figure), the roughness (middle figure) and contour lines (upper figure) of the wind cleaned of any ground influences (see Section 3.2.2). These data are offered in Germany and wide areas of Europe by various data providers (e.g. Risø (European Wind Atlas) and the German Weather Service (DWD)) in a grid smaller than 100 km spacing. In a simplified manner, the author equates the wind freed from ground influence with the geostrophic wind, which is a meteorological term for air flows that dominate at greater heights (500 m to 1000 m) and are only driven by pressure differences and the coriolis force. This wind overarches regions. With this wind one calculates for the site of the wind turbine, taking into account local contour lines, the local roughness length up to a radius of 20 km and the local obstacles, the wind profile and with it the wind velocity at the hub height of the wind turbine.

With Equation (3.3) it becomes clear that the determination of the profile at the site first needs the determination of the local ground roughness z_0 and secondly a reference velocity v_{ref} at h_{ref}. In this procedure the ground roughness at and around the site is determined and the reference velocity is found from the transfer of the reference station.

The sky directions are divided into 12 sectors for the meteorological measuring station, each 30°, beginning in the north with sector zero. For each sector there is available the average wind velocity v_{aver} in m/s, the two parameters of the Weibull frequency distribution A in m/s and k (no unit), as well as the frequency of the wind from this direction in %. The annual energy of the wind turbine is calculated by means of sector-wide calculation of this data with the capacity characteristics (see Section 3.2.9).

The WindPRO program (see [12,13]) works with WAsP and with the above-mentioned method to create energy calculations for individual wind turbines and for a whole wind farm.

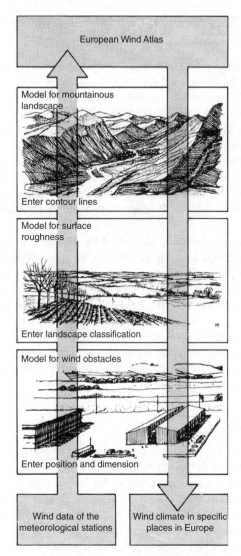

Figure 3.6 Wind calculations according to WAsP European Wind Atlas [2]. Reproduced by permission of DTU Wind Energy

It must be remembered that the average wind velocity is not the same in all years. These fluctuations are described using the wind index. The wind index is an energy index that sets a past year in relationship with the calculated long-term average value of the energy. The German wind index, for instance, can be found directly at the Bundesverband Windenergie (German Wind Energy Association) [15]. A local wind index is suitable for a more accurate estimate. Basically for this and other reasons a safety rebate is deducted for energy calculations; the first estimate is 10%.

Remark: The depictions from the European Wind Atlas [2] were released for copying by Risø DTU National Laboratory for Sustainable Energy at the Technical University of Denmark.

3.2.6   Correlating Wind Potential with Mesoscale Models and Reanalysis Data

Besides the simplified linear models discussed in the previous chapter, increasingly complex mesoscale models are also being used for the correlation of wind potential. The term mesoscale refers to atmospheric phenomena with horizontal scales of a few to several hundred kilometres. Typical phenomena are the land–sea circulation, mountain and valley systems, and larger weather cells. Mesoscale data solve the conservation equations for impulse (movement equations), internal energy (thermodynamic and condensate) and mass (continuity equation) numerically. The processes that cannot be explicitly solved (turbulence, convection, boundary layer) are parameterised. A good overview can be found in [16]. Mesoscale models describe time-dependent phenomena and can simulate daily and annual atmospheric conditions (this is in contrast to the WaSP program that only calculates longstanding average relationships).

In wind energy, mesoscale data are mainly used for the simulation of a map of the average wind velocity (windmapping) and for calculating the time-dependent, three-dimensional conditions of the atmosphere (Wind Atlas: time series for wind velocity, wind direction, temperature, pressure, and so on). Complex atmospheric flows, such as at the crest of a mountain, steep cliffs or in valleys, are often made available with mesoscale simulations for a detailed analysis.

Mesoscale models simulate a portion of the atmosphere for a particular time period. For this, starting and limiting conditions are required. These are usually prescribed by reanalysis data. At the lowest border of the atmosphere the orography and the utilisation of the land describe the character of the Earth's surface.

3.2.6.1   Reanalysis Data

In reanalysis, observation data from the past are prepared with an atmospheric simulation model and interpolated onto a three-dimensional grid. Probably the best known and most used reanalysis data in wind energy are the NCEP/NCAR data [17,18] (see Figure 3.7).

These reanalysis data have been placed on a $2.5 \times 2.5°$ grid at six-hourly intervals since 1948 worldwide and are continually prepared. Observations come from weather stations, ships, aircraft, radio sensors and satellites. The data are relatively consistent because the simulation model is maintained, whereas the observation data-sets vary, for instance, by the introduction of satellite data.

The timely and spatial resolution of the reanalysis system is too coarse for many applications in wind turbine meteorology. Mesoscale models are the tools for calculating wind turbines with higher resolution. This so-called downscaling is used as a statistic–dynamic downscaling for correlating the longstanding average wind potential (windmapping) or for continuous time-dependent simulation (Wind Atlas). Often designations are mixed up, and a simple windmapping is called a Wind Atlas. The differences are described in the following.

3.2.6.2   Windmapping

In statistic–dynamic downscaling it is assumed that the wind climate of a region can be described by means of a suitable set of atmospheric variables (a statistic of weather situations)

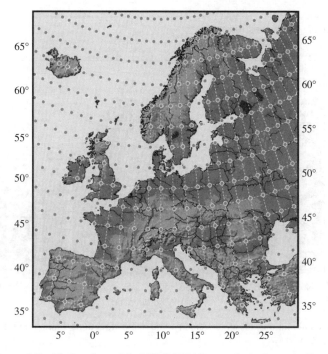

Figure 3.7   Node points of the NCEP/NCAR reanalysis data over Europe

and a description of the Earth's surface characteristics at a corresponding resolution. With reference to the wind climate these are:

- the geostrophic wind (wind velocity and direction);
- the stability of the atmospheric boundary layer;
- the orography;
- the land usage (or roughness of the surface).

The cluster analysis has been shown to be well suited to controlling a statistic of the weather conditions in which a large number of combinations of wind velocities, wind directions and stabilities over a long time period can be divided into clusters. In this division the agreement of the frequency distribution of the original data (time series and reanalysis) is optimised with the distribution of the clusters. Simulations are carried out only for individual clusters, which means a substantial saving in calculation time. With knowledge of the frequency of the individual clusters, it is then possible by means of the combination of wind fields for each individual cluster and the frequency to determine the average wind field. The number of individual results in each cluster determines their respective weightings. An example is shown in Figure 3.8 of all combinations in a time series of the wind velocity and wind direction (dark points) and the clusters derived from them (light squares). The principle of statistic downscaling is shown in Figure 3.9.

   The individual clusters are calculated in a cascade with different spatial resolutions (nested model), because, for instance, the jump from the coarse resolution of the NCEP/NCAR to a

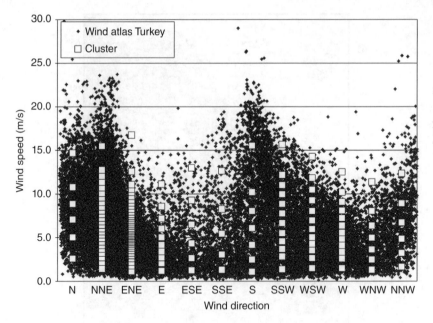

Figure 3.8 All combinations of wind velocity and wind directions (dark points) and the derived clusters (light squares)

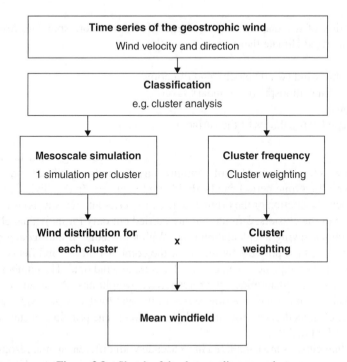

Figure 3.9 Sketch of the downscaling procedure

Figure 3.10 Levels of downscaling

very high resolution of, for instance, some hundreds of metres would be numerically unstable. Figure 3.10 shows the various levels of downscaling down to a resolution of 5 km over Germany. Whilst the reanalysis data have a resolution of around 250 km, the resolution on the first level is 135 km, the second level 45 km, the third level 15 km, and the last level 5 km. The orography used as a lowest limiting condition is shown in Figure 3.11. A comparable picture is given for land use.

The results of such a simulation may be the average field of wind velocity at 100 m height for the time period 1990 to 2006, as shown in Figure 3.12.

3.2.6.3 Wind Atlas

For a Wind Atlas the simulations are carried out in a time-dependent manner. The downscaling occurs in the same way as described previously, by means of embedded simulation areas (see above: nested model). The drive, however, is no longer given by the statistics, but the model is continuously driven by the demands of the limiting conditions. The results are time series of the atmospheric, dynamic and thermodynamic parameters for each of the horizontal and vertical model grid points. In wind energy one also talks of 'virtual measuring masts'. One such simulation is characterised by the simulation area (model domain), the spatial (horizontal and vertical) resolution, the timely resolution (time step of saving the results) and the simulated time period.

Models and methods of these simulations are similar to regional climate modelling and require similarly large calculation capacities and competence in the implementation and analysis of the simulations.

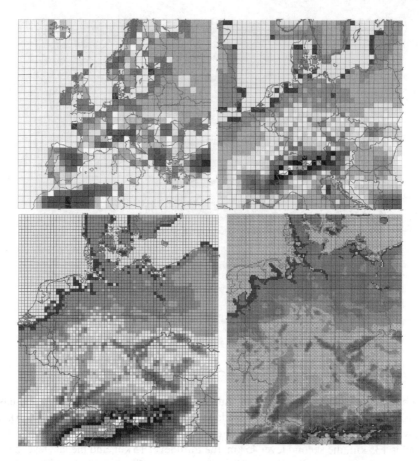

Figure 3.11 Orography for 135 km, 45 km, 15 km and 5 km resolutions

3.2.6.4   Verification and Time Series

The verification of the model results with observations is a constant challenge. For wind energy usage one normally compares the time series of the wind velocity and direction, the frequency distribution and statistical dimensions such as average value, standard deviation, extreme values, and so on. In the evaluation it must always be borne in mind that a model size that is representative of a model grid resolution is compared with a local size (anemometer measurement).

A comparison of the wind velocity and direction for a site in North Poland for a grid spacing of $5 \times 5\,\mathrm{km}^2$ is shown in Figure 3.13. From numerous similar comparisons it can be seen that mesoscale models are quite capable of simulating the time variations in wind velocity and direction in a manner that is close to reality. The agreement in the absolute value of the wind velocity or the wind potential varies very greatly, and is particularly dependent on the complexity of the landscape.

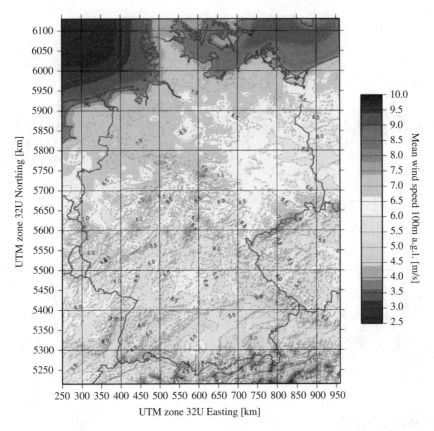

Figure 3.12   Average wind velocity for 100 m height in Germany for the period 1990–2006

3.2.6.5   Areas of Applications

Time series of the wind velocity, wind direction, temperature, and the derived energy production of wind turbines are used for many requirements. Primarily the data are a long-term
source for correlation with short-term wind measurements or with actual production data of
wind turbines. Thus, wind measurements for a 12-month time period are referenced to those
of a 20- or 30-year period and it is possible to take short-term annual fluctuations into
account. When wind turbines have been in operation for a few months, it can be estimated by
comparison with long-term production data what long-term yield can be expected.

Often wind turbines are required to be shut down for particular times due to official limitations, or must be operated with reduced nominal capacities (noise-reduced operation during
night hours, temporary shutdowns due to bird migrations or high air temperatures). The loss
of yield can be calculated in detail by means of analysis of the corresponding production time
series. Figure 3.14 shows the loss of production due to reduced operation because of bat flight
restrictions during the year and per day.

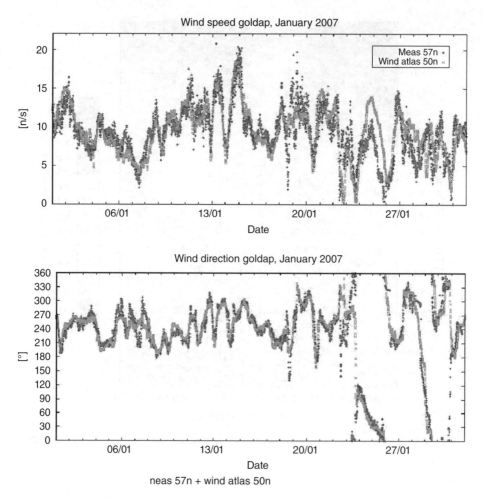

Figure 3.13 Comparison of wind velocity (top) and wind direction (bottom), calculation (light grey) and measurement (dark grey), for a site in North Poland (calculation for 50 m height, measurement at 57 m height)

3.2.7 Wind in the Wind Farm

Wind turbines are mostly erected in wind farms. Both geometric and non-geometric configurations are selected. The distances between the turbines are not unified. As a rule, one can assume that turbines arranged laterally to the main direction of the wind are spaced at least three rotor diameters apart and those in the main wind direction are five rotor diameters apart (see Figure 3.15). At sea greater distances are selected, for instance seven. It is not necessary to position the second row in the gaps of the first row, as the main wind direction does not consist of a single degree value but is often spread over two direction sectors.

The given distances are a good compromise in practice between shadowing the wind (see below) and structural loading of the turbines of the second row due to the turbulent lee

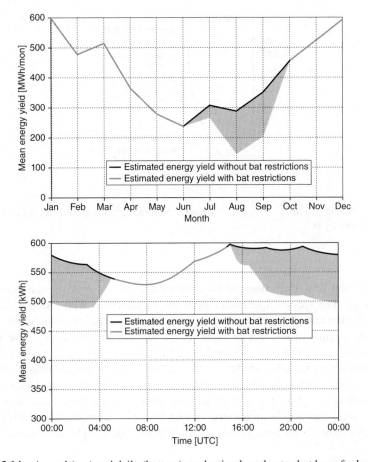

Figure 3.14 Annual (top) and daily (bottom) production loss due to shutdown for bat flights

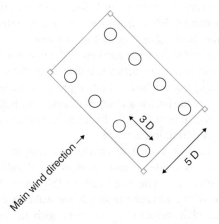

Figure 3.15 Principle of arrangement in a wind farm, where D is the diameter of the rotor

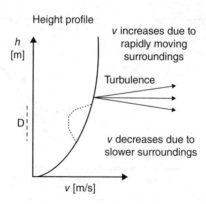

Figure 3.16 Height profile, showing the undisturbed profile, and the disturbed profile due to a wind turbine with diameter D

winds of the first row. Both demand large spacings, and the space requirement demands close spacings. The space requirement must not be confused with space utilisation. The space between the turbines and also beneath the rotor continues to be used for agricultural purposes. The sealing of the surface refers to the foundation, and in this mostly the tower itself, possibly a transformer house, and gravel-covered access paths to the turbines. The space requirement is included in the limited accounted-for wind advantage areas that are suitable for wind turbines. This leads to the clear tendency to select small spacings in wind farms. According to a grid study (2010) for large-area planning, the requirement for surface area per nominal capacity is mentioned as 7 ha/MW (1 hectare is 100×100 m in extent). For real wind farms in an arrangement according to Figure 3.15, more capacity per area is assumed.

As the wind velocity drops in the wake of the plants of the first row, the plants of the second row generate less energy. The efficiency of the farm, when referenced to the annual energy, is normally 90%. This corresponds to a loss of 10% only due to mutual shadowing of the wind. For real wind farms, figures ranging from 6% to 12% are found. Further losses due to technical non-availability of, for instance, 5% and grid losses due to internal farm cabling of, for instance, 3%, are two further noticeable losses of wind farms.

The disturbed profile caused by the wind turbines of the first row recovers downstream with larger distances. The cause of this is the turbulence that mixes the vertical velocity layers until the logarithmic profile is re-established, which represents an equalisation condition. In Figure 3.16 the disturbances of the profile are linked to a plant with diameter D, with turbulence related to direction changes (see Section 3.5). Eventually, kinetic energy from the uppermost layers makes up the velocity deficit in the wake.

The wake in a wind farm is described according to a simple model [13,19,20]. Besides the eddy-viscosity model, this model is also used in the WindFamer program [12]. The undisturbed wind has a velocity v. The wind turbine WT has a diameter D, and there the velocity sinks to the value of v_w, whereby w stands for wake. A cone forms at a distance x behind the plant and the wind velocity v_w recovers with its increasing area expansion (Figure 3.17).

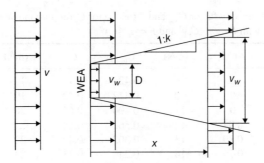

Figure 3.17 Wake behind a wind turbine, showing the velocity decrease at the rotor level with C_t, and recovery of the velocity geometrically at 4° (corresponds to k) [19]

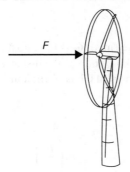

Figure 3.18 Thrust coefficient C_t describes the thrust force F horizontal to the hub

Described mathematically, the wind velocity decays at the rotor level with the thrust coefficient G. The thrust coefficient describes the horizontal thrust on the hub according to Equation (3.12) analogous to the flow rsistance (see Figure 3.18).

$$F = \frac{1}{2}\rho A C_t v^2 \tag{3.12}$$

where F is the shear force, ρ is the density (kg/m³), A is the covered rotor surface (m), C_t is the thrust coefficient, and V is the undisturbed wind velocity at hub height (m/s).

As already mentioned above, C_t is used in order to describe the wind weakening. The reason for this method of description, and not the method of converted energy, is the same as for the Betz laws.

The energy conservation cannot be fully described, but impulse conservation and especially mass conservation are always fulfilled. The reduced velocity at the rotor level is described by means of the impulse that is linked to the thrust force and is used in the following according to the law of the conservation of mass:

$$v_w = v\left[1 - \left(1 - \sqrt{1 - C_t}\right)\left(\frac{D}{D + 2kx}\right)^2\right] \tag{3.13}$$

Table 3.2 Electrical capacity factor $C_{P\text{-el}}$ and thrust factor C_t for all wind velocities for a modern multi-megawatt plant with $P=\frac{1}{2}\rho AC_p v^3$ and $F=\frac{1}{2}\rho AC_t v^2$

v (m/s)	$C_{P\text{-el}}$	C_t	v (m/s)	$C_{P\text{-el}}$	C_t	v (m/s)	$C_{P\text{-el}}$	C_t
0	0	0	9	0.424	0.77	18	0.114	0.15
1	0	0	10	0.422	0.72	19	0.097	0.12
2	0	0	11	0.400	0.65	20	0.083	0.11
3.5	0.162	0.97	12	0.362	0.54	21	0.072	0.09
4	0.258	0.89	13	0.302	0.42	22	0.062	0.08
5	0.369	0.79	14	0.240	0.33	23	0.055	0.07
6	0.393	0.79	15	0.197	0.25	24	0.048	0.06
7	0.413	0.79	16	0.162	0.21	25	0.043	0.06
8	0.420	0.78	17	0.135	0.17	26	0	0

where v_w is the velocity in the wake (m/s), C_t is the thrust coefficient, D is the rotor diameter (m), $k=0.07$ expansion, $4°=\tanh 0.07$, and x is the distance downstream (m).

The above equation for k with $k=0.07$ is sufficient. In general the following applies for k:

$$k = \frac{A}{\ln\left(\dfrac{h}{z_0}\right)} \tag{3.14}$$

where $A=0.5$, h is the hub height (m), and z_0 is the roughness length (m).

In Equation (3.13) the front part of the bracket – which sets the distance x of the first plant to zero ($x=0$) – describes the reduced velocity at the rotor level of the plant that is calculated using the thrust coefficient. For $x>0$ the part $(D/(D+2kx))^2$ describes the expansion of the lee wind as a cone with k, which in the first iteration corresponds to a half-angle of $4°$; $k=0.07=\tan(4°)$ applies. As the lee wind expands, the velocity in the wake v_w recovers and approaches that of the undisturbed velocity v. For more exact information Equation (3.14) can be calculated using the factor k. The influence of the turbulence becomes clear with the roughness length z_0. The greater the ground roughness, the greater the turbulence, and the sooner it equalises itself. From a calculation point of view k becomes larger with greater roughness length, that is, the expansion angle becomes larger and the distance x up to the recovery of the profile becomes shorter.

From the theory it can be seen that with a capacity factor $C_P=16/27$ according to Betz, $C_t=8/9$, the result is that $C_t=3/2C_P$ applies. The comparison with the drag factor of a round disc whose $C_w=1.1$ and whose equation with $F=\frac{1}{2}\rho AC_w v^2$ is the same in structure as that of C_t, making it clear that the rotor with its three rotor blades is only 20% more permeable than a full disc referenced to the force.

For wind turbines the real C_t value is given as well as the electrical C_P value, so that the performance can be calculated with the C_P values and the wake with the C_t values. In Table 3.2 the electrical C_P values and the C_t values for a modern multi-megawatt wind turbine are shown.

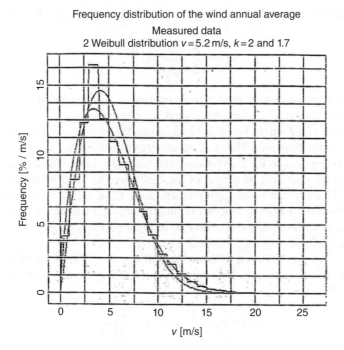

Figure 3.19   Frequency of the wind $p(v)$ over the year as a histogram, as a Rayleigh function and as a Weibull function ($v_m = v_{aver}$ means annual average value, $k=2$ is the Raleigh function, frequency for all three curves in % per m/s, and the bin width = 1 m/s)

3.2.8   Wind Frequency Distribution

The wind is usually measured by means of a vane anemometer over a time period of 10 minutes. The measuring device delivers a velocity coefficient every second. From the 600 data readings, the average value, the maximum, the minimum and the standard deviation are determined. The maximum provides information on the extreme loads, the minimum is less important, and the standard deviation is the statistical dimension of the turbulence (see Section 3.5). The average value of the 10-minute interval will be discussed in the following. This velocity is designated by v.

The wind velocity is measured for a time period of a year meteorologically at a height of 10 m, the wind energy mostly higher, possibly at the level of the hub. After one year, $365 \times 24 \times 6 = 52,560$ velocity values are available, and this is 100%, or statistically 1. The velocities are divided into bins, arranged as a bin width of 1 m/s. The procedure is generally called classification formation. The wind velocities are counted into the bins, so for the velocities 3.666 m/s and 3.667 m/s, Bin 3 is respectively incremented by 1. This technique is used to give meaningful categorical results from data that are collected on a continuous scale. In this example, all velocities between 3 and truly less than 4 are counted into the interval 3. This leads to the term of frequency density, i.e. the frequency per interval. After all the velocity values have been counted into the bins, division by the overall count is carried out and the frequency is obtained in %. The diagram is called a histogram and can be recognised in Figure 3.19 as a cornered curve.

The Rayleigh function is often used as an analytical function to describe the histogram:

$$p(v) = \frac{\pi}{2} \frac{v}{(v_{aver})^2} e^{-\frac{\pi}{4}\left(\frac{v}{v_{aver}}\right)^2} \tag{3.15}$$

where $p(v)$ is the frequency of v, number between 0 and 1, v is the wind velocity (m/s), and the argument of the function v_{aver} is the annual wind velocity (m/s).

An observation of the unit of Equation (3.15) shows the term $v_{(aver)}^2$ generates the unit 1 m/s for p; the other units disappear. Thus p is a number between 0 and 1 that gives the frequency per m/s that corresponds to the above-mentioned bin width 1 m/s. The average annual wind velocity v_{aver} is the only free parameter of the function. The value of the average value is known, and it is easily calculated from the measuring value for the year, here, for instance 5.2 m/s. Therefore the Rayleigh function is fully determined and can be calculated for any velocity v. The result is shown as a solid line in Figure 3.19. As can be seen, the Rayleigh function describes the histogram with modest accuracy.

A greater accuracy is achieved with the Weibull function. This distribution has a further parameter, the form parameter k. It allows a deviation of 2 in the exponent of the e function. The equation is:

$$p(v) = a e^{-\left(\frac{v}{A}\right)} \quad \text{with} \quad a = \frac{k}{A}\left(\frac{v}{A}\right)^{k-1} \tag{3.16}$$

where $p(v)$ is the frequency of v, a number between 0 and 1, v is the wind velocity (m/s), the argument of the function, k is the form factor, A is a scaling factor (similar to the average value but not identical) (m/s), and a is a standardising factor (1 m/s).

The standardising factor a is derived from the exponent in Equation (3.16), as p represents a frequency density. The factor A is a scaling factor similar to the average value but not identical to it. The Weibull function thus has two parameters, k and A. With both together the average value of the wind velocity is obtained as in Equation (3.17) (see [21]):

$$v_{aver} = A\left(0.568 + \frac{0.434}{k}\right)^{1/k} \tag{3.17}$$

The Weibull function is used in preference. The parameters A, v_{aver} and k are given. The parameters A and k describe the Weibull function fully, but v_{aver} is of practical interest.

At many sites the form factor has the value $k=2$, then the Weibull function changes to the Rayleigh function and the knowledge of the average value for describing the frequency distribution is sufficient. Experience has shown that at coastal sites where the wind is steadier, k values higher than 2 occur, but inland they are lower.

If only the annual average wind velocity is available, then the frequency distribution is often calculated only with the Rayleigh function, which represents an acceptable approximation of the Weibull function.

3.2.9   Site Classification and Annual Energy Production

The site evaluation is discussed here, in that a simple method is presented for calculating the annual energy of a wind turbine.

Table 3.3 Annual energy yield

v (m/s)	$p(v)$ (% m/s)	h_i (1/m/s)	$h_i T$	$P(v) = P_i$ (kW)	$h_i T P_i$ (kWh)
0.5	2.24	0.0224	196.5	0.00	0
1.5	6.43	0.0643	562.6	0.00	0
2.5	9.80	0.0980	858.2	0.70	601
3.5	11.98	0.1198	1049.4	5.40	5667
4.5	12.86	0.1286	1126.4	13.60	15,319
5.5	12.54	0.1254	1098.7	26.40	29,005
6.5	11.31	0.1131	990.5	45.20	44,769
7.5	9.51	0.0951	833.3	71.10	59,248
8.5	7.51	0.0751	658.2	103.70	68,260
9.5	5.60	0.0560	490.1	142.60	69,894
10.5	3.94	0.0394	345.0	176.50	60,892
11.5	2.63	0.0263	230.0	196.50	45,198
12.5	1.66	0.0166	145.5	205.30	29,867
13.5	1.00	0.0100	87.4	208.00	18,178
14.5	0.57	0.0057	49.9	208.00	10,380
15.5	0.31	0.0031	27.1	208.00	5639
16.5	0.16	0.0016	14.0	208.00	2916
17.5	0.08	0.0008	6.9	208.00	1436
18.5	0.04	0.0004	3.2	208.00	674
19.5	0.02	0.0002	1.4	208.00	301
20.5	0.01	0.0001	0.6	208.00	129
21.5	0.003	0.0000	0.3	208.00	52
22.5	0.001	0.0000	0.1	208.00	20
23.5	0.0004	0.0000	0.04	208.00	7
24.5	0.0001	0.0000	0.01	208.00	3
25.5	0.00005	0.0000	0.004	0.00	0
Total	100.2	1.002	8776.5		468,457

Firstly, the average wind velocity at the height of the hub must be known. In an example (see Table 3.3) the wind velocity is averaged over the whole year and all direction sectors and is 5.9 m/s. With the Rayleigh function according to Equation (3.12), 1 m/s will be calculated for the bin width and always for the velocity in the middle of the bin (see v in the table in column 1), and the frequency $h_i = p(v_i)$ in 1/m/s (in column 3). The frequencies h_i correspond to a histogram in which the unit p, thus 1 per m/s with reference to the bin width 1 m/s, disappears; one obtains a pure number between 0 and 1. A year has $T = 8760$ hours. The frequency h_i multiplied by T gives the number of hours for which the wind blows with velocities in the limits of this bin (see $h_i T$) in column 4. In column 5 the capacity characteristics P_i of the plant are given; this is the performance in kW for the wind velocity given in the first column. The multiplication of the number of hours in column 4 with the capacity P_i in column 5 gives the energy in column 6 in the unit of kWh for each bin. This sum gives the annual energy production E_{anno} in kWh:

$$E_{anno} = \sum_{i=0}^{n} (h_i T) P_i \quad \text{with} \quad h_i = p(v \text{ index } i) \text{ and unit } 1/(m/s) \text{ for a bin width of 1 m/s} \quad (3.18)$$

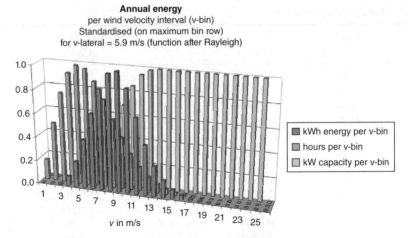

Figure 3.20   Annual energy in kWh is the total of the products of capacity in kW times hours, typically with a maximum in the partial load region below the nominal velocity

The calculation is graphically depicted in Figure 3.20; the rear bar row is the capacity characteristic, the middle one is the Rayleigh function, and the front one is the energy. Each bar corresponds to a bin. The bar series is standardised to 1 for a better overview.

It can be seen that the greatest annual energy is achieved in a velocity range that is in the partial load range just below the velocity at which the nominal capacity is achieved.

From the formation of a maximum in the energy bins it can be surmised that the plant concept of a fixed speed of revolution for all wind velocities is a good concept from the energy point of view. With a fixed speed of revolution there is only one operating point for the optimum design of a plant. This is fixed on the wind velocity at which the described maximum is positioned. Normally this velocity is 8 m/s. At this velocity the plant is just as good as a variable-revolution plant, because 'optimum' cannot be increased. At other velocities the fixed-revolution plant is somewhat at a disadvantage, but the overall deficit compared with a variable-revolution plant is not so great.

Yet variable-revolution plants are favoured. On the one hand, the inverter technology that is of necessity used is very advanced, and on the other hand, variable-revolution plants can vary in their rotation in gusting winds. Impacts of the torque on the drive train are reduced, which makes the building of very large plants possible.

Full-load hours is a measure for comparing plants and sites. For a plant with a nominal capacity of P_{nom}, the annual energy must either be calculated as above, or the measurements for a year must be known. The full-load hours are the quotient of both:

$$\text{Full-load hours} = \frac{E_{anno}}{P_{nom}} \qquad (3.19)$$

where E_{anno} is the annual energy (kWh) and P_{nom} is the nominal capacity (kW). Alternatively the capacity factor is given (8760=hours in a year):

$$\text{Capacity factor} = \frac{\text{Full-load hours}}{8760h} = \frac{P_{aver}}{P_{nom}} \qquad (3.20)$$

Equation (3.20) allows P_{aver} to be given. This is the average annual capacity of the plant. Good values are just over 2000 full hours, which corresponds approximately to a capacity factor of ¼, or in other words, a quarter of the nominal capacity is delivered on average in a year. If the full-load hours fall to below approximately 1500, then the plant becomes uneconomical. If 3000 full-load hours are achieved, it should be considered whether a wind turbine with a different relationship of rotor surface to nominal capacity should be selected in order to reduce the full-load hours, because then the plant becomes more economical. Meanwhile, occasionally, contrary views are to be heard. If the production of the infrastructure, for instance the laying of the cables, is very expensive (e.g. at sea or in little-exploited areas far from power-user centres), it could be cheaper to select the capacity factor of the cables (Equation (3.15) referenced to the transmission capacity of the cable P_{aver}/P_{Nnom}) as high as possible, with the result that the wind turbines must achieve high full-load hours.

This means again the wind turbines must be designed for large rotor surfaces with small generators for high wind areas, such that the relationship of rotor surface to nominal capacity corresponds to weak-wind turbines, but the stability corresponds to strong-wind turbines.

In practice, energy prognoses (see above calculations) are carried out with three extensions. The bin width is 0.5 m/s according to the IEC standard. The energy calculation is carried out and totalled with the frequency distribution for each of the 12 direction sectors. The frequency distribution is carried out for known form factors k according to Weibull.

3.2.10   Reference Yield and Duration of Increased Subsidy

In the style of the technical guidelines TR2, TR5 and TR6 [5], and following the EEG [22], the following describes how the reference yield of a wind turbine and thus the duration of the increased subsidy is calculated. The specifics of the subsidies mentioned in this section apply to current rules in Germany only, and will vary in different countries.

The capacity characteristics of a wind turbine $P_i = P(v_i)$ in bin width steps of 0.5 m/s for $i = 0...n$ must be known. For instance, for the velocity 0 to 25 m/s it is $n = 2.25 + 1 = 51$. For the non-relevant values in which the velocity is below the switch-on velocity of around 3 m/s, or where the velocity is above the switch-off velocity, or for which there are no statistically relevant values from the measurements, which is often above 22 m/s, there are mostly no capacity values, so that n is less than 51. The capacity characteristic for the wind velocity v_{NH} at the height of the hub NH is always given. The wind velocity at the height of the hub is found for the reference yield from the standardised profile according to Section 3.2.2. With Equation (3.3), the average wind velocity at the height of the hub can be found by using:

$$v_{NH} = v_{NH} \ (h = NH, \ z_0 = 0.1 \text{ m/s}, \ v_{ref} = 5.5 \text{ m/s}, \ h_{ref} = 30 \text{ m})$$

In practice the reference yield does not need to be calculated. The value is given with the data of the wind turbine or can be found in the FGW Guidelines [5].

With the Rayleigh function, not given as above as a frequency density, but depicted as a sum function (i.e. the frequency from 0 to v_i), the sum frequency $F(v_i)$ for all n values of v_i can be given as:

$$F(v_i) = 1 - e^{-\frac{\pi}{4}\left(\frac{v_i}{v_{NH}}\right)^2} \tag{3.21}$$

In this way energy yield of each bin AEP(v_i) can be calculated (AEP stands for annual energy production), here per bin i:

$$AEP(v_i) = 8760\,h\ (F(v_i) - F(v_{i-1}))\left(\frac{P_i + P_{i-1}}{2}\right) \tag{3.22}$$

The term $F(v_i) - F(v_i - 1)$ corresponds to the frequency density $p\ ((v_i + v_{i-1})/2)$ in Equation (3.15) but is more accurate because of the non-linearity of $p(v)$. The terms $(P_i + P_{i-1})/2$ of the average yield and the hour count of the annual hours correspond to the scenario in Section 3.2.9. The advantage of this depiction is that the bin width need not be taken into account, and the given supporting values of the capacity characteristics can be used without further ado.

As above, the sum of all bins gives the annual energy AEP:

$$AEP = \sum_{i=1}^{n} AEP(v_i) \tag{3.23}$$

A period of 5 years is used for the reference yield R, there thus applies:

$$R = 5\,AEP \tag{3.24}$$

For the first five years each wind turbine is given an increased subsidy. The amount depends on several circumstances; to a first approximation, 9 cents per kWh (in the year 2011) can be assumed. As a sum, E_5 is taken as the generated energy over the first five years. According to Equation (3.25) the duration of the yield Δ in months results from:

$$\Delta = \left(1.5 - \frac{E_5}{R}\right)\left(\frac{2}{0.0075}\right) \tag{3.25}$$

where Δ is the duration of the increased yield in months, E_5 is the energy yield (kWh) in the first five months after start-up, and R is the reference yield (kWh) according to Equation (3.24).

The duration of the increased subsidy does not fall below zero, $\Delta_{min} = 0$. If the actual yield is equal to the reference yield, $E_5 = R$, then this corresponds to a 100% site; in this, Δ is greater than zero and covers several years. Obviously, and deliberately, strong-wind sites are less preferred because of the shorter duration than weak-wind sites. Thus, weak-wind sites can be used under certain circumstances. Therefore there is an incentive to develop and install plants for weak-wind areas, but the incentive is limited. The limit of economy is seen as the 60% site, that is, $E_5 = 60\%$ of R. For lower E_5, Δ falls to zero. Experience has shown that it makes little sense to use a site at the 60% limit. Because of the low wind, the energy yields are so low that despite the long duration of the increased subsidy, profitability is not achieved.

After the end of the duration of the increased subsidy, the subsidy falls to the standard level. This is also dependent on various circumstances. As a first approximation, one can assume 5 cents per kWh (in 2011).

The level of the increased subsidy is constant over the operational length of the plant, but the level is dependent upon the point in time of the start-up of the wind turbine. The later a wind farm is installed and connected to the grid, the lower is the increased subsidy. But after start-up, no further changes occur. The degression has been changed several times, as a first approximation at 1.5% per year. The degression is a purposeful incentive to make wind energy usage ever more economical.

There are additional subsidies, so-called bonus surcharges. For instance, these affect the system service or the repowering. As a first approximation one can assume 0.5 cents per kWh. In parallel there is also the possibility of direct marketing. There are higher subsidies for off-shore plants; at present 15 cents per kWh is mentioned. The rules for subsidies are changing, but current levels of German subsidies can be obtained from the Bundesverband Windenergie [German Wind Energy Association] (Osnabruck).

3.3 Acoustics

The measurement, calculation and evaluation of noise generated by wind turbines is comprehensively described in the standards and guidelines (see [5,6]). Although basically they are described by the same equations, a distinction is made between *emission*, the noise arising from the wind turbine, and *imission*, the noise that arrives at a place. In this, only the noise that is perceived by the human ear is taken into account, which is described by the acoustic weighting *A*. An overview of the noise based on the standards mentioned is given here. Discussions will involve the dB(A) unit, the source of noise, the attenuation of the noise, standard values, frequency analysis, and impulse surcharge and distance rules.

3.3.1 The dB(A) Unit

Noise is given as acoustic power level L, here with three indices L_{eq}, L_{WA} and L_{pa} in dB(A) units.

Two circumstances must be noted in the decibel (dB) scale. Firstly, it is logarithmic, not linear, so that the normal assumptions one makes for scales do not apply. Secondly, a doubling of the acoustic power means an increase by 3 dB or 3 dB(A) (see below).

For instance, noise is measured using a condenser microphone; for a simple circuit depiction see Figure 3.21. The condenser foil of the microphone reacts to the air pressure p, which changes the capacity of the condenser that then results in a strengthened voltage signal U. The acoustic power is given as the square of the pressure p, analogous to the electrical capacity $P = U I$, which can be converted by Ohm's law to $P = U^2/R$. For noise, U is replaced by p, and R is replaced by the hearing limit p_0. Analogously, one receives not the capacity, but the

Figure 3.21 A condenser microphone reacts to pressure p. Change of pressure $p \rightarrow$ changes capacity $C \rightarrow$ changes the voltage U.

capacity per area, the intensity J, unit W/m^2. The level is the logarithm base 10 multiplied by the factor 10, so that the figures are easier. This is the dB scale:

$$L = 10\log\left(\frac{p^2}{p_0^2}\right) \tag{3.26}$$

where L is the acoustic power level in dB or dB(A), P is the air pressure at the microphone, or more accurately, the effective value alternating pressure (Pa), and p_0 is the standard limit of hearing $= 2 \times 10^{-5}$ Pa.

The microphone is equally sensitive for all frequencies, and this is called linear weighting. The human ear can hear noise from 20 Hz to 20 kHz. The greatest sensitivity is at 1000 Hz. Towards the named limits the sensitivity is reduced continuously. The frequency-dependent sensitivity of the ear is technically called the A weighting, whereby A stands for acoustic weighting.

The hearing threshold is at 0 dB(A), the pain limit is at 130 dB(A), and normal speech is at 55 dB(A). This is the approximate noise level of a wind turbine when one stands at a distance of its construction height (hub height plus rotor radius) on the ground. The noise level at the site of a switched-off plant is often 10 dB(A), lower by 45 dB(A).

For the addition (subtraction) of the actual acoustic power it must be energetically added (subtracted). For this purpose, two acoustic power levels L_1 and L_2 are delogarithmised, added, logarithmised again, and the factor of 10 is taken into account wherever necessary.

$$L_{\text{sum}} = 10\log\left(10^{L_1/10} + 10^{L_2/10}\right) \tag{3.27}$$

L_1, L_2, and L_{sum} are in dB or dB(A), and for the analogue difference L_{diff}, there is a minus instead of a plus in the bracket.

With Equation (3.27) it is easy to calculate that with $L_1 = L_2 = L$ as sum, there follows $L_{\text{Sum}} = L + 3$, units dB or dB(A). A doubling of the acoustic power therefore means an increase by 3 dB or 3 dB(A). This is used in the reflection of the noise from the ground (see below).

It follows further that for large differences between two acoustic powers (e.g. 10 dB), the addition of a first approximation can be ignored. For instance, the result of this is that the noise from a wind turbine that is quieter by 10 dB(A) can be ignored at great distances (for details see [6]).

The calculation of the energetic difference between acoustic levels is analogous to Equation (3.27), where L_{sum} is replaced by L_{diff} and in the bracket + is replaced by –. Use is made of this in the determination of the noise level of the wind turbine. Two measurement series are carried out. In the first the plant is in operation, and the plant and surrounding noises are measured. In the second measurement series the plant is switched off and the surrounding noises are measured. The energetic difference of the noise level described above is the sought-for noise level of the wind turbine.

Measuring conditions are attempted in which both noise levels have a difference of 6 to 10 dB(A) so that the influence of the lower noise level only has a small effect on the result. This means that the surrounding noises that are not always constant only have a small influence on the result. This linkage can be made clear with the examples for calculating the level differences. Thus it is justified to measure the noise levels that are subtracted from each other, not at the same time, but in sequence.

The measuring time is 1 minute or 10 seconds. With the short measuring time for individual measurement, the number of measurements for a fixed overall time of measurement is

Figure 3.22   Noise measurement: microphone on a high-impedance plate

increased. This allows a more accurate statistical analysis of the measurement. The measuring values within 1 minute or within 10 seconds are energetically averaged, analogous to Equation (3.27) to an equivalent value.

$$L_{eq} = \text{energy-equivalent average value}$$

The energy-equivalent average value is therefore, according to its definition and in its importance, equivalent to the effective value of an electric alternating current.

3.3.2   Sources of Noise

For measuring the volume of the noise source of the wind turbine L_{WA}, the microphone is placed on a high-impedance plate on a level surface (see Figure 3.22) at a defined horizontal distance from the foundation of the plant in the lee position, thus downwind. The horizontal distance is equal to the overall height (hub height plus rotor diameter) with a tolerance of 20%. In order to determine the volume of the source noise, the measuring distance is calculated so that the result is independent of the selected distance. The high impedance plate has a reflection grade of 1, so that one assumes a standard ground.

The noise is best transmitted in the direction of the wind so that one measures the loudest noise downwind (worst-case scenario). It is likely that the volume will increase with increasing wind velocity (see Figure 3.23). Background noises also rise with the wind velocity according to $L_{eq}/(\text{dB(A)}) = 2.5\ v/\text{ms} + 27.5$ [9]. The IEC regulates that the noise volume is to be determined at a measured wind velocity of 10 m/s at a height of 10 m. This velocity is also a worst-case scenario. For lesser wind velocity the plant is quieter, while for a greater wind velocity the nominal capacity is reached and the plant switches off, so the noise does not rise any further. This is verified for plants with pitch regulation. Plants with stall regulation are dealt with in the same manner. As an alternative, the IEC permits using the wind turbine itself as a wind measuring device and to measure the volume of noise at 95% of the nominal capacity. For older plants and small wind turbines the reference velocity at which the noise volume was measured is often given as 8 m/s instead of 10 m/s.

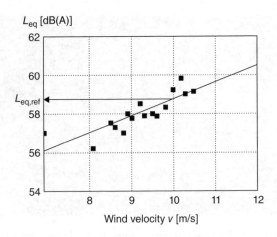

Figure 3.23   Acoustic power level rises with wind velocity, reference value ($L_{eq,ref}$) at 10 m/s

For the purposes of calculating the source volume of the plant L_{WA}, the noise values are measured in operation and when shut down, interpolated to the reference value at 10 m/s at 95% nominal capacity, and energetically subtracted (see above). The pure plant noises $L_{eq,ref}$, are given (see Figure 3.23). The reflection from the high-impedance plate is subtracted and the increase in the noise along the path of the noise by the air from the microphone to the centre of the hub, analogous to Equation (3.29), is added. The acoustic power level of the wind turbine L_{WA} described as a point source is obtained. The level value of the point source is high. However, this value is only a calculation value for the propagation of the noise. Theoretically, one would hear this noise level at a distance $R = 1/\sqrt{(4\pi)} \approx 28$ cm from the centre of the rotor hub; in fact this sound is not heard there. Although the noise is mostly emitted by the outer third of the rotor blades, the noise propagation can be described by the equation of a point source. A sufficiently large distance is already achieved on the ground at a distance equal to the construction height.

The blade tip velocity v_{tip} is entered with a 5th power into the acoustic power to provide an estimate of the point source [8]:

$$L_{WA} = 10 \log\left(v_{tip}^5\right) + 10 \log(D) - 4 \tag{3.28}$$

where L_{WA} is the acoustic power of the wind turbine as a point source (source volume) in dB(A), v_{tip} is the circumferential velocity of the blade tip in m/s, and D is the rotor diameter (m). Values of $L_{WA} \geq 100.0$ dB(A) are normal for large plants. Note that this acoustic power is a source power. The volume is much less even at the foot of the plant.

(Advanced information: the measurement of the high-impedance plate is unusual and it is not immediately obvious that the reflection on the plate must be measured with 6 dB instead of the usual 3 dB. This is caused by the extremely short distance between microphone and plate which leads to a coherent overlaying of the direct and reflected noise wave. For normal reflections with distances greater than the wavelength, the capacity is added, which means a 3 dB rise due to reflection.)

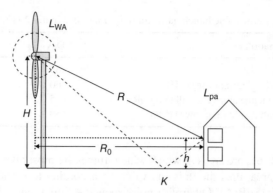

Figure 3.24   Simple noise propagation

3.3.3   Propagation through the Air

According to TA Lärm [23], a simplified noise propagation without taking attenuations into account, especially air attenuation, is permissible as an estimate. The equation used is between units. This means that the measuring and calculation dimensions are to be used in the given units as pure numerical values. The result is a pure numerical value that provides the calculated size in the given unit. The wind turbine emits the acoustic power of L_{WA} mathematically from its hub (see Figure 3.24 and Equation (3.29)).

The level is reduced during its path through the air R; this is the sum on the right side of the equation. The site of imission at a residence is placed at 5 m height; that is the first floor where the bedrooms are usually situated. R is derived from the hub height H, the height of the imission site h, and the horizontal distance R_0 between wind turbine and house. In addition, the noise also reaches the imission site by reflection from the ground.

Again the worst case is assumed, with complete reflection and ignoring the short extra path so that power doubling can be assumed, which means an addition of 3, thus $K = 3\,dB(A)$.

$$L_{pa} = L_{WA} - 10\log\left(4\pi R^2\right) + K \tag{3.29}$$

where L_{pa} is the acoustic power at the imission site in dB(A) (house, height 5 m, first floor), L_{WA} is the acoustic power of the wind turbine in dB(A) (wind turbine as point source, approximately 100 dB(A)), R is the direct distance through the air in metres (wind turbine to house, first floor), and K is the increase in noise volume by reflection from the ground $= 3\,dB(A)$.

In this calculation attenuations are not considered. When the calculated noise value falls below the limiting value, then nore accurate calculations need not be carried out

3.3.4   Imission Site and Benchmarks

According to [23], permissible benchmarks for noise imissions at the imission site for the following areas are as shown in Table 3.4. The borders between day and night are 6:00 and 22:00 hours. The most important benchmark is 45 dB(A), which is to be adhered to at night in

Table 3.4 Permissible benchmarks for noise imissions in dB(A) [23]

Area designation	Day	Night
Business area (BU)	65	50
Mixed area (e.g. farmhouse) (FH)	60	**45**
General residence area (e.g. village) (GR)	55	40
Pure residential area (e.g. town) (PR)	50	35

mixed areas. Mixed areas are areas in which wind turbines are most often erected. If there is a house in the mixed area, then the imission value L_{pa} according to Equation (3.29) must not exceed the value of 45 dB(A). Otherwise noise-reducing limitations will be applied in the approval, or the approval will be refused.

3.3.5   Frequency Analysis, Tone Adjustment and Impulse Adjustment

Frequency analyses are also required for the noise investigation. These can be available within the framework of a certification or a type approval. Both ⅓ octave and narrow-band analyses are to be presented, from which the spectral properties of the emitted noise are derived. Frequency-dependent attenuations can be calculated with the third spectra. If such spectra are not available, then an approximation can be calculated with the 500 Hz frequency. If in the spectrum not only a broad sound but a particularly distinct tone is present, such as that of the tooth-meshing frequency of the drive, then a tone surcharge can be added, in this case an individual tone surcharge of $K_T = 0$ to 6 dB(A). In modern plants, individual tones are no longer the state of the technology, since they no longer occur. The same applies to the impulse adjustment. In the acoustic sense, impulse is a short, sharp rise and fall of the sound volume like a hammer blow, usually from the passage of a rotor blade at the tower. Impulses are also no longer the state of the technology.

3.3.6   Methods of Noise Reduction

Sources of noise in wind turbines are the rotor blades and the gondola. With rotor blades it is the leading edge, the trailing edge, both radials in the outer third of the blade, and to a lesser extent the blade tip, which, together with the turbulent air that they move through, generate a wide aerodynamic noise. An effective countermeasure is a lowering of the blade tip velocity (see Equation (3.28)).

Sources of noise in the gondola are the drive, secondary devices and ventilators with exhaust ducts. In addition, vibrations in the gondola and to the tower are transmitted as mechanical vibrations. Basically, all these sources of noise are avoidable. Measures for this include uncoupling the mechanical vibrations by means of elastomers as support points of drives and generators as well as other components. Design measures include reduced-noise designs of all components and silencers in exhaust ducts. The uncoupling of an elastic, torsionally stiff coupling of the rapid shaft between drive and generator is also beneficial. The state of the technology is low-noise gondolas with weak vibration-transfer inside the gondola and within the tower.

One active measure for reduction of noise is a reduction in the speed of revolution (see Equation (3.28)). This is clearly connected with a reduction in the capacity and is only carried

Table 3.5 Distance regulations and type of utilisation of the imission areas according to the regulations for overall height of wind turbines higher than 70 m

Utilisation of the imission area	Distance
Individual houses and splinter communities (≤ 4 houses)	400 m
Communities in general	800 m
Special areas that serve for recuperation	800 m
Business and industrial areas at the edge of communities	500 m

out for corresponding limitations. The process can only be applied to variable-revolution plants. Three methods are selected for this:

• The nominal capacity is maintained but is only achieved at a higher wind velocity. For this, the speed of revolution is reduced over the whole range of the characteristics.
• The nominal capacity is reduced in that, for instance, a reduced speed of revolution is set. The speed is not reduced in the partial load region.
• A combination of the first and the second.

In the control systems of modern plants, which are occasionally called sound management systems, all three methods are offered for the choice of the customer.

3.3.7   Regulations for Minimum Distances

Wind turbines must adhere to certain spacings from individual houses and communities. According to common circular decrees of two ministries, the regulations given in Table 3.5 apply to Schleswig-Holstein. The distances must be adhered to for plants with an overall height above 70 m. There are special regulations for overall heights to 30 m, to 50 m and to 70 m, and plants below 50 m overall height are also approved according to the building regulations and also BImSchG (see Section 3.8). The overall height H is the hub height NH plus the rotor radius $D/2$. There is a difference between individual houses including splinter communities (< 4 houses), villages, and urban communities. Approximate calculations for noise propagation according to Section 3.2.3, with a typical value for the source of noise L_{WA} and guide values from Section 3.2.4, result in lesser distances. The result of this is that the distance regulations have been cleverly arranged so that a single wind turbine, when adhering to the distances, cannot exceed the guide values for noise. The same applies to the cast shadow. For several plants in a wind farm, calculations must be carefully carried out due to the summation effects.

3.4   Shadow

The calculation of the duration of the periodic cast shadow, thus the sum of the time period in which the position of the sun is behind the rotor, is calculated for the whole year with a resolution of 2 minutes under astronomic conditions (see Figure 3.25). It is assumed that there are no clouds during the whole year and that the wind is positioned so that the rotor is arranged laterally to the point of view, therefore again a worst-case scenario. The guide value of the Länderausschusses für Immissionsschutz (LAI) [State Committee Imission Protection] of 30 hours a year is also determined astronomically. If the astronomically calculated duration of the shadow cast is less than the astronomical guide value, then there are no limitations. If this theoretical guide value is exceeded,

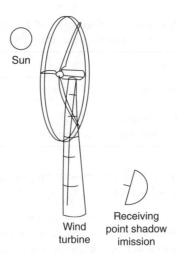

Figure 3.25 Periodic cast shadow: sun behind rotating rotor

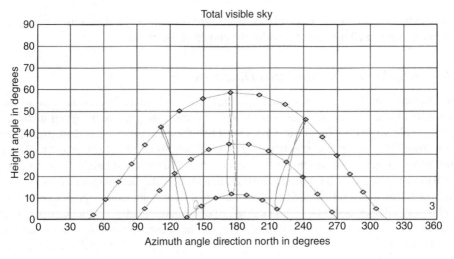

Figure 3.26 Total visible sky, height angles in degrees over azimuth angle pointing north in degrees, for three days of sun in Flensburg in hourly steps, from top: summer sun, to bottom: winter sun

then the real cast shadow must be compared with a real time value. The relationship of the real cast shadow in Germany to the theoretical is approximately ¼. The LAI guide value for real cast shadows is 8 hours per year or, more accurately formulated, per 12 months for all effective wind turbines. A real cast shadow is created with solar radiation with a capacity of more than 120 W/m². The real duration of the cast shadow at an imission point is determined by instrumentation and by calculation. In the case of exceeding the real time value of 8 hours/year, a switch-off module must be installed in the wind turbine system (to control moving shadows). This applies to all imission points. A switch-off module can control several plants or whole wind farms. Control measurements on the part of the approval authorities at critical points are possible.

For the sake of clarifying the astronomical calculation, Figure 3.26 shows the whole sky over Flensburg. The azimuth angle from 0 degrees, which is north, to 360 degrees is shown

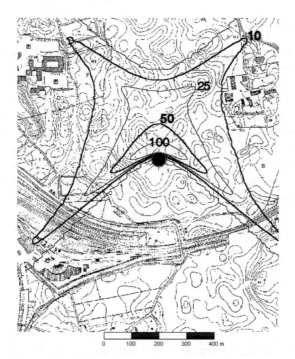

Figure 3.27 Cast-shadow isolines in hours per year. The black circle is the wind turbine, hub height 50 m, rotor diameter 30 m, scale bar: 4 × 100 m

horizontally. The height angle from 0 degrees, the horizon, to 90 degrees, which is the zenith, is shown vertically. The three extreme daily paths of the sun, the summer solstice (top), the winter solstice (bottom) and the day–night equinox. The step widths are hourly. If one views the sky from an observer's point of view one can enter the circular area of the rotor in the depiction and calculate the dwell time of the sun in this circle.

On the one hand the result is a calculation for individual critical imission points such as houses or terraces near houses, and on the other hand, calculations for all points of an area, so that an overview of the average cast-shadow isolines in hours per year is given (see Figure 3.27). The typical butterfly form is caused by the fact that for both solstices the sun rising and setting points at the horizon, thus the azimuth, hardly change from day to day, so that when totalled over the year there is a lot of shadow in this direction. Further results are calendars which show exactly to the minute when each wind turbine will shadow-cast a given number of several critical points with start, duration and end of the cast shadow.

A further astronomical guide value is that 30 minutes cast shadow per day must not be exceeded.

At present in the calculations all wind turbines up to a distance of approximately 2000 m must be taken into account. However, increasing plant heights can require that larger radii are required, so that the periodic cast shadow of a 183 m high wind turbine is 2460 m.

3.5 Turbulence

The turbulence in the natural surroundings and the plant-specific turbulence are described in the following.

3.5.1 Turbulence from Surrounding Environment

As already shown in Section 3.2.2, the wind is turbulent. The overall variation of the wind is not homogeneous over time and is shown by the wind fluctuations. Measurements are taken over a long time period with high resolution, and the amplitude of the fluctuations is shown vertically compared with the frequency of the fluctuations horizontally. Actually, the inverse value of the frequency, thus the duration of the period, is plotted horizontally. In the spectrum, a fluctuation period of around 10 minutes forms a gap and therefore only has small amplitude. When fluctuations with longer periods, up to days or a year occur, then the cause of this is the weather; when fluctuations with shorter periods occur they are referred to as turbulence. For this, see the van der Hoven spectrum, among others shown in the *Wind Atlas* [2]. This gap is the reason why, meteorologically, the horizontal velocity v is measured for 10 minutes. From this 10-minute interval, the average value v_{aver} is evaluated as velocity and the standard deviation $\sigma = \sigma_v$ as turbulence as a statistical dimension. From this the intensity of the turbulence I is:

$$I = \frac{\sigma_v}{v_{aver}} \tag{3.30}$$

where I is the turbulence intensity, σ_v is the standard deviation of the wind velocity horizontal component for a 10-minute measuring time (m/s), and v_{aver} is the average wind velocity horizontal component for a 10-minute measuring time (m/s).

Actually the wind velocity has a constantly fluctuating portion in all three vector components. The two lateral components to v horizontally and vertically are linked to directional changes. According to Prandtl (see [4]), there is an even distribution of the fluctuations of these vector components, so that here the limitation on the fluctuation of v is sufficient for describing the turbulence.

The environmental turbulence, also called natural turbulence, can be described to a first iteration according to the *Wind Atlas* as a function of the ground roughness z_0 and the height:

$$I_{env} = \frac{1}{\ln\left(\dfrac{h}{z_0}\right)} \tag{3.31}$$

where I_{env} is the natural environmental turbulence, h is the height above the ground (m), and z_0 is the roughness length (m).

Analogous to flow over a flat plate (see [4]), this generates the particle size k and here the roughness length z_0, the turbulence. The velocity gradient dv/dh with its rotation can be seen as a generator for turbulence. The velocity gradient falls with rising height h, thus the turbulence becomes less, which is also described with Equation (3.31). Typically for the roughness classification 2 at 80 m height, the mathematical turbulence intensity is 15%. Measurements are more reliable. It is to be noted that for measured turbulence intensities, only those with wind velocities greater than 4 m/s are taken into account, because with smaller wind velocities the turbulence intensity rises steeply according to Equation (3.30), which, however, has little meaning for the alternating loads on the plant because the forces are small, so that the switching-on velocity is hardly ever reached.

3.5.2   Turbulence Attributed to Turbines

Turbulence generates alternating loads on the turbines which are described with Wöhler curves. Taking the alternating loads into account, wind turbines are designed from a stability point of view for turbulence intensities or turbulence classes. Although the definition of turbulence is assumed according to Equation (3.30) for turbulence intensities, yet the turbulence intensity is raised by a factor in order to take the particularly damaging fluctuations after high velocities into account. This leads to several definitions of the turbulence that have various high surcharges. The most important turbulence intensities are characteristic, representative and effective turbulences. The definition and application for forming the turbulence classifications A, B and C are not obvious and are not detailed here. Reference is made to the standards IEC 61400-1 Ed. 2/DIBt and Ed. 3 (see [6]).

A wind turbine in the second row stands downwind of the first and is subject to additional turbulence. Complex processes are used for their calculation. Common to them is that the total turbulence I of the downwind results from the standard deviations of the squared addition of the natural turbulence I_{env} and the turbulence I_{ind} induced by the plant, according to:

$$I = \sqrt{I_{env}^2 + I_{ind}^2} \quad \text{with} \quad I_{ind} = I_{ind}(C_t, D, Z, \lambda, I_{env}) \tag{3.32}$$

The induced turbulence is a function of thrust factor C_t, of the rotor diameter D, the number of blades Z, the tip speed ratio λ and the environmental turbulence I_{env}. Reference is made here to further literature [24–26].

3.6   Two Comprehensive Software Tools for Planning Wind Farms

Two software programs are available with which the complete planning of wind turbine farms can be carried out. These are WindPRO [27] and WindFarmer [12]. Both programs can directly or indirectly access the WAsP program, with which the above-mentioned wind potentials are calculated. Both programs can depict and calculate the energy, and all emissions (shadow, noise, visual changes), the farm cabling and the economics. In addition, wind measurement can be integrated.

Both programs, WindFarmer and WindPRO, are oriented on maps. The wind farm is laid out on a digital topographical map with all its objects and calculation results, whereby, if properly divided, the whole wind farm becomes very clear. The following can be individually processed with these programmes:

- Entry of the base data geography, wind turbines with complete characteristics, for which an almost complete catalogue of standard market installations is available.
- Wind potential calculations: direct application or import of WAsP, WINDSIM or other sources.
- Wind data analyses: multiple analyses of measured wind data for integration into the energy calculation.

- Energy: calculation of the annual energy of the wind farm.
- Optimisation of the energy with consideration of the degree of efficiency and of fog conditions.
- Noise: complete calculation of the acoustic fields including comparison with guide values.
- Shadow: complete calculation of the periodic cast shadow including comparison with guide values.
- Optimisation of the energy with and without the requirements of subsidiary conditions, such as noise restrictions.
- Observation analysis: depiction of the viewability of the plant as a map.
- Composite photo: visualising the planned wind farm in real landscape photos.
- Electrical diagram of the internal farm cabling.
- Financial calculations for the wind farm.
- Turbulence: calculating the natural turbulence and the induced turbulence, calculation of representative or characteristic turbulences in a program as standard.
- RIX analysis of the steepness of the site.
- Analysis of the interruption of civil and military radar stations in a program.

All the information is available under the given Internet addresses and demo versions of the programs can be downloaded.

3.7   Technical Guidelines, FGW Guidelines and IEC Standards

In the following, two standards that apply are named as standards for wind energy. These are the technical guidelines of the FGW and the International Standard IEC 61400. Both sets of standards correspond basically to each other. FGW stands for Fördergesellschaft Windenergie [Society for the Promotion of Wind Energy], IEC for International Electrotechnical Commission, whereby wind turbines are summarised in No. 61400.

The FGW guidelines (TR) can be downloaded, among others, from the FGW at www.wind-fgw.de [5]. Parts mentioned here are:

TR Part 1	Determining noise emission values (available in German only)
TR Part 2	Determining power performance and standardised energy yields
TR Part 3	Determining electrical properties – power quality (EMC)
TR Part 4	Determining electrical properties – power plant behaviour
TR Part 5	Determining and applying the Reference Yield (available in German only)
TR Part 6	Determination of the wind potential and energy yields
TR Part 7	Maintenance of wind farms (available in German only)
TR Part 8	Certification of the electrical characteristics of power generating units and systems in the medium-, high- and highest-voltage grids

IEC 61400 can be obtained from the IEC Central Office Switzerland (www.iec.ch) or from the VDE-Verlag at www.vde-verlag.de [6]. The parts mentioned here are:

IEC 61400-12	Wind turbine power performance testing
IEC 61400-11	Acoustic noise measurement techniques

IEC 61400-13   Measurement of mechanical loads
IEC 61400-14   Declaration of apparent sound power level and tonality values
IEC 61400-21   Measurement and assessment of power quality characteristics of grid-connected wind turbines
IEC 61400-22   Conformity testing and certification
IEC 61400-23   Full-scale structural testing of rotor blades
IEC 61400-24   Lightning protection
IEC 61400-25   Communications power plant components, turbines to actors, Scada
IEC 61400-1   Design requirements
IEC 61400-2   Design requirements for small wind turbines
IEC 61400-3   Design requirements for offshore wind turbines
IEC 61400-4   Gears
IEC 61400-5   Wind turbine rotor blades

3.8   Environmental Influences Bundes-Immissionsschutzgesetz (Federal Imission Control Act) and Approval Process

Just as with human life, with living and working, it is not possible to do anything without the use of energy. Any type of energy conversion into electrical power has an effect on the environment. Environmental effects can occur in the acquisition of raw materials, the conversion into electrical energy, and with dismantling plants or disposing of residues.

This also applies for the use of wind energy, whose expansion is politically desired and offers the following advantages:

- Climate-affecting gases and radioactivity are not released when the plant is operated or during disturbances.
- The Earth's reserves are not used up at the expense of future generations.
- Wind energy is available throughout the world and provides no potential for conflict or attack.
- At a regional level, the use of wind energy provides widely spread income and ensures tens of thousands of German jobs.
- Police protection is not required for wind turbine transport.
- The supply security is increased due to decentralised infeeds.

Besides the abovementioned advantages, also the disadvantages must be mentioned, because where there is sunlight there is also shadow. Because of the newness of this technology, protective mechanisms against long-term effects on human health are not readily available. From government approval and monitoring activities, the jurisprudence and from scientific investigations, it is known that residents in the effectiveness region of wind turbines find themselves affected:

- approximately 16% by hindrance-warning measures characteristics;
- approximately 48% by the periodic cast shadow;
- approximately 58% by noise;
- approximately 64% by movement (not imission).

Feared crop failure in areas in which the turbulent wake from wind turbines touches the ground have not been confirmed. Instead US investigations have shown that turbulent wake flows promote the pollination of crops. In addition it has been shown that with moist weather conditions the formation of mould is prevented due to rapid drying.

Furthermore it has been determined that the community and the individual citizens react more sensitively to environmental loading. For this reason, with the implementation of the 'Article Law' in Germany in 2001, further types of plant to be approved such as wind farms are subject to the comprehensive approval and intervention instruments of the Federal Imission Control Act (BImSchG) [28]. Besides, with foresight, the participation of the public has also been increased.

The purpose of the Federal Imission Control Act is to protect humans, animals and plants, the ground, the water, the atmosphere, as well as cultural and other goods from damaging environmental effects, and to prevent the occurrence of future damaging environmental effects.

3.8.1 German Imission Protection Law (BImSchG)

In the sense of the BImSchG part 3, damaging environmental effects are imissions, which, because of their scope or duration, could cause hazards, substantial disadvantages or substantial annoyance for the community or the neighbourhood. Imissions include air pollution affecting humans, animals and plants, the ground, water, the atmosphere, as well as cultural and other goods, noises, vibrations, light, heat, radiation, and similar environmental influences.

Emissions are the air contamination from a plant: noise, vibrations, light, heat, radiation and similar environmental influences. With reference to wind turbines these are especially:

- noise;
- periodically cast shadows due to the splitting of the sun's rays by rotor blades;
- 'disco' effects due to reflection of the sunlight on rotor blade surfaces;
- turbulence for neighbouring wind turbines and overhead lines of railways and grid operators;
- beaconing the wind turbine as an air traffic hindrance.

In addition, other effects on the environment that are caused by the erection or operation of the wind turbines are to be mentioned. Other laws or sublaws apply to these effects:

- Movement signals due to rotation of the rotor blades.
- Changes to the landscape.
- Ice throw and lightning.
- Devaluation of neighbouring cultural goods.
- Disturbances due to radar and microwave radio relay.
- Grid fluctuations (today hardly a factor).
- Effects on animals.
- Ground warming or ground drying (limited to the surroundings of the earth cable).
- Ground sealing due to the foundation and compaction in the access region.
- Mistaking the hindrance designation lights of wind turbines with traffic installations of marine waterways (e.g. behind the Elbdeich in the neighbourhood of the NOK lock entrance).

3.8.2 Approval Process

For the sake of clarity, the sequence of the approval process according to BImSchG is shown as a block diagram (Figure 3.28). The important elements of the sequence are explained in the following. Because of the limited spatial effect of small wind turbines up to 50 m overall height, they are approved according to the respective state construction laws. A comprehensive legal imission-protection test and approval process for wind turbines is provided by Clause 1.6, Sp.2.4 BImSchV (1997) [29] only for an overall height of more than 50 metres. The approval process of the imission protection law includes construction approval. As a real concession, it applies independently of the operator and can be transferred.

After approval, start-up must take place of within three years or a reasoned application for extension must be made to the approval authorities. This limitation is applied due to the advances in knowledge of imission protection. The process for obtaining an imission protection approval is oriented to the requirements of the ordinance on the approval process [30] for the following:

- erection and operation of a wind turbine;
- important changes in the position, characteristics or operation of the plant;
- obtaining a preliminary decision or a partial approval;
- permission for an earlier start; or
- a subsequent arrangement according to part 17 (1a) BImSchG for fulfilling the duties from the BImSchG or its ordinances.

As far as an approval has not been applied for, and according to part 15 BImSchG, the responsible authorities must be informed in writing of the changes in the approval-required wind turbine at least one month before the changes are started, if the changes can influence the goods affected by the protection. Corresponding documents must be included.

The arrival of the notice must be acknowledged by the authorities. It must also test within one month whether the change requires approval.

The applicant may undertake the change as soon as the responsible authorities have informed him that the change does not require approval, or if they have not responded within one month.

Additionally in the European Union, the environmental impact assessment is anchored in the EIA guideline which was implemented by the member states of the Union by the issue of a determination on the EIA, and also in the Federal Republic of Germany by the law on the environmental impact assessment (EIA) of 12 February 1990 with applicability from 1 August 1990 [31].

3.8.3 Environmental Impact Assessment (EIA)

The environmental impact assessment, also called the environmental impact study or environmental impact investigation, is a legal procedure comprising a systematic testing process with which the immediate and later effects on the environment or goods affected by the decision or intent are determined, described and evaluated. The difficult field of interactions also has to be taken into account.

This environmental impact assessment with further application documents is a key part of the basis for verification by the expert and approval authorities within the framework of the approval process.

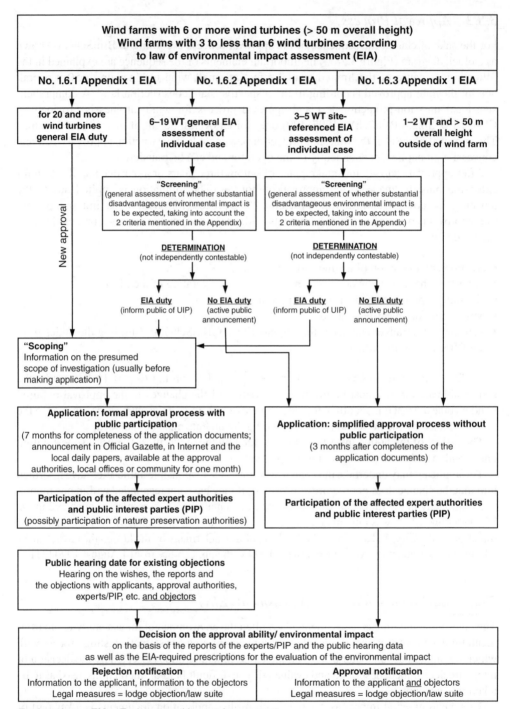

Figure 3.28   Sequence of the BImSchG approval process

3.8.3.1 Screening

Before handing in the application documentation for imission protection approval of a wind turbine, it is to be checked in a preliminary test of the individual case (screening) by means of a cursory test of the necessity of an EIA by the responsible authorities. The criteria for the screening can be found in Appendix 2 of the EIA.

3.8.3.2 Site-Specific Assessment

As can be seen from the wind energy manual of the State Environmental Agency [32], for the site-specific preliminary assessment of the individual case, that only the characteristics of the site are important for the screening. It is to be assessed whether, despite a small number of 3 to 5 wind turbines, just because of the special site conditions, substantial environmental effects for the affected protected area are to be expected.

As a rule there are no substantial negative effects when the distances for the protected areas of the state-specific wind turbine permits are adhered to, or when the wind farm is situated within a specified wind-suited region and there are no new points of view which have not been taken into account in the specific regional area-utilisation or construction plans.

3.8.3.3 General Pre-Approval

In the general pre-approval it should be taken into account how far the threshold value of six wind turbines is exceeded and whether it approaches the dimension for the EIA requirement of 20 plants (wind turbines that were erected before 14 March 1999 were not considered as regards the number of plants).

However, as in the general pre-approval, the size of the project cannot be decisive; the authorities, in their determination of the threshold for the EIA requirement, have already evaluated the effects of a project in a general way, and as a rule only use the necessity of an EIA from 20 wind turbines upwards. For this reason in the general pre-approval of the individual case the site criteria must be included, and from the size of the project and the interaction of the specific project characteristics with the local conditions, they must determine whether there are substantial negative environmental effects. Because of the expert knowledge required for this, it is sensible to include, for the screening, representatives from nature protection authorities and monument preservation and mayors of the affected communities.

In order for these cursory assessments to give comparable and reproducible results, a multi-page screening questionnaire was developed which, after the presentation of the project by the applicant and on the basis of available information and without additional study and investigations, can be answered 'Yes' or 'No' by the representative of the approval authorities.

The number of questions answered with 'Yes' is not decisive here for the question of whether an EIA is to be carried out or not; besides the contextual evaluation, it can only be an indication for the weighting of values.

At the end of the 1–2 hour screening hearing, the responsible representatives of the authorities will determine whether there is a requirement for the application to carry out an environmental impact assessment or not, based on whether substantial disadvantageous effects are to be feared. A court review of this decision is not foreseen.

In the case that, during the screening hearing, a duty for carrying out an environmental impact assessment is determined, a scoping hearing will follow with the participation of public interest parties, and on the basis of a proposal handed in by the applicant, the scope of the EIA will be defined.

3.8.3.4   EIA Audit Criteria

The EIA audit criteria consist of the following parts that are mentioned here as an overview:

1. Introduction
2. Project-relevant planning requirements and framework conditions
3. Description of the available and planned usage at the site and in the affected area
4. Description and evaluation of the environment situation
5. Development prognosis of the condition of the environment without realising the project
6. Description and characteristics of the project
7. Determination and description of the spatial and environmental impacts
8. Proposals and measures for preventing, reducing and ameliorating the environmental impacts
9. Information on occurring difficulties and existing knowledge gaps in the compilation of the information
10. Generally clear summary.

3.8.4   Specific Aspects of the Process

3.8.4.1   Application for Imission Protective Rights Approval

After determination of the environmental impact study and adding it to the other application documents, the BImSchG approval application is complete and can be processed. Now the application for approval of a wind turbine according to part 4 of the Federal Imission Protection Law can be handed into the responsible approval authorities.

If no EIA duty has been determined then the approval application can be made in a simplified process according to part 19 BImSchG.

In order to optimise the times for a star-shaped participation of the public interest parties (PIPs), application documents must be supplied in the following quantities: 3 × complete application sets (applicant, regional or town building department, and approval authorities) as well as 10 × 'thinned out' application sets (for the roads department, flight safety authorities, regional building department, community, monument protection, maritime directorate, coastal protection department, railway department, and so on). Operators of non-overhead operated directional radios are only informed.

3.8.4.2   Issuing the Permit

The imission permit for a wind turbine (>50 m) in a simplified process should be granted within approximately 3 months, and in an EIA formal process after 7 months insofar as:

- the application documentation was complete;
- the building conditions were correct;

- the community agreement was issued; and
- the participating public interest parties have agreed.

Conditions and instructions are taken up in the approval notification and must be reasoned.

3.8.4.3 Difficulties in the Approval Process

The approval process can become more difficult due to the following circumstances:

- Application documents are incomplete and must be supplemented (signatures, lease contracts, reports, equalisation payments, and so on).
- The planning rights permit has not yet been given. The community agreement is missing or been refused, or a height limitation of 100 metres is foreseen (a need-based beaconing can render such limitations pointless).
- The aviation authorities do not approve of the wind turbine proposal (signed reports sometimes assist in this).
- Unforeseen further requirements must be dealt with (report on determining the impact on people, nature, landscape, migrating and wintering birds and bats, as well as fishes, horses, but also for protection from being mistaken for traffic signals).
- An upcoming second approval process forces the first into EIA.
- Insufficient participation of citizens, or for near-border projects possibly border-straddling authorities' participation taking their regulations into account (EIA in Germany from 20 wind turbines $\neq$ EIA in Denmark from 85 m overall height).

In the approval notification for a wind turbine, because of the imission protection and the supply, subsidiary determinations (conditions, limits and information) for the protection of the neighbourhood and the environment are mentioned.

3.8.4.4 Noises are Imissions in the Sense of BImSchG Part 3 [2]

In connection with noise imissions, care must be taken in the evaluation of the plant operation in the framework of the BImSchG approval process or neighbourhood complaints, especially regarding the technical instructions for protection against noise – TA Lärm [Technical instructions, noise] – and also the recommendations of the working group 'Geräusche von Windenergieanlagen' [Noises from wind turbines].

The TA Lärm [23] defines imission guide values and describes the basic evaluation of noise imissions in approval and neighbourhood complaint procedures. Included in the approval process, besides the imission values, the draft DIN ISO-9613-2 of 1997 is of particular importance. Thus, for instance, for residences in outer areas – analogous to the village mixed areas – for the total effects of all plants, the following guide values must not be exceeded: 60 dB(A) by day and 45 dB(A) at night.

Adherence to the guide values is to be prognosticated within the framework of the approval process, and after start-up of the approved wind turbine is to be verified by means of a report of imission measurements 1 m in front of an opened window at the nearest imission site. In the case of exceeding the guide value, the noise emissions are to be reduced by means of stepped capacity reduction or possibly by non-operation at night.

3.8.4.5   Optical Imissions, Strobe Effects and Periodic Shadow Casting

Cyclic strobe effects/disco effects as well as periodic shadow casting are also imissions in the sense of the Federal Imission Protection Law.

Strobe effects are to be prevented by the requirement for medium reflecting colours (e.g. RAL 7035-HR) and matt surfaces, according to DIN 67530/ISO-2813.

Periodic shadow casting depends on the interaction of the wind direction, the sun position and the condition and operation of the wind turbine. Timed switching-off measures for the plants are required in order to reduce periodic shadow casting to a bearable amount. On the basis of scientific investigations of the Psychological Institute of the University of Kiel in 2000, such an effect is usually not very annoying when the astronomical maximum possible shadow duration is no more than 30 min/day and also not more than 30 hr/year (worst case). This corresponds to a real time of 8 hours per year.

If these times are exceeded per residence during the totality of all working wind turbines, then switching-off measures for adherence to the guide values are required. Corresponding programmable switch-off modules are available. Their suitability was proven by a two-year test under real conditions at the test site of Kaiser-Wilhelm-Koogs in Dithmarschen.

3.8.4.6   Turbulences in the Wake of Wind Turbines

Turbulences/wind-shears are formed on the lee side of wind turbines. The over- and under-pressures associated with turbulence are to be evaluated as an environmental impact in the sense of part 3 Abs. 2 BImSchG.

These turbulences can also impact high-tension lines and other wind turbines. This can lead to material fatigue with effects on the length of life of the plants. The operators of neighbouring plants are thus to be included in the approval process. If doubts are cast by them and if the distance is less than five times the rotor diameter of the wind turbine to other plants, then a site-referenced report furnished by the applicant must prove that the distance has no disadvantageous effects for the plants situated on the lee side. Building ordinance requirements remain unaffected by this.

If the distances are less than three times the rotor diameters to high-voltage lines, then these must be equipped with vibration dampers.

3.8.4.7   Characterising wind Turbines as Aviation Obstacles

Due to the development of plants of heights more than 100 m above ground, the portion of wind turbines that are characterised as an aviation obstacle according to the aviation law is rising steadily.

A scientific investigation in this respect by the Psychological Institute Halle-Wittenberg showed that in the visible segment of the wind turbines equipped with obstruction beacons, 18% of local inhabitants considered themselves harassed and what disturbed them most were the xenon strobe lights.

On various levels a solution is being sought in transponders or primary radar systems in order to activate the wind turbine obstacle lights only in the rare case of the approach of an aircraft to the wind turbine. This would increase the acceptance of higher wind turbines at sites and also corresponds to the wishes of the federal Government formulated in its energy report of 2010.

3.8.5 Acceptance

In order to improve acceptance in the neighbourhood, the possibility of citizen participation in wind energy projects should be extended insofar as noise and shadow casting of these plants are concerned, in order to compensate disadvantages with advantages. Community borders are insignificant in this.

3.8.6 Monitoring and Clarifying Plant-Specific Data

On the part of the authorities, the following measuring instruments are available for plant assessment by approval and monitoring authorities, or for clarifying complaints by affected residents:

- **Laser-supported telescope**: For determining the plant height and the rotor diameter as well as the distance between wind turbine and residences.
- **Gloss degree measurement device**: For checking the required degree of gloss of the rotor blades for disco-effect prevention.
- **RAL colour table**: For checking the colours used for medium reflection behaviour.
- **GPS hand device**: For determining the site data in WGS or Gauss-Krüger coordinates.
- **Acoustic power level measuring device**: Calibrated instruments are available for oriented measurements.

In addition, Internet research via satellite-supported aerial photos permits a first view of the conditions at site. Measuring of the site of the approved plant and the noise imission values is carried out in any case by known institutions corresponding to the limitations of the approval notification after erection and start-up. Further plant data can then be made available and can be taken from the construction type approval and application documents.

3.9 Example Problems

1. **Height exponent**
   At a site the average annual wind velocity at 50 m height is 6.0 m/s. The Hellmann height exponent is 0.16. What is the annual wind velocity at 100 m height?
2. **Roughness and profile**
   The average wind velocity over a landscape of roughness classification 2.0 is 5.9 m/s at 50 m height. What is the wind velocity at 80 m height? (Name the intermediate resulting length.)
3. **Convert roughness**
   What roughness length z_0 does the roughness classification 2.1 have?

4. **Calculate roughness**

   What is the roughness length of the following village district: over an area of $10,000\,m^2$ there are 10 houses, each with a height of 5 m and a width of 20 m, and the land between the houses is described with a roughness length of 0.03?

5. **Speed-up on hill**

   The reference velocity in front of a gentle hill, $v_{ref}=11.9\,m/s$, is measured at a reference height $h_{ref}=50\,m$ over ground of roughness length $z_0=0.03\,m$. What is the velocity v_0 at a height $h=25\,m$? The hill has a height $H=100\,m$ and the half-value length is $L=250\,m$. The ground roughness is the same throughout and also at the tip of the hill. How great is the undisturbed velocity v_0 at $h=25\,m$ height? How large is the speed-up factor S? What is the correction factor Cor? What is the velocity v at 25 m above the tip?

6. **Obstacle**

   A house is an obstacle and is 10 m high. At a height of 20 m the undisturbed wind velocity is 6.0 m/s. What is the wind velocity at this height at a distance 250 m behind the obstacle? Estimate the solution using the diagram fig. 3.5 Wind weakening.

7. **Thrust coefficient**

   From the capacity coefficient Betz $C_p=16/27$, derive the equation for the capacity $P=\frac{1}{2}\rho A C_p v_0^3$, the equation for the thrust $F=\frac{1}{2}\rho A C_t v_0^2$, the connection of capacity, force and velocity at the rotor level $P=Fv_R$, the estimate according to Betz for the velocity at the rotor level $v_R=\frac{2}{3}v_0$, the relationship C_t to C_p, and the value for C_t.

8. **Wind weakening in the farm**

   A plant is subject to free wind of 10 m/s at the hub. It has a thrust coefficient of 0.7. To what velocity has the wake flow directly behind the plant sunk? What is the velocity of the wake at a distance of five times the diameter of the plant behind the plant?

9. **Weibull parameter**

   $A=5.8\,m/s$ and $k=2.1$ are known for a Weibull distribution. What is the average wind velocity v_{aver}?

10. **Annual energy production**

    A plant with nominal capacity of 200 kW and an annual average wind velocity of 5.9 m/s at the height of the hub generates an annual energy production of 496,000 kWh.

    a) Calculate with the Rayleigh function the frequencies $p(v)$ for the velocity bin 8–9 m/s. *Information*: calculate with $v=8.5\,m/s$. The distribution provides the frequency for the bin width 1 m/s, convert to %, remain in the standardisation at 1, as the equation provides, and work to 4 decimal places.

    b) Convert the frequency from the above solution into a number of hours in the year (a year has 8760 hours).

    c) At a velocity of 8.5 m/s, the capacity of the plant is 103.7 kW. How much annual energy does the plant generate in this velocity bin?

    d) Give the full-load hours for the plant.

    e) Give the capacity factor that is occasionally given instead of the full-load hours.

11. **Duration of enhanced subsidy**

    In the first five years a wind turbine generates the energy $E5=2,345,000\,kWh$. The reference energy $R=2,422,766\,kWh$. For how many months is the payment of the subsidy extended? For how many years is the increased subsidy paid?

12. **Noise evaluation**

    What noise level is permitted at night in business areas?

13. Noise attenuation

How loud is it at a house (L_{pa}), when the wind turbine has a noise volume $L_{WA} = 100\,\text{dB(A)}$, the distance between the house and the plant through the air is $300\,\text{m}$, and the simplified formula TA-Lärm is to be used?

14. Estimate acoustic power

Estimate the acoustic power of a wind turbine. Use the empirical dependence LEA of the blade tip velocity, which is $63\,\text{m/s}$ and a rotor diameter of $30\,\text{m}$. What is the speed of revolution per minute of the rotor?

15. Noise addition, simple

How many dB(A) is a doubling of the noise? Calculate exactly 10 times $\log_{10}$ (2 times L by L) where L is a noise value that can be abbreviated. What does the (A) behind dB mean?

16. Noise addition

a) Doubling is $3\,\text{dB}$. By means of exact energetic addition (multiple logarithmic and factor 10: $L_3 = 10 \log (10^{(L1/10)} + 10^{(L2/10)})$) that $50\,\text{dB}$ energetic plus $50\,\text{dB}$ equals $53\,\text{dB}$.

b) The stronger wins by $10\,\text{dB}$ difference. Show by exact energetic addition that $60\,\text{dB}$ energetic plus $50\,\text{dB}$ is equal to $60\,\text{dB}$. As an exception, give both results to 3 decimal places. It is then plain that the statement is only approximate, which is practically meaningless because the dB values in front of the decimal point are strong but behind it are weak.

17. Turbulence

What is the connection between standard deviation, turbulence intensity and 10 minutes as regards wind? Over the roughness length $0.1\,\text{m}$ at a height of $30\,\text{m}$ and at a velocity of $5.5\,\text{m/s}$, what is the turbulence intensity and the standard deviation of the wind velocity? What is the formula sign of the standard deviation? What unit does this standard deviation possess? Does the turbulence intensity increase or decrease with increasing height? At the same height, does the turbulence intensity increase or decrease with greater roughness? *Information*: use the iteration formula $I = 1/\ln(h/z_0)$.

18. Acceptance

How can one improve the acceptance in the neighbourhood of wind turbine projects?

3.10   Solutions to the Problems

1. Height exponent $6.7\,\text{m/s}$.
2. Roughness and profile: $0.1\,\text{m}$; $6.346\,\text{m/s}$.
3. Convert roughness: equation: $\ln(z_0/1\,\text{m}) = 1.333 \cdot \text{Kl-}4.9$; $0.1224\,\text{m}$.
4. Calculate roughness: 0.280.
5. Speed-up on hill: $v_0 = 10.79\,\text{m/s}$; $v_0 = 10.79\,\text{m/s}$; $S = 0.444$; $\text{Cor} = 1.444$; $v = 15.58\,\text{m/s}$.
6. Obstacle: 15%; $0.90\,\text{m/s}$; $5.10\,\text{m/s}$.
7. Thrust coefficient: $C_t = 3/2\,\text{CP}$; $C_t = 8/9$.
8. Wind weakening in the farm: $5.477\,\text{m/s}$; at $x = 5D$, $v_{wake} = 8.435\,\text{m/s}$.
9. Weibull parameter: $5.2\,\text{m/s}$.
10. Annual energy production: (a) 0.0751, unit can be left out; (b) $658.2\,\text{h}$; (c) $68260.1\,\text{kWh}$; (d) $2345\,\text{h}$; (e) 26.77%.
11. Duration of enhanced subsidy: $\Delta = 141.9$ months; +5 years: 16.8 years.

12. Noise evaluation: $50\,dB(A)$.

13. Noise attenuation: $L_{pa} = L_{WA}$ minus 10 times log (4 times π times R in m^2) plus 3; 42.5 dB(A).

14. Acoustic power estimate: $L_{WA}/dB(A) = 10\log((v_{tip}/m/s)^5) + 10\log(D/m) - 4$; 100.74 dB(A); 40.11 min^{-1}.

15. Noise addition, simple: 3 dB(A); $10\log(2\,L/L) = 10\log(2) = 3.0103$; acoustically weighted, corresponds to the sensitivity of the human ear.

16. Noise addition: (a) 53.010; (b) 60.414.

17. Turbulence: I = standard deviation/average value of the wind velocity both measured over the time interval of 10 minutes; 17.5%; 0.964 m/s; σ; m/s; decrease; increase.

18. Acceptance: Create possibility of citizen participation in wind turbine project for citizens living in the extended region of noise and shadow, irrespective of community borders.

References

[1] Troen, I. and Petersen, E.L. 1989. *European Windatlas*. Meteorology and Wind Energy Department, Risø National Laboratory, Roskilde.

[2] Troen, I. and Petersen, E.L. 1990. *Europäischer Windatlas*. Meteorology and Wind Energy Department, Risø National Laboratory, Roskilde.

[3] European Windatlas Homepage, Wind Energy Division, Risø, Roskilde DK, available at http:// www.windatlas. dk/Europe/About.html [accessed May 2011].

[4] Oertel, H. 2008. *Prandtl, Führer durch die Strömungslehre*. Vieweg Teubner Verlag, Wiesbaden.

[5] FGW-Fördergesellschaft Windenergie, Berlin, Technische Richtlinien (TR) Teil 1 bis 8, erhältlich, available at http://wind-fgw.de [accessed May 2011].

[6] IEC 61400 International Electrotechnical Commission, Wind Turbines, Part 1, 2, 3, 4, 5, 11, 12, 13, 14, 21, 22, 23, 24, 25. Bezug, IEC Central Office, Geneva. Available at http: //www.iec.ch, auch Bezug, VDE Verlag GmbH, Berlin. Available at http:// www.vde-verlag.de [accessed May 2011].

[7] WindFarmer Homepage, GL Garrad Hassan, Hamburg, Oldenburg, Bristol. Available at http:// www.gl-garradhassan.com/en/software/GHWindFarmer.php [accessed May 2011].

[8] Manwell, J.F. and McGowan, J.G. 2009. *Wind Energy Explained: Theory, Design and Application*. John Wiley and Sons, Ltd., Chichester.

[9] Hau, E. 2008. *Windkraftanlagen*. Springer Verlag, Berlin.

[10] Windpower. Available at http:// www.windpower.org/de/stat/unitsw.htm#roughness [accessed March 2006].

[11] Horns Ref, Anstieg der Turbulenzintensität mit der Windgeschwindigkeit. Available at http://130.226.56.153/rispubl/reports/ris-r.1765.pdf [accessed May 2011].

[12] *WindFarmer Theory Manual, Version 4.2*. 2011. Garrad Hassan and Partners.

[13] Niesen, P., *et al.* 2010. *WindPRO-Handbuch, Version 2.7*. EMD International A/S, Aalborg.

[14] Astrup, P. and Larsen, S.E. 1999. *WAsP Engineering Flow Model for Wind over Land and Sea*. Risø National Laboratory, Roskilde.

[15] Bundesverband Windenergie. Available at http://www.wind-energie.de/infocenter/statistiken [accessed May 2011].

[16] Pielke, R.A. 1984. *Mesoscale Meteorological Modeling*. Academic Press, New York, p. 612.

[17] Kalnay, E., Kanamitsu, M., Kistler, R., *et al.* 1996. The NCEP/NCAR 40-year reanalysis project. *Bulletin of the American Meteorological Society*, 77: 437–71.

[18] Kistler, R., Kalnay, E., Collins, W., *et al.* 2001. The NCEP-NCAR 50-year reanalysis: monthly means CD-ROM and documentation. *Bulletin of the American Meteorological Society*, 82: 247–68.

[19] Jensen, N. 1983. (Article) A note on wind generator interaction. Risø M-2411.

[20] Katic, I., Hoestrup, J. and Jensen, N. 1986. (Article) A simple model for cluster efficiency. EWEC.

[21] Molly, J.-P. 1990. *Windenergie*. Verlag C.F. Müller GmbH, Karlsruhe.

[22] EEG 2009, Erneuerbare-Energien-Gesetz 29, Wind Energie. Available at de/eeg_2009/index.html [accessed May 2011].

[23] TA Lärm. 1998. *Sechste Allgemeine Verwaltungsvorschrift zum Bundes-Immissionsschutz-gesetz* (Technische Anleitung zum Schutz gegen Lärm, TA Lärm).

[24] Burton, T., Sharpe, D., Jenkins, N. and Bossanyi, E. Wind Energy Handbook. John Wiley & Sons, Ltd., Chichester. 2001.

[25] Quarton, D.C. and Ainslie, J.F. 1990. Turbulence in wind turbine wakes. *Journal of Wind Energy*, 14.

[26] Garrad Hassan and Partners. 1989. Characterisation of wind turbine wake turbulence and its implications on wind farm spacing – wake turbulence characterisation. ETSU WN 5096.

[27] WindPRO Homepage, Energi- og Miljødata EMD, Aalborg, DK. Available at http://www.emd.dk/ WindPRO/ Frontpage [accessed May 2011].

[28] BImSchG – (Law) Gesetz zum Schutz vor schädlichen Umwelteinwirkungen durch Luft-verunreinigungen, Geräusche, Erschütterungen und ähnliche Vorgänge (Bundes-Immissionsschutzgesetz – BImSchG) i. t. v. of 26. Sept. 2002 (BGBl. I S. 3830).

[29] BImSchV – (law) Vierte Verordnung zur Durchführung des Bundes-Immissionsschutz-gesetzes (Verordnung über genehmigungsbedürftige Anlagen – 9. BImSchV) in the publication of 19 May 1992 (BGBl. I S. 1001).

[30] BImSchV9 – (Law) Neunte Verordnung zur Durchführung des Bundes-Immissionsschutz-gesetzes (Verordnung über das Genehmigungsverfahren – 9. BImSchV) i. t. v. of the publication of 5 September 2001 (BGBl. I S. 2351).

[31] BGBl.I – Gesetz über die Umweltverträglichkeit in the version of 05.09.2001 (BGBl. I S. 2351).

[32] Herten. 2007. *Windenergie, Handbuch des Staatlichen Umweltamtes Herten.*

4

Aerodynamics and Blade Design

Alois Schaffarczyk

4.1 Summary

The basics of aerodynamics of wind turbines as a quantitative description of the flow around parts of or whole wind turbines or even wind farms are shown. As there is a comprehensive discussion of this field under preparation [1] and space here is limited, we will refer in many places to further studies of the original literature. For that reason an extensive discussion of computer-based methods (CFD) applied to wind turbine flow problems must also be omitted. The interested reader therefore should consult the final report [2] of the IEAwind Task 29, in which a comparison was undertaken by teams from 11 countries to meet the experimental wind tunnel results of Mexico (Model Experiments in Controlled Conditions) [3].

This chapter is divided into three parts: horizontal rotors, vertical rotors and wind-propelled vehicles. In a first overview, Figure 4.1 shows (without wind-propelled vehicles) that c_p is a dimensionless (comparison) capacity and λ, the tip speed ratio, a dimensionless comparison number.

4.2 Horizontal Plants

4.2.1 General

For the mechanical design of wind turbines, it is necessary to refer back to theoretical models, which, in their simplest form, go back to the theories of Rankine [4] and Froude [5]. They were originally developed for ship and airplane propellers and then transferred by Betz [6] and Glauert [7] to wind turbines. Despite all attempts to exceed the tight limits of applicability (for

---

Understanding Wind Power Technology: Theory, Deployment and Optimisation, First Edition.
Edited by Alois Schaffarczyk. Translated by Gunther Roth.
© 2014 John Wiley & Sons, Ltd. Published 2014 by John Wiley & Sons, Ltd.

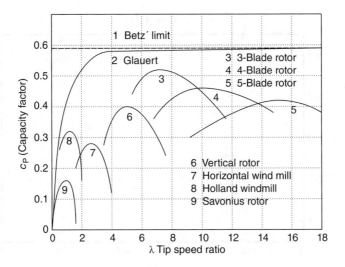

Figure 4.1 Map of the wind turbines

instance with CFD, computational fluid dynamics) this blade element method is still the most used in practice today because, in comparison with low implementation and entry effort, it provides usable results. From many points of view, the development of the aerodynamic theory of wind turbines runs parallel to that of ship's propellers [8] and helicopters [9], whereby naturally there are big differences. Thus, for ship's propellers the cavitation and the associated much lower underpressures on the suction side of the profile are of greater importance, whereas with helicopter rotors the Mach number (= flow velocity/velocity of sound) is situated near 1 and therefore the compressibility of the medium must be included.

4.2.2 Basic Aerodynamic Terminology

If a two-dimensional, aerodynamic profile (Figure 4.2) of depth c has a direct flow v, then the resultant is two force components in the direction of drag (D) and vertical to the wind (lift – L) per metre span width, with the factors:

$$C_D = \frac{D}{\frac{\rho}{2} v^2 c \times 1} \tag{4.1}$$

$$C_L = \frac{T}{\frac{\rho}{2} v^2 c \times 1} \tag{4.2}$$

A typical polar (application of these values against each other or relative to the angle of attack) is shown in Figure 4.3 (AOA is the angle of attack, α is the angle between the incoming direction and the chord).

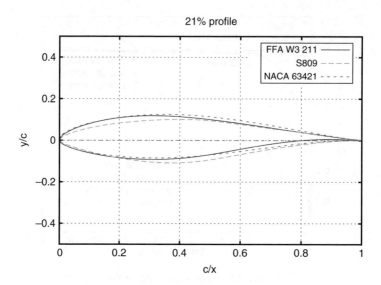

Figure 4.2 Wind turbine profile with 21% thickness

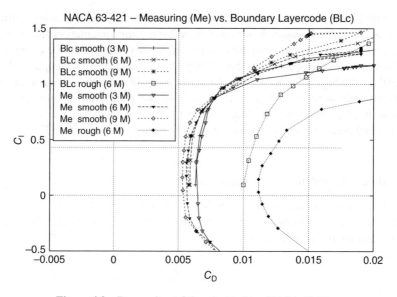

Figure 4.3 Drag polars of the wind turbine NACA 63421

In practice, use is mostly made of tables (e.g. NACA [10] or the Stuttgart Profile Catalogue [11]) or some, partly very complex measurements [12,13] must be carried out when their characteristics are not known.

We will advance step-by-step. For this reason we will first look at systems that are driven by the wind in a purely translational manner. An example of this can be a sailing boat or an ice glider (see Figure 4.4).

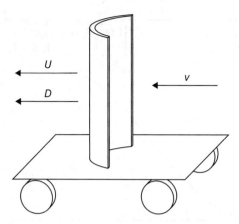

Figure 4.4   Advance using drag or lift forces (*D*)

4.2.2.1   Drag Forces

C_D and C_L according to equations (4.1) or (4.2) are the coefficients of drag or lift. If $A = cl$ is the surface of the sail and U is the wind velocity and v that of the boat, then for the performance there applies:

$$P = \frac{\rho}{2}(U - v)^2 C_D \times cl \times v \tag{4.3}$$

If one refers to a performance coefficient of that of the wind ($P_w = \rho/2 \times U^3$) then the result is:

$$c_P = c_D (1 - v/U)^2 (v/U) \tag{4.4}$$

If one differentiates for maximising according to $a = v/U$, then it is shown that the maximum performance coefficient of such a drag runner cannot be greater than:

$$c_P^{max,D} = \frac{4}{27} c_D \tag{4.5}$$

thus $c_P^{max,D} \approx 0.3$ at $v = \frac{1}{3}U$. Such a sailing boat at wind force 7 (according to Beaufort ≈ 30 knots ≈ 16.2 m/s) travels no faster than 10 knots. With 30 m² sail area it has a capacity of $P = 24$ kW.

4.2.2.2   Lift Forces

In order to increase the efficiency of such vehicles, use is made of the lift forces that in themselves perform no work as they are vertical to the velocity.[1]

However, the geometric interconnections are somewhat more complicated. The wind is now vertical to the direction of travel (half-wind). The performance is $P = Nv$. N is the normal

---

[1] $P = \dot{W} = \vec{F} \cdot \vec{v} = 0 \Leftrightarrow \vec{F} \perp \vec{v}$

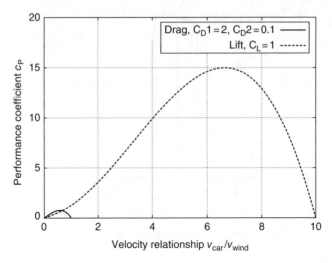

Figure 4.5   Performance coefficients of drag and lift vehicles in comparison

force vertical to the wind and parallel to v. The velocities U and v form a triangle and N is made up of:

$$N = N_L + N_D = L \cos \phi - \sin \phi \tag{4.6}$$

By means of algebraic conversion this leads to:

$$c_P = a \left(\sqrt{1 + a^2} \right) (C_L - C_D \, xa) \tag{4.7}$$

The maximum performance coefficient (see Figure 4.5) is now:

$$c_P^{max,L} = \frac{2}{9} C_L \left(\frac{C_L}{C_D} \right) \sqrt{1 + \frac{4}{9} \left(\frac{C_L}{C_D} \right)^2} \tag{4.8}$$

Note:
a) Now the maximum velocity is greater than the wind and is reached at:

$$a = \frac{v}{U} = \frac{2}{3} \frac{C_L}{C_D} \tag{4.9}$$

In our example, therefore, $v_{max} \approx 6U = 180$ knots.
b) A (not negligible) force T parallel to the wind must be compensated.

In Figure 4.5 both cases for $C_D = 2$ or $C_L = 10$ and $C_L/C_D = 10$ are compared. A somewhat more comprehensive discussion of wind-propelled vehicles is carried out in Section 4.5.

4.3   Integral Momentum Theory

4.3.1   Momentum Theory of Wind Turbines: the Betz Limiting Value

In the simplest case a wind turbine reduces the wind energy only by means of slowing down, which, because of its mass retention, must occur with spreading (see Figure 4.6). This results

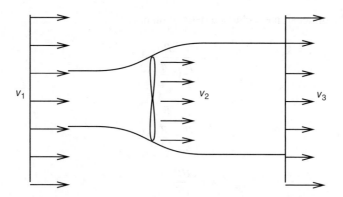

Figure 4.6   Idealised flow relationships for a wind turbine

in a pressure drop behind the turbine, thus a thrust force $\vec{T}$ in the direction of the wind. If one further assumes that only axial (i.e. in wind direction) velocity components are available, then in the overview the result is the following simple derivation:

Energy:	$E = \dfrac{1}{2}mv^2$	(4.10)
Performance:	$P = \dot{E} = \dfrac{1}{2}\dot{m}v^2$	(4.11)
Removed performance:	$P_T = \dot{E} = \dfrac{1}{2}\dot{m}\left(v_1^2 - v_3^2\right)$	(4.12)
Froude's Law:	$v_2 = \dfrac{1}{2}(v_1 + v_3)$	(4.13)
Mass flow:	$\dot{m} = \rho A v_2$	(4.14)
Thus:	$P_T = \dfrac{1}{2}\rho A v_1^3\left(\dfrac{1}{2}\left(1 + \dfrac{v_3}{v_1}\right)\left(1 - \left(\dfrac{v_3}{v_1}\right)^2\right)\right)$	(4.15)
	$c_P(\eta) = \dfrac{1}{2}(1+\eta)(1-\eta^2)$	(4.16)
with:	$\eta = \dfrac{v_3}{v_1}$	(4.17)
with:	$a = \dfrac{v_2}{v_1}$	(4.18)
	$c_T(a) = 4a(1-a)$	(4.19a)
	$c_P(a) = 4a(1-a)^2$	(4.19b)

Now also the dimensionless characteristic numbers P:

$$c_P = \frac{P}{\frac{\rho}{2} v^3 \frac{\pi}{4} D^2} \tag{4.20}$$

$$c_T = \frac{T}{\frac{\rho}{2} v^2 \frac{\pi}{4} D^2} \tag{4.21}$$

$$\lambda = \frac{\Omega R}{v} \tag{4.22}$$

and $$\Omega = 2\pi N/60 \tag{4.23}$$

are known. The result therefore, as a maximum performance, is the Betz (limiting) value:

$$c_P^{max} = \frac{16}{27} = 0.5926 \quad \text{at} \quad a_{max} = \frac{1}{3} \tag{4.24}$$

One can see that far downstream the wind will be reduced by ⅔, to ⅓. Besides this, it can be seen that this derivation loses its validity if the airflow is completely braked ($a = 0.5$) and hence there is no propeller wash behind the turbine.

In order to determine the effects of the force, one can, because of $T = \frac{d}{dt}mv = \dot{m}v$, compare impulse losses within a control volume to:

$$T = \dot{m}(v_1 - v_3) \tag{4.25}$$

In this way a secure upper limit for the capacity of a wind turbine is given, which, despite stubborn attempts by many inventors, cannot be exceeded.[2]

4.3.2 Changes in Air Density with Temperature and Altitude

Often it is necessary to convert the standard density ($\rho_0 = 1.225\,\text{kg/m}^3$) of air to other temperatures (ρ) and other ground heights (H). For this, use can be made of the formulae:

$$\rho(p, T) = \frac{p}{R_i T} \tag{4.26}$$

$$T = 273.15 + \theta \tag{4.27}$$

$$R_i = 287 \tag{4.28}$$

---

[2]Somewhat higher values – to about 0.62 – can be achieved when expanded turbines such as vertical-axle machines – *Darrieus* rotors – are viewed [14]. This was already recognised by Betz [6] (see Section 4.5.2).

$$p(H) = p_0 e^{-H/H_{ref}} \tag{4.29}$$

$$p_0 = 1015\,\text{hPa} \tag{4.30}$$

$$H_{ref} = 8400\,\text{m} \tag{4.31}$$

4.3.3 Influence of the Finite Blade Number

A frequently asked question is about the number of blades. It should be remarked at this stage that there are no aerodynamic reasons for the universally used three-blade form. Much more decisive are the higher production costs (more than three blades) and increase of the aeroelastic loading during yawing (less than three blades).

The lift L can be linked to an important fluid-mechanical quantity, the circulation, by using the theorem of Kutta-Joukovski:

$$\Gamma = \oint \vec{v}\ ds \tag{4.32}$$

$$L = \Gamma \rho v \tag{4.33}$$

Further, this dimension is linked with the local blade depth $c(r)$ and C_L:

$$\Gamma = \frac{B}{2} C_L c\ w \tag{4.34}$$

The product $c(r) \cdot C_L$ thus defines the aerodynamic blade design.

But the blade element and the actuator disk method (see below) assumes implicitly that at the edge of the blade there is no pressure equalisation and therefore a constant circulation ($\approx$ lift $\approx$ driving moment) can remain. This is strictly only given for B (no. of blades) $\to \infty$. For the case of the finite number of blades, Prandtl [15] has already given an approximation for the circulation loss. It is based on a potential theoretical concept according to which the vortex-induced flow about a B-bladed rotor can be depicted on a batch with B plates.

If F (the reduction factor of the circulation at the blade tip) is defined by:

$$\Gamma = F(B)\ \Gamma_\infty \tag{4.35}$$

then it follows:

$$F = \frac{2}{\pi} \arccos(\exp(-f)) \tag{4.36}$$

$$f = \frac{B}{2} \frac{R-r}{r \sin\phi} \tag{4.37}$$

Here ϕ is the flow angle (see Equation (4.52)). The losses are marked especially by $B = 1$ and $B = 2$. The case for $B = 1$, which was built as Monopteros (see Figure 4.7) was controversial for a very long time. It was Okulov [16] who, with the aid of analytical attempts, was able to find a closed solution that could be included harmoniously into the overall picture.

Figure 4.7   Monopteros in the DEWI test field near Wilhelmshaven, Germany. Photo: Alois Schaffarczyk

Recently [17,18] the methods of flow simulation (CFD) were also used in order to investigate the applicability of these assumptions. It seems that the circulation loss – depending on tip form – can be greater than proposed by the Prandtl approximation. However, the doubts that we will mention in Section 4.4.5 on the basis of comparison with measurements will also apply.

4.3.4   Swirl Losses and Local Optimisation of the Blades According to Glauert

In order to include the rotating motion of the rotor, angular momentum is balanced analogous to the case of pure axial flow in order to determine the moment and thus, according to $P = M\omega$, the performance.

$$M = \dot{m} r v_t \tag{4.38}$$

$$dM = d\dot{m} r v_t \tag{4.39}$$

If one divides the rotor surface in increments of length dr and surface $dA = 2\pi r\, dr$, then it follows:

$$dM = 4\pi r^3 v_1 (1-a) a' \omega\, dr \tag{4.40}$$

Here $a' = \omega/2\,\Omega$ is the induction factor for tangential velocity components. For the whole performance there is:

$$P = \int \omega\, dM \tag{4.41}$$

and with:

$$c_P = \frac{8}{\lambda^2} \int_0^\lambda (1-a)a'x^3 \, dx \tag{4.42}$$

whereby $x = \omega r / v_1$ is the local tip speed ratio formed with the hub distance. In comparison with the Betz discussion, there are only two optimising parameters a, a' that are both dependent on x. A further equation, the so-called Glauert orthogonality condition (see [19,20]):

$$a'(1-a')x^2 = a(1-a) \tag{4.43}$$

is necessary in order to close the equation system.[3] According to Glauert the turbine is now locally optimal when the function

$$f(a, a') = a'(1-a) \tag{4.44}$$

under the subsidiary condition Equation (4.43) becomes extreme. Deriving from Equation (4.44) leads to:

$$(1-a)\frac{da'}{da} = a' \tag{4.45}$$

or from Equation (4.43):

$$(1+2a')x^2 \frac{da'}{da} = 1 - 2a \tag{4.46}$$

together thus:

$$a' = \frac{1-3a}{1+4a} \tag{4.47}$$

As a' must be positive, it follows that $0.25 < a < 0.33$. The equations (4.43) and (4.47) determine the course of $a(x)$ and $a'(x)$ (see Table 4.1).

In order to carry out the integration over the blade elements (i.e. over x, see Equation (4.42)) the function $x(a)$ need only be inverted, i.e. $x(a) \rightarrow a(x)$ after a' Equation (4.43) is eliminated, which, however, takes some effort. The results are summarised in Table 4.1 and Figure 4.8.

Recent investigations (see, for instance, [22]) show that this relationship, just as in the rotation-free, axial case, applies strictly only in the border case of weakly loaded rotors, e.g. $C_T \rightarrow 0$. However, $C_T \approx 0.7...0.9$ is typical for wind turbines and is not taken into account in the greater sense.

---

[3] Also here it can be seen that this is strictly valid only for $C_t \rightarrow 0$ (see [21]).

Table 4.1 a, a' from $x = \omega r/v$

a	a'	x
0.25	∞	0.0
0.26	5.500	0.073
0.27	2.375	0.157
0.28	1.333	0.255
0.29	0.812	0.374
0.30	0.500	0.529
0.31	0.292	0.753
0.32	0.143	1.15
0.33	0.031	2.63
0.333	0.003	8.58
1/3	0.00	∞

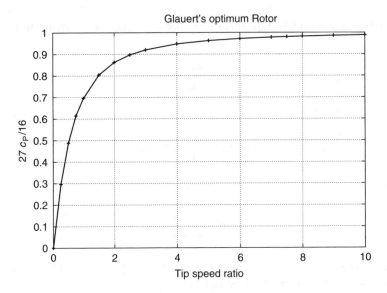

Figure 4.8   Optimum rotor according to Glauert, containing only vortex losses

Glauert's [7] analysis (see Table 4.1) shows that the performance factor is only markedly reduced by vortex losses when λ drops below 2. The integration through x (dimensionless reach) also means that a transfer to the blade sections must be undertaken.

4.3.5   Losses Due to Profile Drag

The last important mechanism for reducing the performance of a wind turbine will now be discussed with the so-called profile losses. A summary of all mechanisms is shown in Figure 4.9. E1 and E2 are two turbines of a German manufacturer and are meant to show the advances in only ten years.

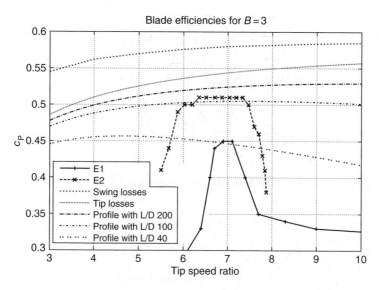

Figure 4.9 Comparison of the mechanisms for a blade number $B = 3$

4.4 Momentum Theory of the Blade Elements

4.4.1 The Formulation

The actual blade element method (Figure 4.10), in connection with the differential momentum method, presents a suitable procedure for making measurable statements about the performance of rotors (propellers and impellers) and loads.

On the assumption of the impulse theory, it is further stated that now the origin described by the forces, lift and drag, is assumed, due to the parameters C_L and C_D (see above). The increments of the moment (dQ) and the axial force (thrust dT) are referenced to a radial increment dr:

$$dT = d\dot{m}(v_1 - v_2) = 4a(1 + a)2\pi r\, dr \frac{\rho}{2} v_1^2 \tag{4.48}$$

$$\frac{Bc(r)}{R_{tip}} = 2\pi \frac{8}{9} \frac{1}{C_L} \frac{1}{\lambda^2 \left(\dfrac{r}{R_{tip}}\right)} \tag{4.49}$$

Here r is the observed distance from the hub, v_1 is the direct axial velocity and $v_2 = (1 - a)v_1$, the velocity at the rotor plane.

Figure 4.11 shows the resultant velocity and force triangles. These are:

$$C_n = C_L \cos\phi + C_D \sin\phi \tag{4.50}$$

$$C_{tan} = C_L \cos\phi - C_D \sin\phi \tag{4.51}$$

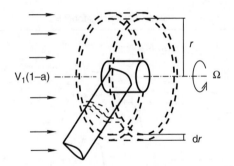

Figure 4.10   Blade elements

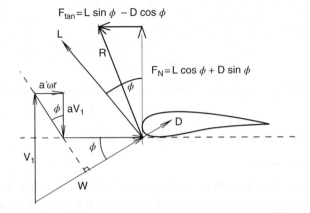

Figure 4.11   Velocity and force triangles

with
$$\tan \phi = \frac{1-a}{1+a'} \frac{v_\infty}{r\Omega}$$
(4.52)

and
$$\phi = \alpha + \theta$$
(4.53)

for the flow angles ϕ and θ of the twist of the blades.

This method has been completed in much detail by means of supplementary assumptions. For instance, the abovementioned losses due to 3D effects at the hub and blade tip have been included.

Thus there is then (w is the overall flow velocity):

$$dT = Bc\frac{\rho}{2} W^2 C_n \, dr$$
(4.54)

$$dQ = Bc\frac{\rho}{2} W^2 C_t \, r dr$$
(4.55)

$$w^2 = (v_1(1-a))^2 + (\omega r(1+a'))^2$$
(4.56)

Analogous to the axial induction factor a is $a' = 2w/\Omega$, as previously mentioned. A Froude's theorem is also assumed for a', according to which the induced components at the propeller level are half as large as in the distant wake. Equation (4.48) applies to the thrust only until $a \leq \frac{1}{3}$ (or at least $\leq \frac{1}{2}$), otherwise empirical extrapolations must be used.

If one presents equations (4.48) and (4.49) with reference to a and a' then:

$$\frac{a}{1-a} = \frac{BcF_t}{8\pi r \sin^2 \phi} \tag{4.57}$$

$$\frac{a'}{1+a'} = \frac{BcF_n}{4\pi r \sin^2 \phi} \tag{4.58}$$

Overall the following iteration scheme emerges:

- Estimate a and a' (at the start $a = a' = 0$).
- (*) Determine φ from Equation (4.52).
- Determine the attack angle α from Equation (4.53).
- Determine C_L and C_D from a table (see Section 4.2.2).
- Determine C_n and C_t from equations (4.50) and (4.51).
- Determine a and a' from equations (4.57) and (4.58).
- Go to (*) and iterate until below a given error.

4.4.2 Example of an Implementation: WT-Perf

WT-Perf is a FORTRAN implementation by M. Buhl of the National Renewable Energy Laboratory (NREL) in Golden, Colorado, US, which has been derived from the legendary PROP Code by S. Walker. The source can be translated by any GNU compiler.

4.4.3 Optimisation and Design Rules for Blades

The external shape of a blade is geometric (aerodynamic), in other words, determined by depth distribution $c(r)$, twist distribution $\rho(r)$ and profile section (= aerodynamic profile). Many so-called optimum designs are derived from the above definitions [7,20,23]. In the simplest case, for the relative depth c/R:

$$\frac{c(r)}{R} = 2\pi B \frac{8}{9} \frac{1}{C_L} \frac{1}{\lambda^2 \left(\dfrac{r}{R}\right)} \tag{4.59}$$

(see, for instance, [23]), and for the twist:

$$\theta = \arctan\left(\frac{3}{2}\lambda \frac{r}{R}\right) \tag{4.60}$$

Results of a local aerodynamic optimisation according to Wilson [20] and de Vries [19] are shown in Figure 4.12 and compared with various real designs. The inclusion of further

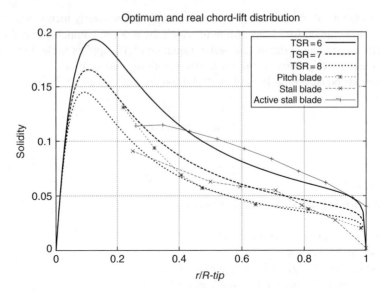

Figure 4.12 Optimum blade design (3 blades) according to Wilson/de Vries; comparison with actual blades

important subsidiary conditions, such as loads, and the importance of optimising the overall costs of the energy thus generated, are discussed in many publications. Reference is made to a current example [24]. An attempt is made in [25] to mirror commercial blade concepts to this process.

4.4.4 Extension of the Blade Element Method: The Differential Formulation

The blade element method is strongly based on integral energy and impulse balances that can apply only at three places (far in front of the rotor, at the rotor itself and far downstream). As it is assumed that at the first and last positions there are no radial variations of the velocity, this is a one- rather than two-dimensional method. The so-called actuator-disk methods represent a fully 2D extension. They can be solved semi-analytically (see [22]). It is mentioned at this point that one can 'downgrade' full 3D methods (see Section 4.5.5) to 2.5D in order to find a middle way with less calculation effort.

The implementation in computational fluid dynamics (CFD) occurs basically via an additional volume force. For this purpose both lift and drag are divided into axial and circumferential directions and distributed $2\pi r dr$ over the surface element. As a volume force this pressure jump must then be referenced to the thickness t of the element. In the following we present a few examples of the effectiveness of the method.

4.4.4.1 Validation by Means of an Example of an Active Stall Blade

A so-called active stall blade ((ARA48, [26]) was used to compare both methods (see Figure 4.13). Deviations in the region of the tip of the blade are clearly seen; as in the CFD formulation, no Prandtl correction has been included.

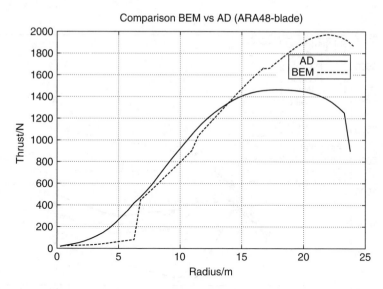

Figure 4.13 Comparison of integral (BEM) and differential (CFD) blade element methodology as an example of the thrust of a 600 kW active stall blade

4.4.4.2 Wind Turbine with Concentrator (Diffuser)

Hansen *et al.* [27] presented the effects of a wind concentrator consisting of a NACA 0015 (ring) profile. Our method [77] was able to choose a suitable blade shape for this combined system, as there was a very strong interaction between rotor and diffuser and this forbade a separate design.

In his dissertation, Phillips [28] investigated many further attempts and parameters. We can show [29] that an integrated blade-element-momentum (BEM) method is unsuitable. A current summary of the state of the theory and experiment can be found in [30].

4.4.4.3 Contra-Rotating Tandem Wind Turbine

An interesting variant that is usefully applied to ship propellers and helicopters is the so-called contra-rotating wind turbine. However, investigations show that the expected yield profit (approx. 5–10%) could be nullified by too great an extra effort, especially in the design of the blade shapes.

4.4.5 Three-Dimensional Computational Fluid Dynamics (CFD)

The use of effective numerical methods for investigating wind turbines has advanced in parallel with the general development of aerodynamic theories. For the past 20 years numerical procedures for solving the fully three-dimensional Navier-Stokes equation – together with empirical turbulence models such as the k-ε or the k-ω model – have been commercially available. Unfortunately these turbulence models are not yet sufficiently matured to provide

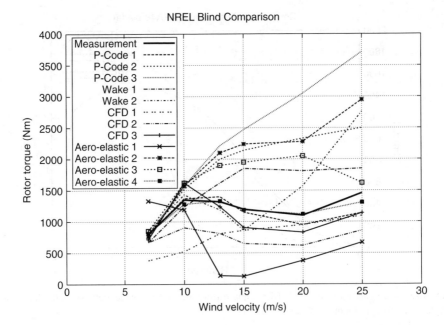

Figure 4.14   Blind comparison of the NASA-AMES-measurements of 2000

sufficiently accurate information. However, the results of many experiments on various detailed questions seem to indicate that CFD (in this sense the numerical solution of the 3D Navier-Stokes equation) seems to be the only possibility for consistent extension of the aerodynamic description of wind turbines.

An impressive example of the results achieved with 3D-CFD is a summary of 'blind comparison' of the NASA Ames wind tunnel experiments in 1999/2000, shown in Figure 4.14. The data rows C2 to C3 represent the CFD calculations. CFD3 is the investigation by Risø (now DTU-Windenergy).

4.4.6   Summary: Horizontal Plants

In this chapter we have presented the most important interconnections and results of the aerodynamics of wind turbines with horizontal axes. Even though work has been carried out in this field for 150 years there still seem to be no final answers to many detailed questions. However, the simple but detailed blade element method presented is sufficient for a first approximation if its applicability limits are not lost sight of.

4.5   Vertical Plants

4.5.1   General

Wind turbine plants with vertical axes of rotation (VAWT) (Figure 4.15) present an alternative to the wind turbine with horizontal axes. They also offer the possibility for use of methods of

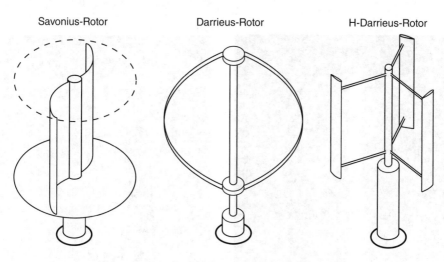

Savonius-Rotor   Darrieus-Rotor   H-Darrieus-Rotor

Figure 4.15   Wind turbines in vertical arrangements

mobile (transient) flow mechanisms. Because of their independence of the direction of the wind, there are some further advantages in their practical realisation compared to the horizontal-axis wind turbine (HAWT). And yet many plants failed – up to the 4 MW capacity and 100 m high ÉOLE-C in Cap-Chat, Canada [71,72] (Figure 4.16a) – because of insufficient attention toward operating loads. Figure 4.16b shows a three-bladed arrangement with 50 kW rated capacity of HEOS Energy.

This section will present important models and methods showing how they compare with the aerodynamics of the horizontal plants. In Figure 4.15 the difference between lift rotors and drag rotor can clearly be seen (Darrieus and Savonius principles).

However, it is also clear that horizontal machines benefited in the past 20 years from an enormous jump in development that was not noticeable in the vertical machines, so that only small plants under 100 kW are developed and produced [31,32]. In Germany, development was mostly carried out by the Dornier companies [33–37] and the Heidelberg Motor Company [38].

In the US the Sandia National Laboratory has carried out numerous experiments whose results are available on the Internet [39,40]. A good overview of these projects can be found in [41].

Worthy of mention is a plant of the Heidelberg Motor Company that provided power in the Antarctic from the beginning of the 1990s to 2008 [42]. Experiments have only recently been renewed [43,44]. A project of a 20 MW vertical-axis wind turbine with a floating support structure (DeepWind) is described in [44].

Assume in the normal manner that v_1 is the wind velocity, $\omega = 2\pi \, \text{rpm}/60$, the wind velocity formed from the revolutions per minute. Then $\lambda = \omega R / v_1$ is the tip speed ratio, an important parameter. As Betz already showed in 1926, this only applies for a wind turbine with a horizontal axis whose effect is described on the wind flow by a (infinitely thin) part of the surface $A = \pi R^2$. Much effort was expended in order to achieve or even exceed $c_p^{\text{max,B}} = 16/27 = 0.593$ (Equation (4.24)).

(a)   (b)

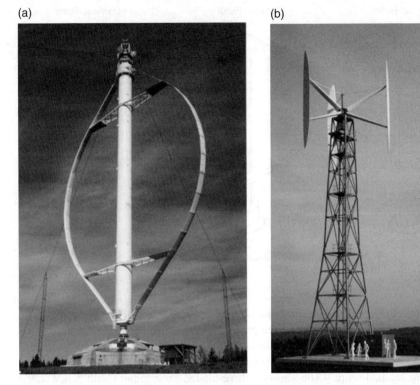

Figure 4.16 (a) The presently largest vertical plant ÉOLE-C in Cap-Chat, Canada, $A=4000\,m^2$, $P=4\,MW$, reproduced with kind permission of Prof. Dr.-Ing. Tamm, Nordakademie Elmshorn; (b) a modern 75 kW plant, reproduced with the kind permission of HEOS Energy GmbH, Chemnitz

The main questions for these rotors are therefore:

- How near can one come to these values?
- What are the operating loads on such a plant?
- Can they be competitive with horizontal axes wind turbines?

Naturally, the last question can only be decided in connection with cost models [45,46].

4.5.2   Aerodynamics of H Rotors

4.5.2.1   Momentum Theory 1: Single Streamtube of Wilson *et al.* (1976)

In order to be able to determine and discuss the capacity factor c_P of a VAWT analogous to a HAWT, Wilson assumed a circumferential circulation linked to the blade with $C_L = 2\pi \sin \alpha$.[4] If one averages over a period, one arrives at a first approximation [47]:

$$c_P(x) = \pi \ x \left(\frac{1}{2} - \frac{4}{3\pi} x + \frac{3}{32} x^2 \right) \tag{4.61}$$

---

[4]α is also here the angle of attack that, strictly speaking, can only be properly measured for 2D direct flow.

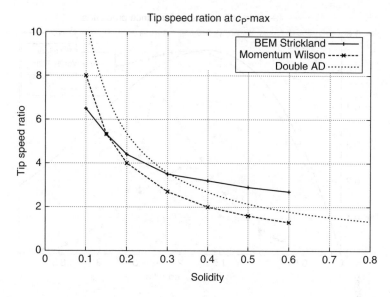

Figure 4.17 Tip speed ratio for maximum C_p according to Wilson (friction-free), Strickland (BEM) and Loth and McCoy (double actuator disk)

with
$$x = \lambda \frac{Bc}{R} \tag{4.62}$$

This model then gives a maximum capacity factor $c_p^{max} = 0.554$ of:

$$a_{max} = \frac{1}{2}\sigma \ \lambda = 0.401 \tag{4.63}$$

Here $\sigma = Bc/R$, with B as the number of blades and c the depth, is the so-called solidity of the rotor.

Remarkably important results can be obtained from this comparatively simple model. For great solidity (such as, for instance, self-starting plants such as the DAWI-10 plant, and others), the optimum tip speed ratio determined from Equation (4.63), $\lambda_{opt} = 0.8/\sigma$, is too small by a factor of around 2,[5] so that very probably, for instance, *dynamic stall* effects must be taken into account, as the angles of attack (AOA) in this case will be too large so that the connection $C_L = 2\pi\alpha$ can no longer be accepted as valid. However, it can be seen from Figure 4.17 that already the inclusion of profile friction clearly increases $\lambda(c_p^{max})$.

4.5.2.2 Momentum Theory 2: Multiple Streamtube Model of Strickland (1975)

In comparison with Wilson, Strickland provided a blade element method, where measured (2D, see above) polars can be included in the determination of the capacity factors. If one refers this model to a realised small wind turbine, then the results of Figure 4.18 are received.

---

[5] The author thanks Mr Görke, Weserwind AG. for this information.

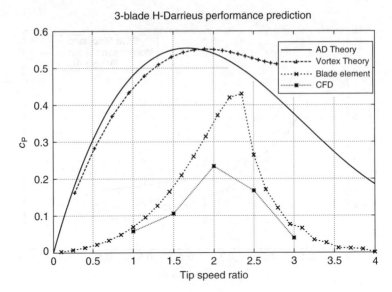

Figure 4.18   Various methods used for performance prediction on a VAWT rotor with solidity $\sigma = 0.48$

The reduction of $c_{\mathrm{P}}^{\mathrm{max}}$ due to profile friction is clearly recognised. The reasons for the further reduction of the $c_{\mathrm{P}}^{\mathrm{max}}$ value within the CFD calculation are discussed further below, but it is important that with all three methods, λ remains approximately the same.

4.5.2.3   Laterally Resolved Streamtube Model

By 1927 Betz noticed that the limiting values named after him could be exceeded if, extended structures were used instead of an infinitely thin effective disk. This idea was worked out in detail by Loth and McCoy [48] for vertical axis machines. With a model with a double actuator disc they obtained (see Exercise 5) a $c_{\mathrm{P}}^{\mathrm{max}} \approx 0.62$. However, in limitation it must be said that, up to present, no prototype has been built that could have explicitly shown these c_{P} values.[6]

4.5.2.4   Vortex Models

(a) Stationary Models
There exists a series of attempts for describing the flow field that, instead of velocity fields, use the vortex strength (vorticity) ($\omega = \bar{\nabla} \times \vec{v}$) as the primary field size because, in addition to the usual ($\vec{v}, p$)-based ones, there are equivalent formulations for the differential equations. These attempts are also able to describe the transient character of the flow field [49–51]. Vortex patch models are especially important here, because they describe the change of the main flow due to the generation, the fanning off and the convection of vorticity at the edges of the blade. The work of Holme [52], was pioneering here as the bound vortex is distributed about a circle (i.e. $B \to \infty$) and thus a stationary but asymmetrical model is obtained. It is

---

[6]The ÉOLE-C had a $c_{\mathrm{p}}^{\mathrm{max}} \approx 0.42$.

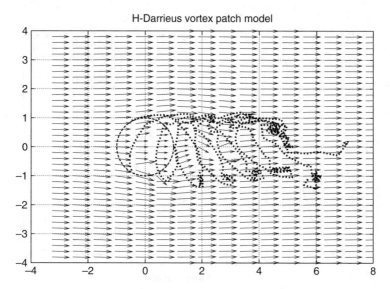

Figure 4.19 Snapshot of the flow field; flow field of a three-blade H-Darrieus rotor after two revolutions

further interesting that Holmes' attempt was able to describe a slipstream deflection perpendicular to the direct flow. Within the framework of an AD formulation, Holmes' model can be implemented in CFD systems, and for this it can be used in its full non-linear formulation [14].

(b) Transient methods
A basic difference in the handling of vertical rotors is the strong periodic change of many quantities during one rotation.
If the blade number is less than 3, then a simple transient vortex model [53] can properly be used for modelling. It could be shown [54,55] that rotational averaged quantities can well be reproduced. Figure 4.19 shows the flow field of a 3-bladed H rotor with solidity $\sigma = 0.3$ and $\lambda = 2$ after two rotations. The interest in such vortex methods has recently re-awoken [56,57]. In addition, one can identify effective load variations per rotation, which are important for the load collective for determining the permanent stability.

(c) Dynamic stall
The phenomenon of the dynamic stall is understood as the marked change of the poles due to dynamic influences, such as oscillation of the blades. A comprehensive discussion with reference to VAWT can be found in Oler *et al.* [58]. It is only remarked here that this effect is especially found for large solidities and lower tip speed ratios, and thus for large angle-of-attack changes.

(d) Remarks on the aerodynamic profile selection for VAWT
In contrast to the horizontal machines in which the circulation $\Gamma \approx c \times C_l$ is an important parameter for determining the blade, for most models for VAWT this function takes on the solidity $\sigma = Bc/R$. For this reason, C_l, and thus the profile, is only included implicitly as in all other concepts either ideally $C_l = 2\pi\alpha$ is to be set, or otherwise polar tables are used.

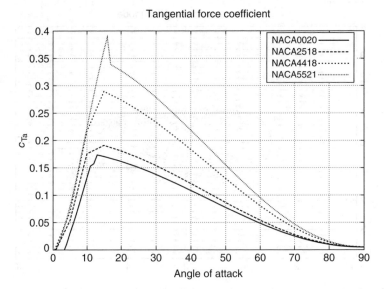

Figure 4.20 Profile comparison $C_t(\alpha)$

If one views the tangential forces generating the driving moment in a resolved manner, then:

$$C_t = C_L \sin\alpha - C_D \cos\alpha \tag{4.64}$$

$$= C_L \left(\sin\alpha - \frac{1}{GZ} \cos\alpha \right) \tag{4.65}$$

with $GZ = c_L/c_D$.

For this reason, local optimising would set a high C_L for larger glide ratio.[7] However, more effective is global optimising, i.e. the maximum of:

$$\frac{1}{2\pi} \langle C_t \rangle = \int_0^{2\pi} (C_t / \phi) \, d\phi \rightarrow \max. \tag{4.66}$$

whereby one is then confronted with a variation problem. Figure 4.20 shows the dependency of the tangential force on the angle of attack for various four-number NACA profiles. It can therefore be clearly recognised that symmetrical profiles for small angles of attack (<7°) and subcritical (Re < 500 k) Reynolds numbers can develop negative tangential forces over wide regions (up to 40°) of the azimuth angle.

Paraschivoiu [41] as well as Duetting [59] and Kirke [60] in their investigations are more detailed on this question, whereby [59] concentrates especially on the starting torque behaviour. Experimental investigations were carried out on the DAWI-10 plant by Meier *et al.* [61]. The DAWI-10 possesses extremely high solidity of σ=0.69. It is shown that measured C_1

---

[7] In [41], p. 248, Paraschivoiu presents a profile that for the same glide ratio (GZ=75) gives almost double the lift factor $C_L = 1.0$, Re=3 M.

sequences dependent on the circumferential angle deviate substantially from those that were theoretically determined.

We would like to close this section with the following recommendations for profile selection:

- The effect of camber and attack (nose outwards) is equivalent to an increase in the depth.
- The effects of dynamic stall are to be included in the aerodynamic modelling if, because of the high solidity, no tip speed ratios occur.
- Modern symmetrical (see [62], and for discussion NACA 00tt vs. NACA 63ctt[8]) profiles can be developed, for instance, with the EPPLER code or generic optimising algorithms (see [63]).

4.5.3   Aeroelastics of Vertical Axis Rotors

The structural design of a wind turbine plant is determined by the loads originating from the aerodynamics. Both extreme (100-year gusts) as well as operational loads (nominal wind including turbulent fluctuations) determine the dimensioning of the material. The specific aerodynamic model formation thus has effects far beyond the pure aerodynamics. Unfortunately there are at present no, or only a few, systems available that can model VAWTs. An exception is the GAROS system developed by A. Vollan [64, 65], with first investigations carried out in [66]. An example of a stability diagram (a Campbell diagram) is reproduced in Figure 4.21.

It can be seen that below the nominal speed of revolution (in this case 120 rpm = 2 Hz) and between 1P and 3P ($nP = n$ times the rotor revolutions), there are indications that a number of resonant frequencies can be found. The overall damping of the system, however, is still negative

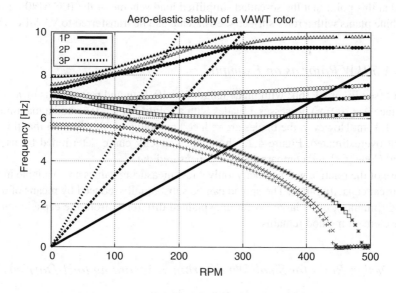

Figure 4.21   Campbell diagram of a vertical rotor

---

[8]Miglore [62] talks of a 20% yield increase.

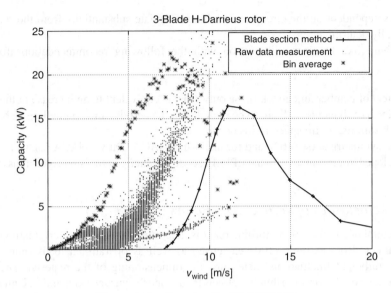

Figure 4.22   Capacity of a 50 kW VAWT

so no aerodynamic divergence is to be expected. The large tower head masses of about 650 kg (corresponding to 130 kg/kW) seem to be especially responsible for the relatively small value of the resonant frequencies. However, one must take into account that because $v \sim m^{-\frac{1}{2}}$, a halving of the tower head mass with only a 30% increase in the frequencies emerges. It is to be remarked at this point that the so-called simplified load scheme of the IEC 61400-2 for small wind turbine plants with a rotor surface $< 20\,m^2$ can also be transferred to VAWTs [67].

4.5.4   A 50 kW Rotor as an Example

This plant (see Figure 4.16b) with blade length $H = 10.21$ m and rotor radius of $R - 5$ m (i.e. rotor surface $A = 100\,m^2$) possesses 3 blades with a solidity of $\sigma = 0.68$ and a maximum depth of $C_{max} = 1.31$ m. However, the blades are reduced markedly at the edges so that this value is somewhat overestimated. Figure 4.22 shows the capacity curve determined by us with the Strickland Code [68] and compared with the measured data.

   Because of the relative wind velocity, only a few measuring values are available in the relative electrical capacity range. The spread can be substantially reduced by means of a suitable averaging (binning), but a large, so-far unexplained difference from the capacity curve with the blade element method remains.

4.5.5   Design Rules for Small Wind Turbines According to H-Darrieus Type A

- An overall concept must be created before any design.
- If the plant is to be self-starting, i.e. $C_m(0) > 0$, then the torque-RPM characteristic of the electrical generator determines the aerodynamic design.
- A three-bladed plant with cambered, not too thick, profiles often seems to be suitable.

- It is possible to develop [69] and use new profile families. They represent an improvement over the four-digit profiles of the NACA family.
- The optimum tip speed ratio is determined by the solidity $\sigma = B_C/R$. It should be below 3 so that the maximum angle of attack does not become too great and require dynamic stall effects to be considered.[9]
- An aeroelastic simulation model (e.g. with GAROS) must be created in the early development stage so that the specific tower head mass can be kept in a sensible region clearly below 100 kg/kW.
- An attempted type certification is designed to support the seriousness of the undertaking.

4.5.6   Summary: Vertical Rotors

4.5.6.1   Basics

First, a short overview of the currently developed aerodynamic models was presented. It is shown that many older semi-analytical methods (e.g. the vortex ring (cylinder model)) can be integrated into existing CFD systems without problems. With this it is possible to carry out stationary (i.e. circumferential-averaged) investigations in a short time period. The most important results are: it seems to be impossible to exceed the Betz limiting value (= 0.59) by small amounts; an optimum tip speed ratio can be determined from Equation (4.63) or Figure 4.17.

4.5.6.2   Profiles

Plant-optimised aerodynamic profiles for VAWTs are not as widely available as for HAWT. An exception is [69]. This is shown not least in the continuing use of four-digit NACA profiles that were developed in the 1930s. The development of individual profiles, as was the case for HAWT in Delft or Risø, is still totally lacking.

4.5.6.3   Starting Behaviour

The possibility of self-starting vertical plants $(c_m(\lambda = 0) > 0)$ was already investigated in the 1980s in various dissertations [13,16]. The results of these investigations were plant concepts with a high degree of solidity (>0.4) and cambered profiles such as NACA 4418 for example. These concept recommendations are used by some new plant types. Mechanical and electrical torque characteristics should be tuned to each other.

4.6   Wind-Driven Vehicles with a Rotor

4.6.1   Introduction

In Section 4.2.2, vehicles with rigid sails were briefly discussed and it was shown that it is advantageous to use lift forces for moving forward. In this connection it is interesting to discuss so-called headwind vehicles in a bit more detail. The concrete reason for this is that in 2008 an international competition, *Racing Aeolus*, was proposed and held in Den Helder,

---

[9] Paraschivoiu ([41], p. 177) takes the dynamic stall below λ 4!

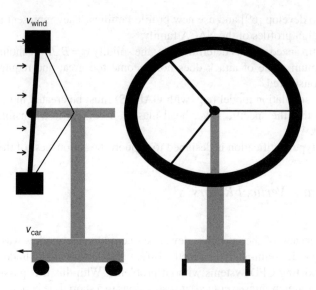

Figure 4.23   Principles of a wind car

Netherlands, for the first time. This was continued in 2009 and 2010 in Stauning, Denmark, and then in 2011 and 2012 took place in Den Helder again.

From the group of all participants the Danish [70], the Stuttgart [71,72] and also the Netherlands [73] team have published the details of their design criteria, so that it seems that this is also generally used for the vehicles from Kiel [74].

Figure 4.23 shows the principle of the wind car. A wind turbine converts the kinetic energy of the wind into power that can be used by the drive train to move the vehicle.

4.6.2   On the Theory of Wind-Driven Vehicles

The concepts for converting the kinetic energy of the wind by means of a wind turbine into mechanically useful energy are discussed in detail in Section 4.2. Further information can be found, for instance, in [19, 25]. An important criterion in this case is that the rotor is designed in such a manner as that shown:

$$c_P = \frac{P}{\frac{1}{2}\rho v^3 \frac{\pi}{4} D^2} \tag{4.67}$$

is maximised. It is easily shown that the thrust (= force on the turbine in the direction of the wind) is given by:

$$c_T = \frac{T}{\frac{1}{2}\rho v^2 \frac{\pi}{4} D^2} \tag{4.68}$$

with $c_T^{Betz} = 8/9 = 0.89$.

Thus, a wind turbine optimally designed for capacity removal provides a large amount of thrust. The dimensionless relationship c_P/c_T is 2/3.

4.6.3   Numerical Example

In the competition the covered rotor surface is limited to $2 \times 2 = 4\,m^2$. If a rotor with a horizontal axis is used, then the surface area is reduced to $\pi = 3.14\,m^2$. If one assumes a wind velocity of $7\,m/s$ and a vehicle velocity of $3.5\,m/s$, then numerically we have $P = 1320\,W$ and $T = 188\,N$. If no other forces were to be overcome, then the transport of the thrust alone would need the capacity of $P_{thrust} = 660\,W$.

This is the idealised case. In reality a well-designed rotor has a $c_p = 0.36$ or only about 60% of the named capacity $(=780\,W)$. A vehicle with $50\,N$ roll friction requires an additional capacity of $50 \times 3.5 = 175\,W$. Therefore the vehicle does not move against the wind. For this reason it is sensible to design the blade in such a manner that $v_{car}/v_{wind} \rightarrow$ max. This occurs within the framework of the so-called blade element method (see Section 4.3).

Two optimising strategies for creating blade geometry have already been described:

- Optimising the local direct flow angle φ and selecting $c(r)$ conventional (Stuttgart).
- Optimising according to $c(r)$ and select the angle of attack so that the profile works for the largest glide ratio c_L/c_D (Copenhagen).

The Stuttgart design is so detailed that a comparison can be made: [71,72] designed the (2008) rotor is such a way that $\lambda = 5$, $c_p = 0.24$ and $c_T = 0.23$. For $v_{wind} = 8\,m/s$, $v_{car}^{max} = 5\,m/s$ and $P = 250\,W$. In contrast, the Danish design [42] is summarised in the following equation:

$$C_{PROPF,loc} = \eta_P \eta_T \left(1 + \frac{1}{v/v_{wind}}\right) C_{P,loc} - C_{T,loc} \rightarrow max. \tag{4.69}$$

4.6.4   The Kiel Design Method

We start from an approach from de Vries [75] and begin with:

$$C_T = \frac{8}{\lambda^2} \int_0^\lambda (1 - aF)aF \left(1 + \frac{\tan\phi}{GZ}\right) x \, dx \tag{4.70}$$

$$C_P = \frac{8}{\lambda^2} \int_0^\lambda (1 - aF)aF \left(\tan\phi - \frac{1}{GZ}\right) x^2 \, dx \tag{4.71}$$

Here GZ is the glide ratio of the profile, $x = \lambda r/R$, the local tip speed ratio, and F the so-called 'tip factor' according to Prandtl [25] (see also equations (4.35) to (4.37)).

The optimising task, Equation (4.69):

$$C_{PROPF,loc} \rightarrow max. \tag{4.72}$$

is solved within a short FORTRAN program with the aid of a duplex binary search process. Results may be found in Table 4.2.

Table 4.2 Results of optimising for a wind-driven vehicle

K	a	v_c/v_{wind}	c_P	c_T	c_P/c_T	$\eta_{drivetrain}$
0.013	0.05	2.0	0.16	0.18	0.89	>0.8
0.06	0.1	1.0	0.29	0.35	0.83	0.7
0.14	0.13	0.75	0.37	0.47	0.79	0.7
0.33	0.18	0.5	0.44	0.60	0.73	0.7
1.0	0.24	0.25	0.50	0.74	0.68	0.7
3.1	0.29	0.1	0.51	0.82	0.62	0.7
36	0.31	0.01	0.52	0.86	0.60	0.7

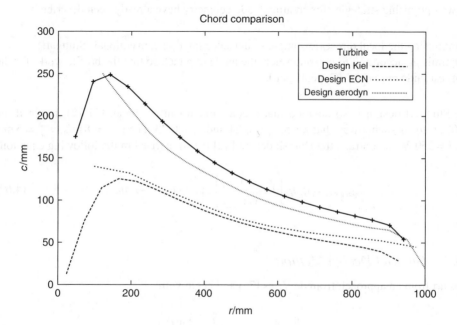

Figure 4.24 Optimum depth distributions of blades for wind vehicles

4.6.5   Evaluation

Results of the above given optimising task is given, as usual in mechanics, in the form of a dimensionless drag coefficient for the vehicle:

$$K = C_D \frac{A_V}{A_R} \tag{4.73}$$

C_D is the drag coefficient of the vehicle which is defined by $D = c_D \, \rho/2v^2 \times A_V$.

A_V and A_R are the projected surfaces of the rotor and the vehicle (see also Figure 4.23). The problem with this attempt is the inclusion of a constant roll resistance.

An important result that can be seen is that basically there is no upper limit for the vehicle speed. Special blade geometries (depth and twist path) can also be derived (see Figures 4.24 and 4.25).

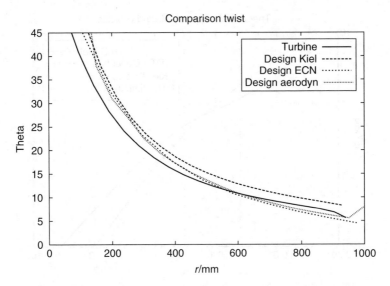

Figure 4.25   Optimum twist distribution for blades of wind vehicles

The design parameters of our concept are: $B = 3$, $\lambda = 5.5$, design lift $c_L = 1.0$ and glide ratio = 80. The rotor radius is 900 mm and the overall drive train efficiency is assumed to be 70%. It is not excluded that the results of this special blade geometry can also be reproduced by feathering commercially available (wind dynamics) blades. Figure 4.26 compares our results with the optimisation by Sørensen [75].

It should be noted that our values, because of additionally included quantities (c_D and blade tip correction), should provide basically lower speeds. As can be seen in Figure 4.26, two additional upper K limits are drawn in that exclude unrealistically high c_p values. In addition, it can be seen that the publicised values from [71] can only be achieved with K values below 0.005. If in 2009 only one vehicle with the $v_{car}/v_{wind} \geq 0.5$ received the ECN award (see Figure 4.28), then the condition has been tightened to $v_{car}/v_{wind} \geq 1$.

In the meantime a team from Canada (Chinook) has reached 73.2% in 2012 and one from Denmark as high as 74.5% in 2011. Our analysis (Figure 4.26) shows that this makes high demands of all components.

4.6.6   Completed Vehicles

Starting with the first sobering experience with a VAWT rotor and electrical drive train, a radically simplified concept was created and carried out. A diffuser ring of GFK was built and was easily erected (12 kg mass). Both yaw and pitch systems were dispensed with, and a simple four-gear moped drive (gear ratios of 6, 8, 10 and 17 to 1) was used. The rear axle had two hinged shafts with idling and a chain. Wheels from a wheelchair were used.

This was successful: in the overall weighting Kiel 2009 was in second place (see Table 4.3 and Figure 4.27). The speeds reached from 2010 to 2012 (no summarised report was issued in 2009) are shown in Figure 4.28.

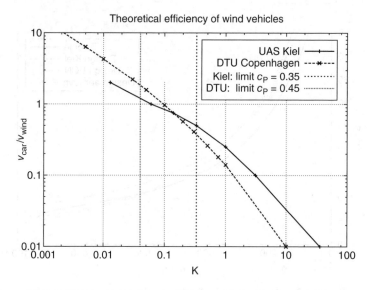

Figure 4.26   Optimum design for wind-driven vehicles from Copenhagen and Kiel

Table 4.3   Results for wind vehicle racing competition Aeolus 2010

Vehicle	Speed relationship	At v_{wind} (m/s)
Amsterdam 2	0.657	4.7
Amsterdam 1	0.495	4.7
Kiel	0.474	4.8
DTU 1	0.427	3.6
Stuttgart	0.402	4.51
DTU 2	0.249	4.67
Flensburg	0.161	5.47
Anemo	0.154	5.4
Bristol	0.049	4.3

4.6.7   Summary: Wind Vehicles

In order to maximise the available performance for the drive of a wind-driven vehicle, a different optimising criterion than for the design of a classical wind turbine must be used. In any case, the loading of the wind turbine whether described by a or c_T must be reduced. A further interesting discussion of the differences and similarities of propeller and turbine operations is found in [70].

It is shown that appreciable speeds ($v_{car}/v_{wind} > 0.5$) can only be reached when the overall resistance of the vehicle (without thrust) has a K value less than 1. This value is substantially reduced ($K > 0.1$) when more realistic c_p values than those required by the optimisation ($c_p \approx 0.35$) are used.

The Kiel *Baltic Thunder 3* wind vehicle was able to reach second place in Stauning Denmark in 2010. This was achieved by minimising all parasitic resistances and by a marked reduction

Figure 4.27   The Kiel wind vehicle *Baltic Thunder*, 2009 to 2012. Photo: Alois Schaffarczyk

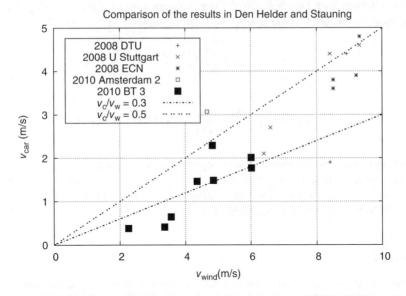

Figure 4.28   Achieved speed relationships in the races 2010 to 2012

in the mass (<150 kg). For the 2011 race in Den Helder, the team developed its own black, as did the team from Stuttgart ECN DTU (see Figures 4.24 and 4.25), was carried out for the race in 2011 in Den Helder.

4.7   Exercises

1. What is the annual yield of a wind turbine with $D = 80\,\mathrm{m}$, $c_P = 0.5$, $v_{nom} = 11\,\mathrm{m/s}$ and with 2000 full-load hours? What is the annual amount (in Euro) of a subsidy of 9 eurocent per kWh?

2. Using equations (4.59) and (4.60), determine the maximum relative thickness c/R of a blade at $r/R = 0.25$ and $C_L = 1.0$. How many meters is this for a blade in question 1?

3. How great is the (relative) subsidy difference for the plant in question 1 if it is operated:
   (a) in winter ($\phi = -20°C$) at sea-level; and
   (b) in summer ($\phi = 30\ °C$) in the mountains ($H = 1500$ meters)?

4. Show, if one assumes a Rayleigh distribution of the velocity:

$$P(v) = \frac{\pi}{2} \frac{v}{\overline{v}} \exp\left(-\frac{\pi}{4} \left(\frac{v}{\overline{v}} \right)^2 \right)$$

then the maximum annual yield according to the 1-2-3 formula can be described as:

$$\overline{P} = \rho^1 \left(\frac{2}{3} D \right)^2 \overline{v}^3$$

What average $\overline{c}_p$ does this correspond to? Compare this with question 1.

5. Do you consider a double adaptor disk as a model of a wind turbine with two contra-rotating rotors or as a Darrieus rotor? On the assumption of a fully developed propeller wash, show the blades 1 as the inflow condition of blades 2 $\left(v_2^{in} = v_1(1-2a) \right)$:

$$c_{P2} = \frac{16}{27}(1-2a)^3$$

and

$$c_P^{max} = c_{P1} + c_{P2} = \frac{64}{125} + \frac{16}{125} = 0.512 + 0.128 = 0.64$$

Further, determine the respective loading c_{T1} and c_{T2} as well as c_T.

6. Design a wind vehicle for $v_{wind} = 8\,m/s$ and $v_{car} = 4\,m/s$. The rotor surface is $3\,m^2$ and K (Equation (4.73)) is 0.1. Discuss your results for dependency of the drive train efficiency and in the light of Figure 4.26 and Table 4.2.

References

[1] Schaffarczyk, A.P. 2014. *Introduction to Wind Turbine Aerodynamics*. Springer-Verlag, Berlin Heidelberg.

[2] Schepers, J.G., Boorsma, K., Cho, T., *et al.* 2012. Final report of IEAwind task 29, Mexnext (Phase 1): Analysis of Mexico wind tunnel measurements, ECN-E-12-004. Petten, The Netherlands.

[3] Schepers, J.G. and Snel, H. 2007. Model Experiments in Controlled Conditions, Final Report, ECN-07-042, Petten, The Netherlands.

[4] Rankine, W.J. 1865. On the mechanical principles of the action of propellers. *Trans. Inst. Nav. Arch.*, **13**.

[5] Froude, R.E. 1889. On the part played in propulsion by differences in fluid pressure. *Trans. Inst. Nav. Arch.*, **390**.

[6] Betz, A. 1926. *Wind-Energie und ihre Ausnützung durch Windmühlen*. Vandenhoeck & Ruprecht, Göttingen.

[7] Glauert, H. 1993. *The Elements of Airfoil and Airscrew Theory*, 2nd edn. Cambridge University Press, Cambridge.

[8] Breslin, J.P. and Andersen, P. 1986. *Hydrodynamics of Ship Propellers*. Cambridge University Press, Cambridge.

[9] Leishman, J.G. 2000. *Principles of Helicopter Aerodynamics*. Cambridge University Press, Cambridge.

[10] Abbot, I.H. and von Doenhoff, A.E. 1959. *Theory of Wing Sections*. Dover Publications Inc., New York.

[11] Althaus, D. 1996. *Niedriggeschwindigkeitsprofile*. Vieweg, Braunschweig.

[12] Freudenreich, K., Kaiser, K., Schaffarczyk, A.P., *et al.* 2004. Reynolds number and roughness effects on thick airfoils for wind turbines. *Wind Engineering*, 5: 529–46.

[13] Timmer, W.A. and Schaffarczyk, A.P. 2004. The effect of roughness on the performance of a 30% thick wind turbine airfoil at high Reynolds numbers. *Wind Energy*, pp. 295–307.

[14] Schaffarczyk, A.P. 2006. *New Aerodynamical Modeling of a Vertical Axis Wind-Turbines with Application to Flow Conditions with Rapid Directional Changes*. Proc. Dewek, Bremen, Germany.

[15] Prandtl, L. 1919. Zusatz zu: A. Betz: Schraubenpropeller mit geringstem Energieverlust, Nachrichten der Kgl. Ges. d.Wiss.,Math.-phys. Klasse, Berlin.

[16] Okulov, V.L. and Sørenson, J.N. 2007. Optimum operating regimes for the ideal wind turbine. *Journal of Physics, Conf. Series* **71**, 1–9.

[17] Hansen, M.O.L. and Johannson, J. 2004. *Tip Studies using CFD and comparison with Tip Loss Models*. Proc. EAWE Conference 'The Science of making Torque from Wind', Delft, The Netherlands.

[18] Shen, W.S., Sørensen, J.N. and Bak, C. 2005. Tip loss corrections for wind turbine computations. *Wind Energy*, **4**: 457–75.

[19] de Vries, O. 1979. *Fluid Dynamics Aspects of Wind Energy Conversion*, AGARDograph, No. 242. Neuilly-sur-Seine, France.

[20] Wilson, R.E., Lissaman, P.B.S. and Walker, S.N. 1976. *Aerodynamic Performance of Wind Turbines*. Oregon State University Report.

[21] Sørensen, J.D. and Sørensen, J.N. 2011. *Wind Energy Systems*. Woodhead Publishing Ltd, Oxford.

[22] Conway, J.T. and Schaffarczyk, A.P. 2003. *Comparison of Actuator Disk Theory with Navier-Stokes Calculation for a Yawed Actuator Disk*. Proc. 8th CASI Meeting, Montreal, Canada.

[23] Gasch, R. and J. Twele, J. (eds) 2010. *Windkraftanlagen*, 6. Auflage, Vieweg + Teubner, Stuttgart.

[24] Xudong, W., Shen, W.Z., Zhu, W.J., *et al.* 2009. Shape optimization of wind turbine blades. *Wind Energy*, **12**: 781–803.

[25] Schaffarczyk, A.P. 2010. Aerodynamics and aero-elastics of wind turbines, Chapter 3 in *Wind Power Generation and Wind Turbine Design* (ed. W. Tong). WIT Press, Southampton.

[26] Schaffarczyk, A.P. 1998. Vergleich verschiedener Numerischer Strömungssimulationsverfahren an einem Aktiv-Stall-Blatt und Schlussfolgerungen für eine aerodynamische Optimierung, 4. DeutscheWindenergiekonferenz, pp. 356–360, Wilhelmshaven.

[27] Hansen, M.O.L., Sørensen, N.N. and Flay, R.G.J. 1999. *Effect of Placing a Diffuser around a Wind Turbine*. Proc. EWEC, Nice, France.

[28] Phillips, D.G. 2003. *An Investigation on Diffuser Augmented Wind Turbine Design*. PhD thesis, University of Auckland, New Zealand.

[29] Schaffarczyk, A.P. and Phillips, D. 2001. *Design Principles for a Diffusor Augmented Wind-Turbine Blade*. Proc. EWEC, Copenhagen.

[30] van Bussel, G.J.W. 2007. *The science of making more torque from wind: diffuser experiments and theory revisited*, Proceedings of 'The Science of making Torque from Wind', Lyngby, Denmark.

[31] Mertens, S., van Kuik, G. and van Bussel, G. 2003. Performance of an H-Darrieus in the skewed flow on a roof. *J. Sol. En. Eng.*, 433–40.

[32] Solum, A., Deglaire, P. and Erikson, S., 2006. *Design of a 12kW Vertical Axis Wind-Turbine equipped with a direct driven PM synchronous generator*, Proc. EWEC, Athens.

[33] Bankwitz, H., Fritzsche, A., Schmelle, J., *et al.* 1975, 1978, 1982. Entwicklung einer Windkraftanlage mit vertikaler Achse (Phase I–III), Friedrichshafen.

[34] Eckert, L. *et al.* 1990. *Analyse und Nachweis der 50kW-Windenergieanlage (Typ Darrieus)*. Interner Bericht, Dornier GmbH, MEB 55/90.

[35] Fritzsche, A. *et al.* 1991. Auslegung einer Windenergieanlage mit senkrechter Drehachse im Leistungsbereich 350–500 kW. Dornier GmbH, Friedrichshafen.

[36] Henseler, H. 1990. Eole-D Abschlussbericht, Friedrichshafen.

[37] Lieferung und Montage einer 2,25MW Darrieus-Windenergienanlage EOLE-D, Interner Bericht, Dornier GmbH, 1990.

[38] Heidelberg, G. and Krömer, J. 1993. *Windkraftanlage H-Rotor: Erfahrungen, Aktuelles und Ausblick*, HusumerWindenergietage, Husum, Germany.

[39] FloWind Corporation. 1996. Final Project Report: High-Energy Rotor Development, Test and Evaluation, Sandia National Laboratories, Sand96-2205. FloWind Corporation, Albuquerque, NM.

[40] Soler, A. and Clever, H.G. 1991. Bau, Aufstellung und Erprobung einer 50 kW-Darriues-Windkraftanlage, Friedrichshafen.

[41] Paraschivoiu, I. 2002. *Wind Turbine Design –with Emphasis on Darrieus Concept*. Polytechnic International Press, Montreal.

[42] Zastrow, F. 1992. *Entwicklung von Windkraftanlagen für den Einsatz in der Antarktis*, Abschlussbericht, BMFT, POL 0041, Bremerhaven.

[43] Lobitz, D.W. and Ashwill, T.D. 1986. *Aeroelastic Effects in the Structural Dynamic Analysis of Vertical Axis Wind-Turbine*, Sandia Report, SAND85-0957, Albuquerque, NM.

[44] Vita, L., Paulsen, U.S. and Pedersen, T.F. 2010. *A Novel Floating Offshore Wind Turbine Concept: New Development*. Proc. EWEC, Warszaw, Poland.

[45] Haris, A. 1991. *The variation in cost of energy with size and rated power for Vertical Axis Wind Turbine*. Proc. Brit. Wind Energy Conference.

[46] Harrison, R., Hau, E. and Snel, H. 2000. *Large Wind Turbines – Design and Economics*. John Wiley and Sons, Ltd., Chichester.

[47] Wilson, R.E., Lissaman, P.B.S. and Walker, S.N. 1976. Aerodynamic performance of wind turbines, Chapter 4 in *Aerodynamics of the Darrieus Rotor*. Corvallis, Oregon, US.

[48] Loth, J.L. and McCoy, H. 1983. Optimization of Darrieus turbines with an upwind and downwind momentum model. *J. Energy*, 313–18.

[49] Katz, J. and Plotkin, A. 2001. *Low-Speed Aerodynamics*. Cambridge University Press, Cambridge.

[50] Thwaites, B. (ed.) 1960. *Incompressible Aerodynamics*. Oxford University Press, Oxford.

[51] White, F.M. 1991. *Viscous Fluid Flow*, 2nd edn. McGraw-Hill, New York.

[52] Holme, O. 1976. *A Contribution to the Aerodynamic Theory of the Vertical-Axis Wind Turbine*, Paper C4, Proc. Int. Symp. on Wind Energy Systems, Cambridge, UK.

[53] Strickland, J.H., Webster, B.T. and Nguyen, T. 1979. *A Vortex Model of the Darrieus Turbine: An Analytical and Experimental Study*. Proc. Ann. Winter Meeting, ASME, New York.

[54] de Vries, O. 1979. *Fluid Dynamic Aspects of Wind Energy Conversion; Chap. 4.5: Vertical-axis turbines*, AGARD-AG-242, Neuilly-sur-Seine, France.

[55] Wilson, R.E. and Walker, S.N. 1983. Fixed wake theory for vertical axis wind-turbines. *J. Fl. Eng*, 389–93.

[56] Ferreira, C.S. 2009. *The near wake of the VAWT*. PhD thesis, Technical University Delft, The Netherlands.

[57] Mikkelsen, R. 2006. Private communication, and Clausen, R.S., Sønderby, I.B. and J. A. Anderkjaer, J.A. 2005. Eksperimentel og Numerisk Undersøgelse af en Gyro Turbine, Master's thesis, DTU, Lyngby.

[58] Oler, J.W., Strickland, J.H., Im, B.J. and Graham, G.H. 1983. *Dynamic Stall Regulation of the Darrieus Turbine*, Sandia Report, SAND83-7029, Albuquerque, NM.

[59] Dütting, J. 1987. *Untersuchungen über das Startverhalten von Windrotoren mit vertikaler Achse*. Dissertation, Universität Bremen, Germany.

[60] Kirke, B.K. 1988. *Evaluation of Self-Starting Vertical Axis Wind Turbines for Stand-Alone Applications*. PhD thesis, Griffith University, Australia.

[61] Meier, H., Schneider, J.-D. and Richter, B. 1988. *Messungen an der Windkraftanlage DAWI 10 und Vergleich mit theoretischen Untersuchungen*, Germanischer Lloyd, WE-4/88, Hamburg.

[62] Migliore, P.G. 1983. Comparison of NACA 6-Series and 4-Digit Airfoil for Darrieus Wind Turbines. *J. Energy*, **4**: 291–2.

[63] Bourguet, R., Martinat, G., Harran, G. and Braza, M. 2007. *Aerodynamic Multi-Criteria Shape Optimization of VAWT Blade Profile by Viscous Approach*, Proc. Euromech Colloquium Wind Energy, Oldenburg.

[64] Vollan, A. 1977. *Aeroelastic Stability Analysis of a Vertical Axis Wind Energy Converter*, Bericht EMSB-44/77, Dornier System, Immenstaad.

[65] Vollan, A. 1978. *The Aeroelastic Behaviour of Large Darrieus-Type Wind Energy Converters derived from the Behaviour of a 5.5 m Rotor*, Paper C5, Proc. Int. Symp. on Wind Energy Systems, Amsterdam.

[66] Kleinmann, D. 2007. *Aeroelastische Analyse einer 5 kW H-Darrieus Anlage mit GAROS*, Studienarbeit, FH Kiel.

[67] Heym, C. 2010. *Entwicklung eines Berechnungsprogramms für den allgemeinen Sicherheitsnachweis von Kleinwindanlagen mit vertikaler Rotorachse nach IEC 61400-2:2004*, Master's thesis, FH Kiel, Kiel.

[68] Strickland, J.H. 1975. *The Darrieus Turbine: A Performance Prediction Model Using Multiple Streamtube*. Sandia National Laboratories, Sand75-0431, Albuquerque, NM.

[69] Claessens, M.C. 2006. *The Design and Testing of Airfoils in Small Vertical Axis Wind Turbines*, MSc thesis, Technical University of Delft, Delft, The Netherland.

[70] Gaunå, M., Øye, S. and Mikkelsen, R. 2009. *Theory and Design of Flow Driven Vehicles Using Rotors for Energy Conversion*. Proc. EWEC, Marseille, France.

[71] Lehmann, J., Miller, A., Capellaro, M. and Kühn, M. 2008. *Aerodynamic Calculation of the Rotor for a Wind Driven Vehicle*. Proc. DEWEK, Bremen, Germany.

[72] Lehmann, J. and Kühn, M. 2009. Mit dem Wind gegen den Wind. Das Windfahrzeug In Ventus Ventomobil. *Physik in unserer Zeit*, 176–81.

[73] Boorsma, K., Machielse, L. and Snel, H. 2010. *Performance Analysis of a Shrouded Rotor for a Wind Powered Vehicle*. Proc. TORQUE 2010, Crete, Greece.

[74] Schaffarczyk, A.P. 2010. *Zur aerodynamischen Auslegung der Kieler Windautos*, Bericht Nr. 66 des Labors für Numerische Mechanik, Kiel.

[75] Sørensen, J.N. No date. *Aero-mekanisk model for vindmølledrevet køretøj, unveröff*. Bericht, Kopenhagen, Dänemark.

5

Rotor Blades

Lothar Dannenberg

5.1 Introduction

The rotor blade is a particularly important element of a wind turbine (WT). Its shape is decisive for the capacity yield. It is optimised according to aerodynamic criteria as described, for instance, in Chapter 4. In order to achieve as low a weight as possible for the rotor blade, according to the aerodynamically determined dimensions, an optimising of its strength, that is, the cross-sectional dimensions and the material used, while still retaining the aerodynamic characteristics, is necessary, whereby the weight again has effects on the strength. In this way the use of material and thus the costs of production of the rotor blade are minimised. After the tower and the foundations, the rotor blade is the second most expensive part of the wind turbine; these costs are approximately 16–18% of the overall costs.

Rotor diameters have become ever larger over the past 25 years, from an average of 20 m diameter in 1985 to 120 m today. At present prototypes are being built with a diameter of more than 160 m (status at the end 2011). Whereas for smaller rotors the wind loads dominate the design for strength, for larger diameters the bending moments and the inherent weight are becoming more decisive. The weights of the rotor blades increase exponentially with the blade length (see Figure 5.1).

The materials for rotor blades can be steel, aluminium, wood and FRP (fibre-reinforced plastic). In modern large rotor blades, fibre-reinforced plastics have gained the upper hand due to their good strength for relatively low material costs.

For very large rotor blades in the category of 5 MW or more, and only on those because of the strength at particularly critical places, carbon and carbon fibre (CF) has been used even though

Understanding Wind Power Technology: Theory, Deployment and Optimisation, First Edition.
Edited by Alois Schaffarczyk. Translated by Gunther Roth.

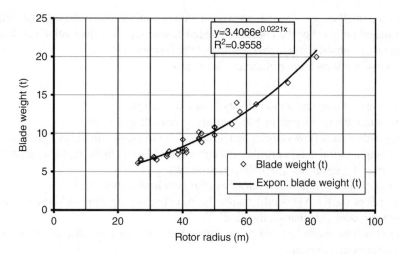

Figure 5.1   Rotor blade weights as a factor of radius, according to [1]

the costs of carbon fibres are substantially greater than for glass fibres. Thus, glass-fibre laminates cost approximately 7 Euro/kg compared with carbon fibre at approximately 70 Euro/kg. The epoxy resins used for fibre laminates cost approximately 7.5 Euro/kg in comparison to steel at approximately 0.15 Euro/kg and aluminium at approximately 0.5 Euro/kg (status middle of 2011). With the weight of a rotor blade of the 5 MW class at approximately 16 tonnes with a fibre volume content of 50–60%, the material costs are therefore a substantial cost factor. The laminate resins used are almost solely epoxy resins (EP resins) although these are almost three times as expensive as the equally suitable unsaturated polyester (UP) resins (PU resins).

In this chapter, the design of static as well as dynamic loads as well as the methods for analysing strength and vibrations of rotor blades will be described. In this, the derivation of the equations used will be largely dispensed with as these can be studied in the basic works on technical mechanics or strength of materials, for instance, [2–4]. Furthermore, the properties of the most common fibre-reinforced materials used in rotor production and their calculation, as well as the manufacture of rotor blades, will be discussed.

5.2   Loads on Rotor Blades

5.2.1   Types of Loads

The loads on rotor blades created by the wind have been discussed in the previous chapter. They will be summarised once again here in order to describe their short- and long-term strength. Depending on the type, the load can be applied once, periodically (i.e. during every rotation of the rotor) or stochastically. In view of the change over time, a distinction must be made between static or quasi-static and dynamic loads. Dynamic loads are seen as all loads that change over time and that can induce fluctuations in the viewed system. When the eigenfrequencies of the structures are clearly smaller or larger than the induced frequencies, then the time-varying loads can be seen as quasi-static, as generally they can be induced to fluctuate only to a very small extent. They are then dealt with as static loads. If the induced frequencies lie in the neighbourhood of the component eigen-frequency, then fluctuation resonances can occur that cause substantially higher loading of the component.

The types of loads on a rotor blade are the result of:

- Wind pressure at constant wind velocity (constant, static).
- Extreme wind due to 50-year or 100-year gusts (one-off quasi-static).
- Changes in the direct flow velocity and direction due to the distribution of the wind velocity over the height (wind height profile); the upper rotor blade is in the region of a greater wind velocity than the lower one (periodic, quasi-static).
- Short-term changes in the wind velocity (gusts, stochastic, dynamic).
- Turbulences of the wind generally and/or due to neighbouring wind turbines in a wind farm (turbulence models, stochastic, dynamic).
- Passing of a rotor blade in front of the tower, thus changing the direct-flow velocity and direction (periodic, dynamic).
- Change of tensile and compressive loading due to the inherent weight during the rotation of the blade (periodic, quasi-static).
- Acceleration forces of the inherent weight of the blade during constant rotation (constant, rotation dependency).
- Change of the set sign of the bending moments in the direction of the fluctuations during every rotation of the rotor (periodic, quasi-static).
- Loading due to relatively rapid changes in the operating conditions of the WT (braking of the rotor, wind-direction following, dynamic).
- Imbalances due to rotor bearings, drives and/or generator (periodic).
- Uneven ice deposits due to corresponding weather (changes of the rotor blade weight and the direct flow (periodic).
- For offshore wind turbines, the rotor blades can additionally be induced to vibrate by wave forces and vortex generation at the foundations or at the tower.

The above types of loading occur both for so-called windward runs (the wind first strikes the rotor and then the tower) as also for lee runs (the wind first strikes the tower and then the rotor). In its guidelines for wind turbines, the GL (Germanischer Lloyd [5]) describes approximately 35 different load cases that must be investigated.

In this chapter, only rotors with horizontal axes will be discussed; vertical rotors such as Darrieus rotors are not dealt with here, although many of the methods described can also apply to them.

As the rotor blades are asymmetrical and the lines of their elastic centres of gravity (see below) are mostly neither straight nor arranged exactly in the rotor blade level, the above-mentioned loads cause a spatial load in the rotor blade with normal and lateral forces, as well as bending and torque moments with the resulting stresses and deformations. In linear elasticity theory, the stresses and deformations resulting from the loading can be individually calculated for each type of load and can then be superimposed (superposition principle). In order that the stresses and deformations, including their prefixes, are clearly defined, the conditions for this will be explained in the following.

5.2.2   Fundamentals of the Strength Calculations

5.2.2.1   Coordinate System, Prefix Rules

A rotor blade is loaded in all three directions. For this reason it is important to introduce an effective coordinate system and to use it consistently for all calculations, including the resulting set signs for the loadings, section dimensions, stresses and deformations.

For a beam in mechanics, use is usually made of a right-handed Cartesian system (right-hand rule) with the longitudinal beam direction as the x axis. The two other axes are at right angles to it, as shown in Figure 5.2b. The Germanic Lloyd uses a right-handed coordinate system for the design of rotor blades in which, however, the longitudinal rotor axis is in the longitudinal or z direction, the direct flow direction of the blade is in the y direction (towards the profile end) and the x direction points to the suction side of the blade (see Figure 5a). In this chapter, for simplified understanding, the normal coordinate system used for the strength calculations is applied, that is, the x axis points in the blade direction, the direct flow direction is the y axis and the suction side of the blade is the z axis.

The set sign rules are: the external (loads) and the internal forces (shear forces) are taken as positive if they act in the direction of the respective positive axes. The external moments (loads) and internal moments (cutting torques) are then positive when their sense of rotation acts positively to the direction of the axis (see Figure 5.3).

5.2.2.2   Static Loads (Static Forces and Cutting Torques)

The shear forces are the result of the stresses in a cross-section; they are obtained by the integration of the stresses through the cross-section surface A (the indicators show the axial directions of the forces).

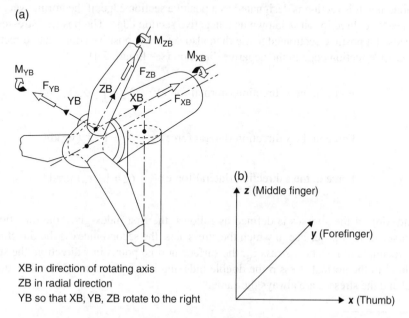

(a)

M_{ZB}

M_{XB}

M_{YB}

F_{YB} F_{ZB}

YB   ZB   XB

F_{XB}

(b)

z (Middle finger)

y (Forefinger)

x (Thumb)

XB in direction of rotating axis
ZB in radial direction
YB so that XB, YB, ZB rotate to the right

Figure 5.2   (a) Coordinate system of the GL [5]; (b) standard coordinate system (right-hand rule)

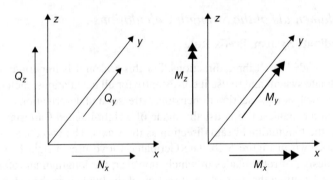

Figure 5.3   Direction of the positive forces and moments

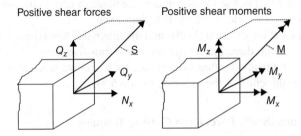

Figure 5.4   Direction of the positive shear forces and shear moments (at the positive section edge)

If the surface normal to a particular section points from the body outwards in the positive direction, then this section is designated as a positive section edge; if the normal vector points in the negative direction, it is known as a negative section edge. The positive stresses or section loads at a positive section edge are then also set in the positive coordinate direction, and for a negative section edge in the negative direction (see Figure 5.4).

$$\text{Force in the } x \text{ direction (normal force):} \quad N_x = \int_{(A)} \sigma_x(y,z)\mathrm{d}A \tag{5.1a}$$

$$\text{Force in the } y \text{ direction (lateral force):} \quad Q_x = \int_{(A)} \tau_x(y,z)\mathrm{d}A \tag{5.1b}$$

$$\text{Force in the } z \text{ direction (lateral force):} \quad Q_z = \int_{(A)} \tau_{xz}(y,z)\mathrm{d}A \tag{5.1c}$$

The indexing of the stresses is defined as follows: the first index gives the direction of the normal vector for the surface in which the stress acts; the second index is the direction of the stress. Example of the shear stress τ_{xz}: the surface normal points in x direction, the stress in z direction.[1] In the normal stress σ the double indexing is not necessary as the direction of the normal and the stresses are always the same.

---

[1] In UK literature the sequence of the indexing is usually reversed.

The section moments are the resultants of the stresses multiplied by the respective levers; they are found by the integration over the cross-section according to the following equations. The indexes give the direction of the moments, i.e. the rotation about the respective axes:

Moment about the x axis (torsion): $\quad M_x = M_T \int_{(A)} (y\tau_{xz} - z\tau_{xy})\,dA$ \hfill (5.2a)

Moment about the y axis (bending): $\quad M_y = \int_{(A)} z\sigma_x\,dA$ \hfill (5.2b)

Moment about the z axis (bending): $\quad M_z = -\int_{(A)} y\sigma_x\,dA$ \hfill (5.2c)

5.2.3   Cross-Sectional Values of Rotor Blades

Rotor blades are thin-walled, non-symmetrical hollow sections with generally different material properties in the cross-sections and in the longitudinal structure (e.g. differing sandwich structure, partial use of sandwich design, and so on). This must be taken into account in the calculation of the centres of gravity and inertial torques and bending, expansion, thrust and torsional stiffness.

In a cross-section of a homogeneous material (density, E-modulus, shear modulus, etc., are the same throughout the section), the surfaces, masses and the so-called 'elastic' centre of gravity lie on the same point. If the cross-section is simply symmetrical, they lie on the symmetry line; if it is double-symmetrical, i.e. it has two symmetry lines, then all centres of gravity are at the crossing point of the symmetry lines. Here, symmetry means that, besides the dimensions, the material properties are also the same.

This is no longer the case with the cross-sections of rotor blades. Density, E-modulus, shear modulus, and so on, can be different throughout the cross-section and over the length of the blade. Then the individual centres of gravity must be viewed separately and calculated with consideration of the corresponding material sizes. At best, use is made of an 'appropriate' coordinate system, for example, with the origin at the middle of the blade connection to the rotor hub, and calculating the reference coordinates of the centres of gravity for the individual cross-sections.

As the complete descriptions of the rotor blade profile (contours, wall thicknesses, material values) are mostly not available in analytical form, one divides the sections of the profiles into appropriate N small rectangles in which the material values are constant. The lengths of the rectangles are referenced to the curves of the profile contour; for large curves, the rectangles are short, for small curves they are longer. The end points are designated with (i) and (k) for the corresponding y- and z-coordinates (see Figure 5.5). End points are to be arranged at the change points of thickness, E-modulus and density, as well as at places at which two cross-sections intersect each other. If parts of the sections consist of circular segments, ellipses, parabolas, and so on, then their surface parts can be calculated directly and used in the subsequent summations, but in view of the later calculations, especially for the determination of the torsion and shear stiffness as well as the shear stresses (see below), they must also be converted into appropriate rectangles. For sandwich components it is necessary that the core and the two shell layers are dealt with as separate rectangles.

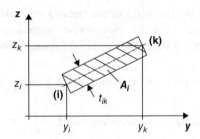

Figure 5.5 Small rectangle for calculating cross-sectional values

In the observed rectangles (index i) the E-modulus and specific density are constant. One obtains:

For the surface of the rectangle: $\quad A_i = t_{ik}\sqrt{(y_k - y_i)^2 + (z_k - z_i)^2}$ $\quad$ (5.3a)

Coordinates of the surface centre of gravity: $\quad e_{y_i} = \dfrac{(y_i + y_k)}{2}; \quad e_{z_i} = \dfrac{(z_i + z_k)}{2}$ $\quad$ (5.3b)

The coordinates to be determined of the overall centres of gravity are (number of rectangles = N):

Surface centre of gravity of the section (at this point all the bending stresses are equal to zero, when the same E-modulus is present everywhere in the section):

$$y \text{ coordinates:} \quad e_{y_F} = \sum_{i=1}^{N} A_i e_{y_i} \Big/ \sum_{i=1}^{N} A_i \qquad (5.4a)$$

$$z \text{ coordinates:} \quad e_{z_F} = \sum_{i=1}^{N} A_i e_{z_i} \Big/ \sum_{i=1}^{N} A_i \qquad (5.4b)$$

Mass or weight centre of gravity (the mass forces act in this point):

$$y \text{ coordinates:} \quad e_{y_M} = \sum_{i=1}^{N} A_i \rho_i e_{y_i} \Big/ \sum_{i=1}^{N} A_i \rho_i \qquad (5.5a)$$

$$z \text{ coordinates:} \quad e_{z_M} = \sum_{i=1}^{N} A_i \rho_i e_{z_i} \Big/ \sum_{i=1}^{N} A_i \rho_i \qquad (5.5b)$$

The mass per unit length of the rotor blade section at point x is:

$$m(x) = \sum_{i=1}^{N} A_i(x)\rho_i(x) \qquad (5.5c)$$

Elastic centre (*EZ*, at this point the bending stresses are equal to zero):

$$y \text{ coordinates:} \quad e_{y_{EZ}} = \sum_{i=1}^{N} A_i E_i e_{y_i} \Big/ \sum_{i=1}^{N} A_i E_i \qquad (5.6a)$$

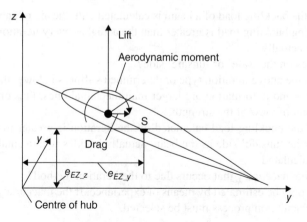

Figure 5.6 Aerodynamic load of a profile

$$z \text{ coordinates:} \quad e_{z_{EZ}} = \sum_{i=1}^{N} A_i E_i e_{z_i} \Big/ \sum_{i=1}^{N} A_i E_i \tag{5.6b}$$

Surface and weight centres of gravity in a section are then in the same place when the material of the section has the same specific density everywhere. The elastic centre is at the same point as the surface centre of gravity when the same E-modulus exists everywhere in the section.

The shear centre of gravity is the point through which all lateral forces must pass so that no additional torsion moments are generated through them at the x axis. Its coordinates are designated by e_{ys} and e_{zs}. The knowledge of the shear centre of gravity is necessary for thin-walled hollow sections in order to be able to determine the shear stress in the section due to the lateral forces and the torsion moments. However, its calculation will not be carried out here as it will exceed the scope of this introduction. The shear centre of gravity must be determined with the aid of the theory of camber force torsion for thin-walled multi-celled hollow cross-sections (see, for instance, [6] or [7]).

The aerodynamic centre is the point at which the aerodynamic forces of lift and drag act on the profile. As this point changes with the direct flow on the profile, a point on the chord of the profile, which is ¼ of the chord length from the leading edge of the profile, is often assumed. The coordinates of the aerodynamic centre are designated with e_{yw} and e_{zw}. Due to the displacement of the force attack site of lift and drag in this point, the resulting torsion moment must be taken into account in the strength calculation (see Figure 5.6). The lift and drag forces are always transformed independently of the direction of the direct flow in the z or y direction.

General remarks on simplified calculation models or iteration process

If iteration processes are used for calculating stresses, deformations, frequencies, and so on, as described above, then it must always be estimated what influence this simplification will have on the accuracy of the results. The following questions must be answered:

- Is the result of the simplification greater or less than the actual?
  Example 1: The actual length of the profile contour is greater (and therefore also the area) than that approximated by the rectangle, as the greater part of the contour is convex. In this way the calculated stiffnesses of the rotor blades (see below) are lower than actual.

Example 2: If the buckling load of a beam is calculated with the aid of an iteration function, then the resulting buckling load is greater than the actual as every iteration makes the beam 'stiffer' than it actually is.

- Are these values on the 'safe' or 'unsafe' side?
  Example 1: As the above iteration type of determined stiffness is lower than the actual one, then the stresses and deformation are larger than the actual ones, thus one is on the 'safe' side from a point of view of the strength.
  Example 2: If the buckling load calculated by the iteration is greater than the actual one, then one is on the 'unsafe' side as the beam actually buckles with a smaller buckling load than the one calculated.

- How great is the inaccuracy that occurs due to the iteration method?
  This can often only be estimated by means of experience. If the inaccuracy is too great, then a more exact calculation process must be selected.

For the calculation of the stresses from the section loads, use is made of the following cross-sectional values, and it can be determined again with the aid of the partition of the blade profile into N small rectangles with constant thickness, E and shear modulus as well as Poisson's ratios. Because of the changes of the profile cross-sections as well as the laminate structures in the longitudinal direction of the blade, they are in general dependent on the x coordinates. For this purpose for the calculation of the sheet, the blade length is divided into M profile sections ($m = 1,2,3,\ldots, M$).

Expansion stiffness for the profile section with the length coordinate x_m for calculating the normal stresses from the normal forces:

$$D^{(m)}(x_m) = \sum_{i=1}^{N} A_i^{(m)} E_i^{(m)} \tag{5.7}$$

Bending strength for the section (m):

$$B_y^{(m)}(x_m) = \sum_{i=1}^{N} \left[E_i^{(m)} \left(A_i^{(m)} \left(z_i^{(m)} - e_{z_{EZ}}^{(m)} \right)^2 + I_{yeigen,i}^{(m)} \right) \right] > 0 \tag{5.8a}$$

$$B_z^{(m)}(x_m) = \sum_{i=1}^{N} \left[E_i^{(m)} \left(A_i^{(m)} \left(y_i^{(m)} - e_{y_{EZ}}^{(m)} \right)^2 + I_{zeigen,i}^{(m)} \right) \right] > 0 \tag{5.8b}$$

$$B_{yz}^{(m)} = \sum_{i=1}^{N} \left[E_i^{(m)} \left(A_i^{(m)} \left(z_i^{(m)} - e_{z_{EZ}}^{(m)} \right) \left(y_i^{(m)} - e_{y_{EZ}}^{(m)} \right) + I_{xyeigen,i}^{(m)} \right) \right] \begin{cases} > 0 \\ = 0 \\ < 0 \end{cases} \tag{5.8c}$$

For a rectangle in any direction, the eigen moment of inertia is $I_{xyeigen}^{(m)}$, that is, referenced to its centre of gravity which is always zero as the rectangle is symmetrical.

If the E-modulus is the same in all part-rectangles, then it can be left out of equations (5.7) and (5.8) and, in place of the stiffness of the surfaces, one obtains the surface moment of inertia $I_y^{(m)}, I_z^{(m)}$ as well as the deviation moments $I_{yz}^{(m)}$.

$$A^{(m)}(x_m) = \sum_{i=1}^{N} A_i^{(m)} \tag{5.9}$$

$$I_y^{(m)}(x_m) = \sum_{i=1}^{N} \left[A_i^{(m)} \left(z_i^{(m)} - e_{z_{EZ}}^{(m)} \right)^2 + I_{y\text{eigen},i}^{(m)} \right] > 0 \tag{5.10a}$$

$$I_z^{(m)}(x_m) = \sum_{i=1}^{N} \left[A_i^{(m)} \left(\left(y_i^{(m)} - e_{y_{EZ}}^{(m)} \right)^2 + I_{z\text{eigen},i}^{(m)} \right) \right] > 0 \tag{5.10b}$$

$$I_{yz}^{(m)} = \sum_{i=1}^{N} \left[A_i^{(m)} \left(z_i^{(m)} - e_{z_{EZ}}^{(m)} \right) \left(y_i^{(m)} - e_{y_{EZ}}^{(m)} \right) + I_{xy\text{eigen},i}^{(m)} \right] \begin{cases} > 0 \\ = 0 \\ < 0 \end{cases} \tag{5.10c}$$

It is also possible to work with the main coordinate system. For this purpose, first the main moment of inertia and the bending stiffness as well as the main direction must be determined from the bending stiffness and the surface moment of inertia. The main directions (1) and (2) are characterised in that the deviation moment or the deviation bending stiffness becomes zero in a cross-section. They are:

$$B_1 = \frac{B_y + B_z}{2} + \sqrt{\left(\frac{B_y - B_z}{2} \right)^2 + B_{yz}^2} \tag{5.11a}$$

$$B_2 = \frac{B_y + B_z}{2} - \sqrt{\left(\frac{B_y - B_z}{2} \right)^2 + B_{yz}^2} \tag{5.11b}$$

The axis of the main inertial system is rotated about the angle α in the y-z plane. The angle is obtained from:

$$\alpha = 0.5 \arctan\left(\frac{-2B_{yz}}{B_y - B_z} \right) \tag{5.11c}$$

Analogously, the surface moments of inertia and the main direction for constant E-modulus in the cross-section are:

$$I_1 = \frac{I_y + I_z}{2} + \sqrt{\left(\frac{I_y - I_z}{2} \right)^2 + I_{yz}^2} \tag{5.12a}$$

$$I_2 = \frac{I_y + I_z}{2} - \sqrt{\left(\frac{I_y - I_z}{2} \right)^2 + I_{yz}^2} \tag{5.12b}$$

$$\alpha = 0.5 \arctan\left(\frac{-2I_{yz}}{I_y - I_z} \right) \tag{5.12c}$$

An advantage of the calculations in the main axis system is the simple stress determination (see below), but a disadvantage is that the main axis systems must be determined for each section and the bending moments as well as the lateral forces must be converted to the directions of the main axes.

5.2.4   Stresses and Deformations

The relationship between the stresses (stress vector $\underline{\sigma}$) and the deformation (deformation vector $\underline{\varepsilon}$) of an orthotropic material (the material properties are different depending on the direction, as is the case for most fibre-reinforced materials) and is described by means of the linear elasticity law (Hooke's law). As we are dealing with rotor blades with thin-walled crosssections, an even two-axes stress condition is assumed, that is, the stresses in the direction of the thickness σ_z and $r_{xz} = r_{zx}$ and the deformation vertical to the wall thicknesses are ignored.

The two-dimensional elasticity law of the even stress condition for orthotropic materials is:

$$\begin{bmatrix} \sigma_x \\ \sigma_y \\ \tau_{xy} \end{bmatrix} = \begin{bmatrix} \dfrac{E_x}{1-v_{xy}v_{yx}} & \dfrac{v_{yx}E_y}{1-v_{xy}v_{yx}} & 0 \\ \dfrac{E_x}{1-v_{xy}v_{yx}} & \dfrac{v_{yx}E_y}{1-v_{xy}v_{yx}} & 0 \\ 0 & 0 & G_{xy}=G_{yx} \end{bmatrix} \begin{bmatrix} \varepsilon_x \\ \varepsilon_y \\ \gamma_{xy}=\gamma_{yx} \end{bmatrix} \Rightarrow \underline{\sigma}=\underline{D}\underline{\varepsilon} \tag{5.13a}$$

or in the component form:

$$\sigma_x = \frac{1}{1-v_{xy}v_{yx}}(E_x\varepsilon_x + v_{yx}E_y\varepsilon_y)$$

$$\sigma_y = \frac{1}{1-v_{xy}v_{yx}}(E_y\varepsilon_y + v_{xy}E_x\varepsilon_x) \tag{5.13b}$$

$$\tau_{xy} = \tau_{yx} = \frac{G_{xy}}{\gamma_{xy}} = \frac{G_{yx}}{\gamma_{yx}}$$

with: $v_{xy}=v_{yx}\cdot E_y/E_x$; $v_{yx}=v_{xy}\cdot E_x/E_y$.

The reverse of this relationship, that is, the determination of the deformation from the stresses, is:

$$\begin{bmatrix} \varepsilon_x \\ \varepsilon_y \\ \gamma_{xy} \end{bmatrix} = \begin{bmatrix} \dfrac{1}{E_x} & -\dfrac{v_{xy}}{E_y} & 0 \\ -\dfrac{v_{yx}}{E_x} & \dfrac{1}{E_y} & 0 \\ 0 & 0 & \dfrac{1}{G_{xy}} \end{bmatrix} \begin{bmatrix} \sigma_x \\ \sigma_y \\ \tau_{xy} \end{bmatrix} \Rightarrow \underline{\varepsilon}=\underline{N}\underline{\sigma} \tag{5.14a}$$

or in the component form:

$$\varepsilon_x = \frac{\sigma_x}{E_x} - v_{xy}\frac{\sigma_y}{E_y}$$

$$\varepsilon_y = \frac{\sigma_y}{E_y} - v_{yx}\frac{\sigma_x}{E_x}$$

$$\gamma_{xy} = \frac{\tau_{xy}}{G_{xy}} = \gamma_{yx}$$

(5.14b)

with normal stresses σ_x in the x direction and σ_y in the y direction, shear stresses τ_{xy} in the x-y plane, expansion in the ε_x direction and ε_y in the y direction; shear angle y_{xy} in the x-y plane; Poisson's ratios v_{xy} (Poisson's ratio in the x direction by means of a normal stress in the y direction) and v_{yx} (Poisson's ratio in the y direction as a result of a normal stress in the x direction[2]; σ is the stress vector, ε the deformation vector, $\underline{S}$ the stiffness matrix and $\underline{N}$ the yield matrix. The stiffness matrix is the inverse of the yield matrix, i.e. $\underline{D}\cdot\underline{N} = \underline{E}$ ($\underline{E}$ = unit matrix).

From this, among others, one can recognise that the equations for isotropic materials (the same material properties in all directions) represent a special case of the relationship for orthotropic materials.

The respective engineering constants E_x (elasticity modulus in x direction), E_y (E modulus in y direction, G_{yz} (shear modulus), v_{yx} (Poisson's ratio in x direction) and v_{xy} (Poisson's ratio in y direction) can be numerically determined for fibre-reinforced materials (see below). However, they must be experimentally determined for stiffness verification, as the actual values due to the production conditions, types of weaves and reinforcing fibres can deviate strongly from those determined mathematically.

The normal stresses resulting from normal forces in the section x_m are dependent on their position and coordinates y and z in the cross-section as well as the E-moduli available there.

$$\sigma_{xN}(x_m, y, z) = \frac{N(x_m)}{D(x_m)}E(x_m, y, z)$$

(5.15a)

If the E-modulus of the observed section is constant, then the normal stress is also constant and Equation (5.15a) is simplified to:

$$\sigma_N(x_m, y, z) = \frac{N(x_m)}{A(x_m)}$$

(5.15b)

In the determination of the bending normal stresses, it must be considered that rotor blades have non-symmetrical cross-sections, so the stresses are to be calculated out of the 'slanted bending'. The normal stresses resulting from the bending moments M_y and M_z in the cross-section x_m are obtained according to Equation (5.16a). In this case the stresses are linearly distributed in parts and are equal to zero in the elastic centre (see Figure 5.7).

---

[2] In the UK the sequence of the indexing is usually reversed.

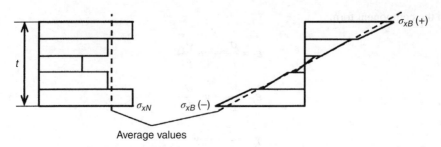

Average values

Figure 5.7 Stress distribution in a multilayer laminate with different elasticity moduli; left: as a result of normal force; right: as a result of bending moments

$$\sigma_{xB}(x_m, y, z) = \frac{E(x_m, y, z)}{B_y B_z - B_{yz}^2} \left[\left\{ M_y^{(m)}(x_m) B_z^{(m)} + M_z^{(m)}(x_m) B_{yz}^{(m)} \right\} z \right.$$
$$\left. - \left\{ M_z^{(m)}(x_m) B_y^{(m)} + M_y^{(m)}(x_m) B_{yz}^{(m)} \right\} y \right] \tag{5.16a}$$

If the E-modulus in the observed cross-section is constant, then Equation (5.16a) becomes:

$$\sigma_{xB}(x_m, y, z) = \frac{1}{I_y I_z - I_{yz}^2} \left[\left\{ M_y^{(m)}(x_m) I_z^{(m)} + M_z^{(m)}(x_m) I_{yz}^{(m)} \right\} z \right.$$
$$\left. - \left\{ M_z^{(m)}(x_m) I_y^{(m)} + M_y^{(m)}(x_m) I_{yz}^{(m)} \right\} y \right] \tag{5.16b}$$

In this case the bending stresses are distributed linearly over y as well as z and equal to zero in the surface centre of gravity.

The normal stresses from the normal force and the bending moments can be superimposed according to their prefixes ('superposition principle of the linear elasticity theory').

In the use of the main coordinate system the determination of the bending normal stresses simplifies to:

$$\sigma_{xB}(x_m, y, z) = \frac{M_1^{(m)}(x_m)}{I_1^{(m)}} z - \frac{M_2^{(m)}(x_m)}{I_2^{(m)}} y \tag{5.17a}$$

and

$$\sigma_{xB}(x_m, y, z) = E(x_m, y, z) \left(\frac{M_1^{(m)}(x_m)}{B_1^{(m)}} z - \frac{M_2^{(m)}(x_m)}{B_2^{(m)}} y \right) \tag{5.17b}$$

The shear stresses resulting from the lateral forces and torsion moments can only be found with the theory of non-uniform torsion. This will not be further dealt with here, as has already been mentioned. However, the shear stresses resulting from the lateral forces can be estimated in the following manner:

$$\tau_{maxxz}^{(m)} = k_z \frac{Q_z}{A}; \quad \tau_{maxxy}^{(m)} = k_y \frac{Q_y}{A} \tag{5.18}$$

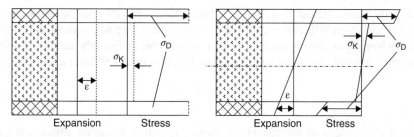

Figure 5.8   Normal stress distribution in a sandwich component with core (K) and shell layer (D); left: resulting from normal forces; right: resulting from bending moments

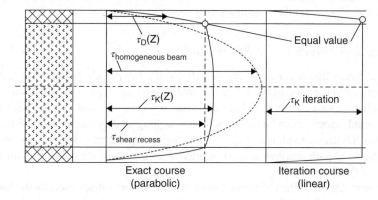

Figure 5.9   Shear stress distribution in a sandwich component resulting from the lateral forces; left: exact; right: approximation-type distribution.

The factors of the shear stress excess increase k_x and k_y can be different for thin-walled hollow sections according to their shapes. For instance, they are:

$$k = \begin{cases} \dfrac{3}{2} & \text{for a rectangle} \\[2mm] \dfrac{4}{3} & \text{for a full circle} \\[2mm] \approx 2 & \text{for a thin-walled tube} \\[2mm] \approx 2.5 - 3.5 & \text{for thin-walled hollow cross-sections} \end{cases}$$

In sandwich components the normal stresses are distributed according to Equation (5.15a) (see Figure 5.8) and the shear stresses according to Equation (5.19) (see Figure 5.9).

The shear stresses in a sandwich component resulting from lateral forces are obtained according to the following relationship ($z_0 =$ observed z coordinates, $b =$ unit width of the cross-section $= 1$, $h =$ height of the sandwich cross section):

$$\tau_{xy}(x, z_0) = \frac{Q_z(x)}{\displaystyle\int_{-\frac{h}{2}}^{\frac{h}{2}} E(x, z)z^2 \, dz} \int_{-\frac{h}{2}}^{z_0} E(x, z)z \, dz \tag{5.19a}$$

Because of the very different E-moduli in the shell layers and the core of sandwich components as well as the small thickness of the shell layers when compared with the core thickness, the iteration formula according to Equation (5.19b) provides sufficiently accurate results (with $\bar{h}$ = core thickness plus half shell layer thickness top, plus half shell layer thickness bottom):

$$\tau_{xz}(x) = \frac{Q_z(x)}{1\bar{h}} \tag{5.19b}$$

Remarks: The set signs of the calculated stresses (positive = tension stresses, negative = compression stresses) are independent of the selected coordinate system. However, the set signs of the section loads (normal and lateral loads, bending and torsion moments) of the support reaction (supporting reactions, restraint moment) as well as the bending and inclination are dependent upon the coordinate system.

5.2.5   Section Forces in the Rotor Blade

The external forces that occur in the cross-sections of a rotor blade generate six section loads (three forces and three moments) due to the spatial loading. Here, only the loads caused by wind, weight and rotation of the blade are dealt with. Forces that change with time (e.g. changes in wind velocity) can be included in this, as long as they can be dealt with in a quasi-static manner. Dynamic loadings, such as those due to rapid changes in the operating conditions like the wind path (coriolis forces) or braking of the rotor, are not dealt with within the framework of this introduction.

   In the determination of the moments it must be remembered that generally the longitudinal axis of the blade is not exactly vertical to the hub but is bent. Further it should be noted that the centre of mass, elastic centre, shear middle point and aerodynamic centre can have different coordinates. This causes additional bending and torsion moments in the rotor blade.

   The forces taken into account are:

- Lift forces as a line load in the positive z direction: $\rightarrow L(x)$; constant during a rotation: application in the aerodynamic centre with the coordinates $e_{y_{W(x)}}$ and $e_{z_{W(x)}}$.
- Drag forces as line loads in the positive y direction: $\rightarrow W(x)$; constant during a revolution; application in the aerodynamic centre.
- Eigen weight of the rotor blade as line load in the positive x direction: for the rotating position of the blade $\beta = 0$ (top) $\rightarrow -m(x) \cdot g$, at $\beta = 180°$ (bottom) $\rightarrow + m(x) \cdot g$, changeable during a rotation, application in the centre of mass with the coordinates $e_{y_{M(x)}}$ and $e_{z_{M(x)}}$.
- Centrifugal forces as line load in the positive x direction $\rightarrow m(x) \cdot x \cdot \omega^2$, constant during a rotation; application in the centre of mass.

These forces cause the following section loads:

- Normal forces N at the coordinates x_i of the rotor blade and the rotor blade setting β. Parts: eigen weight (vertically effective), centrifugal forces (in x direction), both applied to the mass centre of gravity:

$$N_x(x_i, \beta) = N_z(x_i) + N_G(x_i, \beta) = \int_{x_i}^{L_B} m(x)x\omega^2 \, \mathrm{d}x - \cos\beta \int_{x_i}^{L_B} m(x)g \, \mathrm{d}z \tag{5.20}$$

where $N(x_i, \beta)$ is the normal force at the point x_i with rotor blade setting β, $m(x)$ is the cross-section mass per unit length, ω is the circular frequency of the rotor rotation = rpm/(120π), and LB is the length of the blade.

- Lateral forces Q_z in z or Q_y in y directions at the coordinates x_i of the rotor blade. Parts for Q_z: lift forces L, normal forces N_G (depending upon angle β). Parts for Q_y: drag forces W.

$$Q_z(x_i, \beta) = -\int_{x_i}^{L_B} L(x)\,dx - \sin\beta \int_{x_i}^{L_B} m(x)g\,dx \qquad (5.21)$$

$$Q_y(x_i) = -\int_{x_i}^{L_B} W(x)\,dx \qquad (5.22)$$

where $L(x)$ is the lift force per unit length, and $W(x)$ is the drag force per unit length.

- Bending moments M_y and M_z at the coordinate x_i of the rotor blade (they must be referenced to the elastic centre of gravity with the constants $e_{y_{EZ(x)}}$ and $e_{z_{W(x)}}$).
- The centrifugal forces cause bending moments when the position of the centre of mass deviates from the elastic centre of gravity.Parts for M_y: lateral force Q_z, application in the aerodynamic centre; normal force N_x, application in the centre of mass. Parts for

$$M_y(x_i, \beta) = \int_{x_i}^{L_B} Q_z(x, \beta)x\,dx - \int_{x_i}^{L_B} N_x(x, \beta)\left[y_{z_{EZ}}(x) - e_{z_M}(x_i)\right]dx \qquad (5.23)$$

Parts for M_z: lateral force Q_y, application in the aerodynamic centre; normal force N_x, application in the centre of mass:

$$M_z(x_i, \beta) = \int_{x_i}^{L_B} Q_y(x)x\,dx - \int_{z_i}^{L_B} N_x(x, \beta)\left[y_{z_{EZ}}(x_i) - e_{z_M}(x)\right]dx \qquad (5.24)$$

- Torsion moment M_x about the z axis; parts through:
  Lift L and drag W in the aerodynamic centre application;
  Normal force N_x in the centre of mass application (the forces referenced to the shear centre of gravity with the coordinates $e_{y_{S(x)}}$ and $e_{z_{S(x)}}$):

$$M_x(x_i, \beta) = \int_{x_i}^{L_B} L(x)\left[e_{z_w}(x) - e_{z_S}(x_i)\right]dz + \int_{x_i}^{L_B} W(x)\left[e_{y_w}(x) - e_{z_S}(x_i)\right]dz$$
$$+ \int_{x_i}^{L_B} N_x(x, \beta)(x, \beta)\sqrt{e_{yM}^2(x) - e_{zS}^2(x_i)}\,dx + M_W \qquad (5.25)$$

where M_W is the aerodynamic moment.

In this way the six section forces in the rotor blade are defined in terms of the coordinate x_i. The values of the section loads all start with zero at the rotor blade point $(x = L_B)$. The normal and shear stress distributions can be calculated at the rotor blade with the section loads, and with this it is possible to minimise the weight of the rotor blade while making use of the maximum permissible stresses.

The camber moments that occur due to torsion for torsion moment sections changing with x can be ignored for rotor blades, as they are closed cross-sections, and the torsion moments and cross-sectional values change continuously. In this way the contribution of the camber forces and moments to the stresses (camber normal stress and camber shear stress) are only very small.

As the loadings are generally present only numerically, the section loads are numerically integrated according to the trapezoidal rule. The accuracy of the trapezoidal rule is generally sufficient since both the sequence of the section loads as well as the cross-sectional values can only be approximated by means of interpolation.

5.2.6   Bending and Inclination

In a first iteration, the rotor blade can be seen as clamped to the hub, just like a cantilever with asymmetrical cross-sections and E-moduli that change in the longitudinal direction. This forms a 'statically determined' system in which the lateral force and bending moment sequences can be calculated from the given outer loadings with the aid of the balance relationships.

With calculations according to beam theory, the 'linear elasticity theory' is assumed, such that linear material behaviour and small deformations are a condition. In this way the deformations of the beam can be calculated independently of each other through the loadings in the various directions. In this, the displacement or the bending in the x direction is designated with u, in the y direction with v, and in the z direction with w.

The differential equations of the bending lines for $w(x)$ in the z direction and for $v(x)$ in the y direction are then in the linearised form (linear, usual 4th order differential equation):

$$[E(x)\,I_y(z)\,w(x)'']'' = q_z(x) \quad [E(z)\,I_z(z)\,v(z)'']' = q_y(z) \tag{5.26}$$

$$[E(x)\,I_y(x)\,w(x)'']' = -Q_z(x) \quad [E(z)\,I_z(z)\,v(z)'']' = -Q_y(z) \tag{5.27}$$

$$E(x)\,I_y(x)\,w(x)'' = -M_y(x) \quad E(z)\,I_z(z)\,v(z)'' = -M_y(z) \tag{5.28}$$

As the rotor blade is a statically determined system, a second-order differential equation according to Equation (5.29) can be assumed in order to calculate the deformation of the rotor blade as a result of the bending moment:

$$w''(x) = \frac{M_y(x)}{E(x)I_y(x)} \tag{5.29a}$$

$$v''(x) = \frac{M_z(x)}{E(x)I_z(x)} \tag{5.29b}$$

The calculation of the bending moments M_y and M_z has been explained in the previous section. The 2 × integration of equations (5.29a) and (5.29b) with the respective bending moment sequences and the inclusion of the limiting conditions (bending and inclination at the clamping position are zero) provide the bending conditions $w(x)$ and $v(x)$.

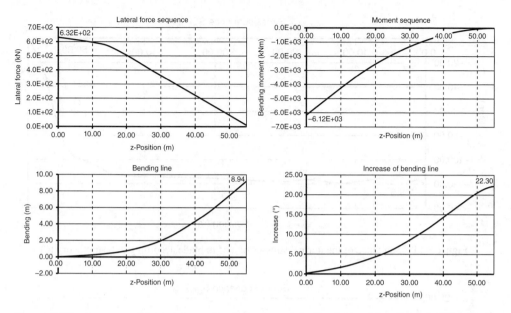

Figure 5.10 Sequence of lateral force, bending moment, inclination and bending of a 55 m blade according to beam theory [1]

Because with rotor blades the sequences for the bending moments $M(x)$, moment of inertia $I(x)$ and E-modulus $E(x,y,z)$ are mostly not available in analytical form, a numerical integration must be carried out. This takes place in parts according to the existing profile sections (see above).

For the determination of the deformation by means of lateral forces (shear bending) and torsion moments (twisting or torsion) for thin-walled, hollow sections, the shear and torsion stiffnesses of multicelled hollow cross-sections are required. These can only be determined with the already-mentioned theory of non-uniform torsion. If the rotor blade is produced only using solid laminates, then the shear bending due to lateral forces can be ignored. If large sections of the blade are manufactured from sandwich laminates, then the shear bending makes a substantial contribution to the overall bending since sandwich laminates, due to the low E-modulus of the usual core material, are 'shear-soft'.

5.2.7 Results According to Beam Theory

The calculation of the cross-section values, internal forces, stresses and deformations can be shown very well in tabular form. Especially in an early project stage, when the final dimensions and material values are not yet available, temporary optimising of the rotor blade cross-sections can be undertaken in this way. The results of such a tabular calculation are shown in Figures 5.10 and 5.11. As a comparison, Figure 5.12 shows the stress sequence of a FEM calculation for a 55 m blade.

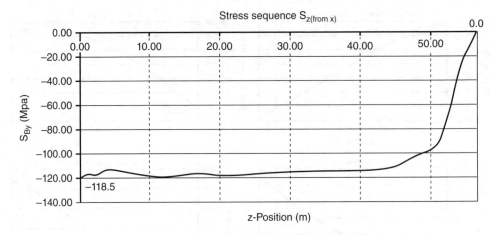

Figure 5.11 Sequence of the normal stresses of a 55 m blade according to beam theory [1]

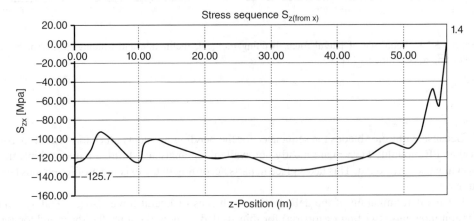

Figure 5.12 Sequence of the normal stresses in a 55 m blade according to FEM calculations [1]

5.3 Vibrations and Buckling

5.3.1 Vibrations

In order to evaluate the vibration behaviour of a rotor blade, knowledge of the eigen-frequencies, the damping behaviour and the frequencies of the excitations due to external forces are necessary.

For damping, one differentiates between material, friction, design or production-conditioned dampings on the one hand, and aerodynamic damping on the other hand. The material, friction and production-conditioned dampings are mostly frequency-dependent (exception: material damping with rubber and plastics such as laminating resins for FRP). Aerodynamic damping is seen as the resistance of a moving body in a medium such as air or water. It is the square of the velocity.

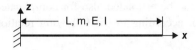

Figure 5.13 Rotor blade as a cantilever with constant cross-sectional values

For damping values, care must be taken of the type of information such as logarithmic decrements, Lehr's damping ratio, loss factors, and so on; they have different values.

For material damping, the logarithmic decrement for steel is 0.005–0.02; for GRP [glass fibre reinforced plastic] it is 0.06–0.1, thus 4 to 5 times as great; for CRP [carbon fibre reinforced plastic] 0.03–0.05; and for rubber 0.2–0.5 depending on the hardness of the rubber. Welded structures have a much lower damping than screwed or riveted ones.

The connection between the two main types of damping information used, the logarithmic decrement A and the Lehr damping ratio D, is:

$$D = \frac{\Lambda}{\sqrt{4\pi^2 + \Lambda^2}} \tag{5.30}$$

Resonances will occur if the excitation frequencies are near to the eigen-frequencies. In this case the vibration amplitudes and thus the deformations and stresses can become larger than is permissible (enlargement function), and could lead to early failure of the components. When the differences between the excitation and the eigen-frequencies are less than ±30% –40%, then vibration resonances must be calculated. The amount of the differences is dependent upon, among others, the damping of the components.

If the eigen and excitation frequencies as well as the damping are known, then the so-called structural answer, that is, the reaction of the rotor blades to the excitation, can be determined by means of the timely changes of external forces. In this way the occurring vibration amplitudes and thus the stresses in the rotor blades due to dynamic loading are obtained. From this, together with the stresses from the quasi-static loads, the load-collective (stress regions and associated load cycles) for the length of life calculation can be determined. This is not within the scope of this introduction; reference is made to advanced literature such as [8] or [9].

An estimate of the blade eigen-frequencies can be carried out by means of strongly simplified models. The simplest model is to calculate the vibration behaviour of the rotor blade as a clamped beam (cantilever) with a constant cross-section and constant E-modulus (Figure 5.13). The eigen-frequencies of such an undamped beam are:

$$\omega_i = \lambda^2 \sqrt{\frac{EI}{mL^4}} (s^{-1}) \quad \text{with } \lambda_i = 1.875;\ 4.694;\ 7.855;\ 10.996;\ 14.137 \quad (j = 1, K, 5) \tag{5.31}$$

Vibration periods:

$$T_j = \frac{2\pi}{\omega_j} (s) \tag{5.32}$$

The first eigen-frequencies can be estimated also for complicated structures such as blades with changing cross-section, according to the following relationships with the aid of the Rayleigh quotient (RQ):

$$E + U = \text{constant} = E_{max} = U_{max} \tag{5.33a}$$

$$E_{max} = U_{max} = \omega^2 \overline{E} \quad (\overline{E} = \text{referenced kinetic energy}) \tag{5.33b}$$

$$\text{Rayleigh quotient RQ} = \omega^2 = \frac{U_{max}}{\overline{E}} \tag{5.34}$$

The referenced kinetic energy $\overline{E}$ and the potential energy U are described with the aid of an iteration function $\overline{\omega}(x)$ for the bending of a beam, taking the border conditions into account. $\overline{E}$ and U_{max} are obtained according to:

$$\overline{E} = \frac{1}{2} \int_{(L)} m(x)\, \overline{\omega}^2(x)\, dx \tag{5.35}$$

$$U_{max} = \frac{1}{2} \int_{(L)} E(x) I(x) (\overline{\omega'})^2(x) dx \tag{5.36}$$

Example: a cantilever with constant cross-section ($m(x) = m$ and $I(x) = I$).
The border conditions for $x = 0$ for clamped cantilevers are: bending at $x = 0 \rightarrow w(0) = 0$; inclination at $x = 0 \rightarrow w'(0) = 0$; bending moment at $x = L \rightarrow w''(L) = 0$, lateral force at $x = L$ $\rightarrow w'''(L) = 0$.

The following iteration can be used for the bending (bending line cantilever):

$$\overline{\omega}(x) = x^2 \left[6 - 4\frac{x}{L} + \left(\frac{x}{L}\right)^2 \right]$$

For the first eigen-frequency with I and E = constant, one obtains with RQ:

$$RQ = \omega^2 = \frac{\int_{(L)} E(x) I(x) \left(\overline{\omega'}\right)^2 (x)\, dx}{\int_{(I)} m(x)\, \overline{\omega}^2(x)\, dx} \tag{5.37}$$

$$\rightarrow \quad \omega_1 = \sqrt{\frac{162 \times E \times I}{13 \times m \times L^4}} \, (s^{-1}) \tag{5.38}$$

Basically the determined frequencies with the iteration process are greater than the actual ones, as iteration processes make the systems stiffer. The smallest eigen-frequency resulting from the different iteration attempts for the vibrations and eigen-forms is the nearest to the actual value. If the exact eigen-form is selected as the iteration attempt, then the exact eigen-frequency is also obtained.

The absolute dimension of the vibration bending or amplitude remains undefined (characteristic of homogeneous equations); for instance, in the Rayleigh quotients the amplitudes are shortened out.

Due to the material and structural dampings in a rotor blade of approximately $\Lambda \approx 0.1$, the vibration frequencies are reduced by approximately 1.5% and this effect can be ignored. This can be estimated according to the relationship of the frequencies for a linear single mass oscillator using Equation (5.39) according to Hauger, among others [10]:

$$\omega_d = \omega_0 \sqrt{1 - D^2} \tag{5.39}$$

where ω_d is the frequency of the damped vibrations, ω_0 is the frequency of the undamped vibrations, and D is the Lehr damping ratio.

The rotor blades prevail over the aerodynamic damping. Their dimensions can only be determined by means of CFD [computational fluid dynamics] calculations as they change with time. It depends on the direct flow of the blade during the vibrations and on the square of the vibration velocity.

The exact eigen-frequencies, especially the higher ones, for instance with changing cross-sections $I(x)$, mass occupations $m(x)$ and line loads $q(x)$ as they occur with real rotor blades, can only be determined with the aid of finite-element modelling (FEM) calculations.

The number of eigen-frequencies depends upon the degrees of freedom of the system; a continuum has an infinite number of degrees of freedom, and thus just as many eigen-frequencies. A FE model has as many degrees of freedom or eigen-frequencies as the equation system has unknowns to be solved. However, we are normally only interested in the approximately 20 lowest eigen-frequencies.

Each eigen-frequency of a system capable of vibrations (elastic continuum) has exactly one 'eigen-form', which is the form in which the system at this eigen-frequency vibrates. As long as the time period of the vibration excitations, as well as the damping behaviour, are unknown, the actual vibration effect cannot be determined but only the vibration or eigen-form.

The lowest eigen-frequencies of rotor blades are in the region of 0.3–2 Hz and the values apply to the relatively flexible direction of impact (from the rotor blades outwards). The eigen-frequencies are clearly higher for the stiff angle of rotation of the blades (in the rotor level). As can be seen in Equation (5.38), they are dependent upon the bending stiffness, among others.

5.3.2   Buckling and Stability Calculations

In the case that compressive stresses occur in the component, it must be checked whether the affected components could fail due to loss of stability. This can occur due to kinking in rod-shaped components or due to buckling for larger surface components. With rotor blades it is particularly the large and long, slightly curved areas in the rear part of the profiles that are endangered.

With most rotor blades there is more chance of a failure of stability (buckling) than a failure of stress (exceeding the bearable stresses of the material). For this reason, measures are taken with large blades to increase the buckling strength. This is firstly with the use of sandwich laminates (increasing the thickness by the use of light core material) by which the bending strength in the endangered regions is increased without much increase in weight. Another possibility is the installation of additional stringers (strengthening ribs) with which the size of the buckling fields is reduced. The results of both these measures are that the weight and the production costs are increased.

With rotor blades, the analytical determination of the critical compression stresses designated as σ_{crit}, above which the blade fails, is very difficult. These are thin-walled surfaces of orthotropic materials in which the size of the 'buckling fields' (length and width) and the support of their edges can only be coarsely determined, and in addition the fields are curved in part. Methods for rough determination of the critical stresses can be found, for instance, in [11].

For sandwich components, besides the failure of the whole structure (kinking as a rod or buckling of the plate or shell), local stability failure of the compressive layers, the shell layers or crinkling of the shell layer or the core (shear buckling) can occur. The cause of this can be either too low a bending stiffness or thickness of the shell layer, or too soft a core (E-modulus or shear modulus too low). The effects can be estimated by the following formulae:

$$\text{Shell layer buckling:}\quad \sigma_{crit} = 0.5 \times \sqrt[3]{E_D \times E_K \times G_K} \tag{5.40a}$$

$$\text{Shell layer crinkling:}\quad \sigma_{crit} = 0.82 \times E_D \times \sqrt{\frac{E_K\, t_D}{E_D\, t_k}} \tag{5.40b}$$

where σ_{crit} is the maximum bearable compressive stress (for stability investigations usually the compressive strength is used as a positive); E_D is the E-modulus of the shell layer; E_K is the E-modulus of the sandwich core; G_K is the shear modulus of the core; t_D is the thickness of the shell layer; and t_K is the thickness of the core.

5.4 Finite Element Calculations

5.4.1 Stress Calculations

The finite element method (FEM) is a further possibility for strength, vibration and buckling calculations of rotor blades. In the advanced project stage, when the dimensions and material data of the rotor blade are basically defined, this method is necessary for final

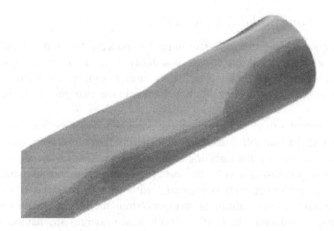

Figure 5.14   FEM stress plot of a 55 m blade at the connection region at the hub [1]

design and as verification, for instance for the approval authorities, since with analytical methods such as beam theory only estimates of the occurring stresses can be provided. In particular, complicated three-dimensional components can be calculated with this method in which local effects such as the introduction of forces, stiffness jumps, stress concentrations, force deviations, and so on, can occur that cannot generally be acquired by analytical methods.

Although the available FEM programs have implemented elements with which laminates can be calculated, for sandwich laminates in which the shell layers and cores have very different properties, such elements lead to large deviations in the determined stresses, especially in the cores. The bearable stresses of the core material are approximately two orders of magnitude less than the laminate in the shell layers, so that even small inaccuracies in the determined stresses can lead to damage of the core.

A possibility for bypassing this problem is a corresponding fine elementing of the shell layers and the core. However, this leads to a large number of elements and very large equation systems, for whose solutions correspondingly capable computers are then necessary.

The FEM results depend strongly on how well the FE model is adapted to the actual conditions, such as support and load application. A large influence on the results will also be the type of element selected, such as triangle or quadrilateral elements for two-dimensional components (plates, disks), tetrahedral or hexahedral elements for three-dimensional components, and also the size of the element. For this reason it is always necessary to check the results of a FEM calculation for plausibility. This can often occur by means of estimates with analytical methods.

5.4.2   FEM Buckling Calculations

The calculation of buckling behaviour of rotor blades in the sandwich method with the FEM process is mainly limited to dividing the blade into shell elements that extend over the whole thickness of the shell. In order to calculate the local effects in a sandwich, such as shell layer buckling or crinkling, several elements must be distributed over the thickness of the shell layer and sandwich core. This would lead to a large number of elements with correspondingly long calculation times for their solution. For this reason the local stability effects are mostly investigated separately. The stresses on the upper and lower sides of the profile surfaces are known from the global stress calculation. With these stresses and material values of the sandwich, they can be compared, for instance, with the critical stresses according to Equation 5.40.

A buckling calculation with FEM programs is carried out in two steps. In the first step a 'normal' stress calculation with the respective loads is carried out, and in the second step the actual buckling calculation. The result is a 'buckling factor' that indicates how much higher the bearable loading to buckling is for this load case.

Figures 5.15 and 5.16 show buckling-endangered regions of the rotor blade (with highly increased scale of the buckling figure), the first with a design with a pure GRP laminate, and the two following ones with the sandwich insert in the endangered regions. The buckling factors in the examples for the same loading are 1.011 (Figure 5.15) and 1.40 (Figure 5.16). For stability calculations by means of iteration, as is usual with the FE method, they are analogous to the eigen-frequency calculations, generally higher than the actual values.

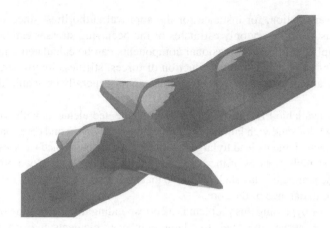

Figure 5.15   FEM plot of the buckling behaviour of pure GFK material (according to [1])

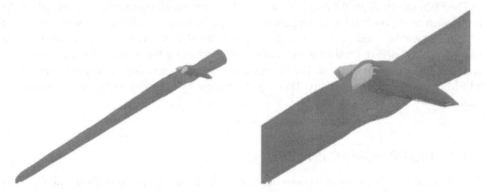

Figure 5.16   FEM plot of the buckling behaviour for sandwich insert in the endangered region. Left: overall view; right: section, according to [1]

5.4.3   FEM Vibration Calculations

With analytical methods only, the first eigen-frequency of rotor blades can be closely determined. The exact calculations, especially for the higher frequencies, are relatively easily carried out with FE methods.

In the determination of the vibration behaviour of rotor blades by means of FEM calculations, it must be noted that all forms of vibrations such as bending in the impact direction, bending in the swing direction and torsion vibrations (about the longitudinal axis) are permitted by the FE model. Combinations of the different forms of vibration can also occur. Expansion vibrations (longitudinal vibrations) are generally not of interest as they only occur at very much higher frequencies because of the high expansion stiffness (on a scale of frequencies ≈ acoustic frequencies in the blade material/blade length [Hz]).

The implementation of FE eigen-frequency calculations is analogous to that of FE buckling calculations (see Section 5.4.2). Just as with the buckling stresses, the frequencies are generally higher than the actual ones. In Figures 5.17 and 5.18 the eigen-forms of vibrations associated with the first and third eigen-frequencies are depicted in the impact direction.

Figure 5.17 Form of vibration of a 55 m rotor blade (first eigen-frequency = 0.737 Hz in direction of impact) [1]

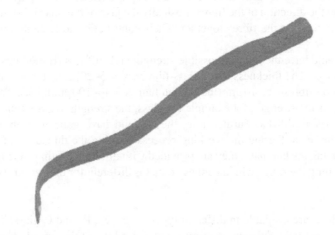

Figure 5.18 Form of vibration of a 55 m rotor blade (third eigen-frequency = 7.79 Hz in direction of impact) [1]

5.5 Fibre-Reinforced Plastics

5.5.1 Introduction

Fibre-reinforced plastics (FRP) are excellent light construction materials as they possess substantially lower specific densities in comparison with metals for high strengths (GRP laminate ≈ 1.8, CRP laminate ≈ 3 1.8 g/cm³, CRP ≈ 1.6).

The tensile strength of GRP is ≈ 250–450 MPa (1 megapascal = 1 N/mm²) and of CRP laminates ≈ 800–1750 MPa. The elasticity moduli of the laminates are: for GRP ≈ 20,000–35,000 MPa, and for CRP ≈ 75,000–200,000 MPa. A further great advantage compared with isotropic metals is that the required material thicknesses can be substantially reduced because of the quantity of fibres (percentage portion of the laminate) and the arrangement of the main portion of the fibre quantity in the direction of the greatest loading. This leads to better material utilisation compared with isotropic materials. Furthermore, the corrosion resistance of FRP is very high. The operational strength of GRP is somewhat lower than that of steel, but better than that of aluminium, and that of CRP is better than that of steel.

A disadvantage is the low fracture expansion of FRP (GRP $\lesssim$ 2%, CRP $\lesssim$ 1%) in comparison with metals (steel $\approx$ 15–40%, aluminium $\approx$ 5–15%). This means that FRPs are much more sensitive to stress peaks and only have a very low flowing property with excess loads; they are less 'good-natured' than, for instance, steel. FRPs are less tolerant to errors.

A further disadvantage is that the material itself must first be produced at the manufacturers of FRP components. For this reason the spread of the material data is much greater than for metals that are used as semi-finisheds.

5.5.2   Materials (Fibres, Resins, Additives, Sandwich Materials)

5.5.2.1   Fibres

Strength and stiffness of FRP are mainly determined by the types of fibres, their proportion in the laminate, and the direction of the fibres. Basically the task of the 'matrix' (laminating resin plus additives) is to keep the fibres together. Their contribution to strength and stiffness is small.

The fibres should have the highest possible strength and stiffness (E-modulus) in comparison with the matrix. The thicknesses for glass fibres are $\approx$ 5–20 μm (1 μm = 1 micrometer = 10^{-6} m), for carbon fibres $\approx$ 5–10 μm (the human hair is $\approx$ 40–100 μm thick). The thinner the fibre, the greater is the strength; for example for glass, the strength of glass fibre $\approx$ 3000 MPa, but for solid glass $\approx$ 55 MPa. Furthermore, they should have good workability and good adhesion with the resin. During the working process (pulling the thread, creating the semi-finisheds, producing the laminate) they are provided with different coatings: the finishing. The last step is providing the fibres with an adhesive that is different for the various types of fibres and resins.

Glass fibres: These are available in different types, as E-, S-, R- and C-glass. They are relatively cheap, electrical insulators, isotropic, and differ little in their mechanical properties, but differ in their areas of application. For rotor blades, mostly E-glass fibres are used, with E-modulus values of $\approx$ 76 GPa (1 gigapascal = 10^9 pascal = 10^3 N/mm²), shear modulus $\approx$ 30 GPa, Poisson's ratio $\approx$ 0.24, tensile strength $\approx$ 1500 MPa, density $\approx$ 2.6 g/cm³, and elongation at fracture $\approx$ 3.5%.

Carbon fibres: These fibres have substantially higher E-moduli, are much more expensive, electrically conducting and strongly anisotropic, that is, their mechanical properties are strongly direction-dependent and they are more difficult to work than glass fibres. They are only used for semi-finisheds that must achieve a high strength and stiffness.

Thus, for instance, mats of carbon fibres are not sensible as these achieve only low strength. The fibres and the components made from them are sensitive to impacts and knocks. There are different types of carbon fibres such as normal fibres (E-modulus $\approx$ 200 GPa, strength $\approx$ 1800 MPa, elongation at fracture $\approx$ 1.3%), high-strength fibres (E-modulus $\approx$ 350 GPa, strength $\approx$ 3500 MPa, elongation at fracture $\approx$ 1.3%), and ultra-high-strength fibres (E-modulus $\approx$ 800 GPa, strength $\approx$ 2000 MPa, elongation at fracture $\approx$ 0.5%).

A temperature expansion for warming in the longitudinal direction of the fibres is negative and in the lateral direction positive (see Table 5.1).

Aramid fibres: aramid fibres (e.g. Kevlar) have special applications such as for knock- or impact-sensitive components. Their mechanical properties are between those of glass and carbon fibres, yet they have big differences. Thus the tensile strength is ≈ 2500 MPa but the compressive strength only ≈ 500 MPa, so they are very sensitive to compression. Furthermore they have a relative high energy absorption capability (impact strength). The temperature behaviour is similar to that of carbon fibres.

The individual fibres are worked into yarns with mostly 64 or 128 fibres. These yarns are used for various semi-finisheds such as mats (non-directional fibres), weaves in various binding types such as screens, body, atlas, etc., non-woven fabrics, rovings (fibre strings), tapes (flat fibre strings) and knits. Depending on the semi-finished, the laminates produced have different strengths for the same fibre proportion. Thus, laminates made of mats have the lowest strength but those made of rovings the greatest. The fibre proportion in longitudinal and lateral directions can be different, from 50% in each direction to almost 100% in one direction (rovings and tapes). In order to produce thick laminates with low production effort, non-woven fabrics are best suited for slightly curved components. The maximum thickness of multi-axial glass-fibre non-woven fabrics is approximately 1 mm, corresponding to approximately 2.6 kg/m², and for carbon fibre non-woven fabrics approximately 0.2 mm, corresponding to approximately 0.35 kg/m². (see Fig 5.19).

Table 5.1 Properties of the technically most important fibres

Properties	Glass fibres	Carbon fibres	Aramid fibres
E-modulus in direction of fibre (GPa)	71–85	200–800	75–130
E-modulus lateral to direction of fibre (GPa)	71–85	6–15	5
Shear modulus (GPa)	28–34	10–15	5–10
Poisson's ratio long/lateral	0.24/0.24	0.2/0.3	0.2/0.34
Density (g/cm³)	2.5–2.6	1.7–1.8	1.4–1.45
Tensile strength (aged) (MPa)	1500–2200	2000–3500	2500–3000
Elongation at fracture (%)	3.5–5.0	0.5–2.0	3–4
Temperature expansion long/lateral (10^{-6}/°K)	5	−0.5/10.0	−5/55

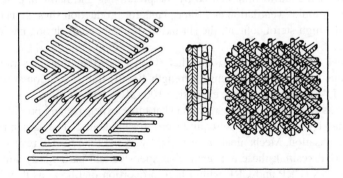

Figure 5.19   Structure of a multi-axial non-woven fabric with 0/90/+45/−45° layers [12]

For laminates with high requirements for strength, evenness of components and rapid processing, use is made of so-called prepregs (pre-impregnated fibres). They are already produced and impregnated with epoxy-resin. In order that the resin does not harden, they are stored in a frozen condition and thawed out before processing. They are then slightly adhesive and can also be processed overhead. Then the components are hardened in a warming oven (autoclave).

Detailed descriptions of the properties of fibres and their semi-finisheds can be found, for instance, in [13]. E-glass fibres are used almost exclusively for rotor blades as they are comparatively cheap and normally provide sufficient strength. Carbon fibres are only used for very long blades at especially highly loaded places. Aramid fibres are not used for rotor blades.

5.5.2.2 Resins

The tasks of the resins or the matrix (resin and additives) are to fix the fibres spatially, to transmit the loads to the individual fibres, to support the fibres for compressive loads, and to protect them from the effects of the environment. The matrix is isotropic; it should have a substantially higher elongation at fracture than the fibres due to the fact that in the matrix, through local effects in the direct surroundings of the fibres, there are larger expansions than in the middle expansion of the laminate. The expansion increases can be greater than three times the middle expansion; the greater the fibre volume proportion, the greater is the expansion increase.

The resins used for the production of laminates are:

Unsaturated polyester resins (UP resins): they are cheap, easy to work, and have a relatively low viscosity (thin-flowing), thus ensuring a good saturation of the fibres. The disadvantage of the UP resins is that the laminates produced with them achieve a relatively low strength, including operational strength, and age faster than other resins. They harden due to polymerisation. Styrol is necessary for hardening the UP resins, and a proportion of 30–40% is required. In order to achieve an even and complete hardening, a higher sterol proportion is added, since at the surface of a component the sterol can evaporate faster that the hardening process takes place. It is also added in order to reduce the viscosity of the matrix. The largest part of the surplus sterol evaporates slowly and leads to the typical sterol smell. In addition, the environmental loading caused by the processing and hardening of free sterol is great, so that complex ventilation and filter plants for production sites are necessary. Also, because of the high sterol content, the shrinking of components during hardening is large and can reach 6%.

   Mechanical values: isotropic, E-modulus 3000–3500 MPa; Poisson's ratio 0.34–0.35; shear modulus 1100–1300 MPa; tensile strength 40–80 MPa; elongation at fracture 4–6%; density 1.2–1.4 g/cm^3; temperature expansion $100–210 \times 10^{-6}/°K$.

Vinylester resins (VE resins): they are somewhat more expensive and provide higher-grade laminates than UP resins, but also require sterol for hardening. The hardening takes place due to polymerisation. Mechanical values: slightly higher than for UP resins.

Epoxy resins (EP resins): these are relatively expensive and particularly suited for highly loaded laminates (GRP and CRP with a high proportion of fibres) as they have the best mechanical values and the best fatigue behaviour of the laminates produced with them.

Their viscosity is substantially greater than that of the UP and VE resins, and therefore thinners must be added in order to achieve good flowing behaviour. Hardening takes place by means of polyaddition; the mixing behaviour of resin and hardeners must be exactly adhered to in order to achieve complete hardening. The shrinking of components during hardening is up to 4%.

Mechanical values: isotropic, E-modulus 3500–4500 MPa; Poisson's ratio 0.38–0.40; shear modulus 1400–1600 MPa; tensile strength 50–100 MPa; elongation at fracture, 3–6%; density 1.1–1.4 g/cm^3; temperature expansion $\approx 60 \times 10^{-6}/°$K.

The abovementioned laminate resins are also suitable as adhesives and they are only slightly modified for this purpose. The epoxy adhesives can also be used for joining components that have been laminated with other resins. Detailed information on the different resins can be found, for instance, in [14].

The processing times of cold-hardening resins (pot times) can be controlled to a large extent by the addition of delaying and accelerating agents: for UP and VE resins between 5 minutes and 4 hours, and for EP resins between 2 minutes and 20 hours.

Epoxy resins are used almost exclusively for rotor blades because of the good mechanical properties, the fatigue behaviour and the wide range of pot times.

5.5.2.3 Additives

In the case of additives that are added to the resins and that together form the matrix, a differentiation is made between additives and fillers. The fillers are meant to 'stretch' the matrix, to make it cheaper or result in a thicker laminate for the same amount of resin. Materials that are cheaper than the resins are added to it, such as cotton flakes or quartz powder, without negatively affecting the important properties of the resin such as viscosity, adhesion and hardening. Such materials are not used for highly loaded laminates.

The tasks of the additives are to influence the properties of the matrix precisely, depending on its application. These are viscosity, lamination process, hardening times, colour of the laminate, fire resistance, wearing strength, workability such as thixotropic behaviour (the matrix must not drain off vertical walls), and so on. Materials used are additives such as fire protection materials, colorants, thinners and surface-builders (only for UP and VE resins), thixotropic substances and light stabilisers. They must be suited to the respective resin and be able to mix well with them. Furthermore, they must not negatively affect the properties of the resin.

5.5.2.4 Sandwich Materials

Sandwich laminates are particularly suitable if a high degree of bending stiffness for relatively low weight is necessary (see Table 5.2). They consist of two cover layers with high E-modulus and strength, such as glass fibre or carbon fibre laminate, and a core of light material as 'spacers' for the cover layers. The strength requirements for the cores are low, but it must be able to withstand the shear stresses. Core materials can be balsa wood, foamed material and web structures. Balsa wood and foamed materials are used almost exclusively for rotor blades.

Table 5.2 Comparison of the bending characteristics of solid and sandwich laminates

	Solid laminate	Sandwich laminate	Sandwich laminate
Thickness of the cover layer	t	t	t
Thickness of the core	0	t	3t
Thickness of the laminate	t	2t	4t
Relative bending strength	1.0	7.0	37.0
Relative bending strength	1.0	3.5	9.25
Relative weight	1.0	Approx 10.5*	Approx 10.5*

*Dependent on the density of the core material.

Balsa wood: balsa wood consists of glued-together blocks; it has a relatively high shear strength and stiffness in comparison to weight, and is easy to work. However, the open pores of the face absorb a lot of resin, thus increasing the weight. It must be protected against rotting and fungal infestation.

Density $\approx$ 0.11–0.2 g/cm^3; shear strength $\approx$ 2–3 MPa; shear modulus $\approx$ 120–160 MPa; tensile/compressive strength $\approx$ 8–12 MPa (the lower values apply to the light balsa wood).

Foamed material: in the foamed materials a distinction is made between polyvinyl chloride (PVC), polymethacrilimide (PMI) and syntactic foams. The PMI foams have much better strength characteristics but are much more expensive than PVC foams. For this reason, almost only PVC foams are used for rotor blades. Syntactic foams (resin mixed with hollow micro-spheres) are not used for rotor blades because of their substantially greater densities.

Densities of the PVC foams for rotor blades: 0.08–0.15 g/cm^3; E-modulus $\approx$ 50–120 MPa; shear strength $\approx$ 1–1.5 MPa; shear modulus $\approx$ 12–50 MPa; tensile/compressive strength $\approx$ 1– 3.5 MPa, the lower values applying to foams with lower densities.

Honeycomb core: honeycomb cores are not used for rotor blades because of their low compressive strength and relatively high prices.

Detailed information on the different resins can be found, for instance, in [13].

5.5.3   Laminates and Laminate Properties

The elastic and shear modulus as well as Poisson's ratios are mainly dependent on the structure of the laminate, such as the type of fibres, fibre content and orientation. There are various methods for calculating these values, such as the software 'mixing rule', the 'Chamis' or 'modified Puck rule'. The Puck rule provides the values that agree best with the experiments.

For a single cover layer (unidirectional, all fibres in the same direction) the relationships according to Puck are:

$$E_1 = E_{Fl}\phi + E_M(1-\phi) \tag{5.41a}$$

$$E_2 = \frac{E_M^*(1+0.85\times\phi^2)}{1.85\phi\times E_M^*/E_{F2} + (1-\phi)^{1.25}} \tag{5.41b}$$

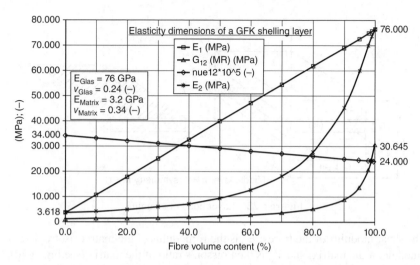

Figure 5.20   Elasticity dimensions of a GFK cover layer dependent on the volume proportion

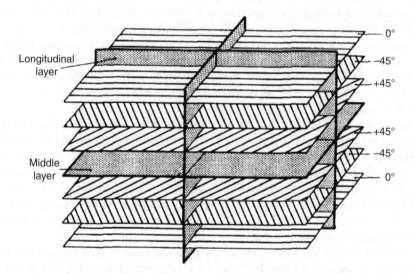

Figure 5.21   Symmetrical structure of a laminate of individual layers [12]

$$G_{12} = G_{21} = \frac{G_M(1 + 0.6 \times \phi^{0.5})}{1.6\phi \times G_M/G_F + (1-\phi)^{1.25}}$$ (5.41c)

$$\text{with } E_M^* = \frac{E_M}{\left(1 - v_M^2\right)}$$ (5.41d)

where E_1 is the E-modulus of the laminate in the direction of the fibres; E_2 is the E-modulus normal to them; G_{12} is the shear modulus of the laminate; ϕ is the fibre volume content; E_{F1} is the E-modulus of the fibres in their longitudinal direction; E_{F2} is the E-modulus lateral to this;

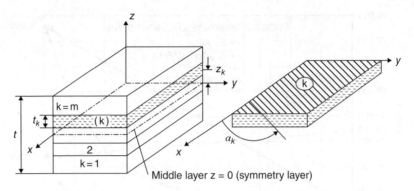

Figure 5.22   Layer bond of a laminate

G_F is the shear modulus of the fibres; E_M is the E-modulus of the matrix (isotropic); G_M is the shear module of the matrix; and V_M is the Poisson's ratio of the matrix (see Fig. 5.20).

Decisive for the mechanical data are the E-moduli of the fibres in the laminate and their orientation, as can be seen from Equation (5.41). The fibre volume proportion ranges from approximately 15% in the use of materials up to 70% for extruded profiles. Volume proportions of approximately 40–60% are used for the production of rotor blades.

The largest proportion of fibres is normally laid in the main loading direction (first direction under 0°) and a smaller proportion vertically to this (second direction). If larger shear stresses are to be transferred by the component, then fibres must also be arranged at angles other than 0° or 90°, optimally below ±45° as fibres in the two former directions cannot transfer shear and are thus 'shear-soft'.

If a laminate is built up of several individual layers, then care must be taken that the layer arrangement is symmetrical about the middle layer, i.e. the topmost and the lowest layers are to be identical (thickness, type of fibre, direction of fibre, volume portion); the layers below or above these can be different but must again be symmetrical (see Fig. 5.21). If the laminate is not symmetrical, then for a level load the component will be additionally curved (coupling effect between normal forces and bending moments). In the case of dominating bending loads of the laminate, the outermost layers should possess the greatest strength, as the bending stresses are largest there (see Fig. 5.22).

The fibre volume content is used in the mathematical estimation of the material data such as E-modulus, strength, and so on. However, for manufacture, the weight % is decisive as the weight of the fibres is easier to determine than their volume. The differences are due to the different densities of fibres and resins. The conversion from volume proportion to weight proportion and back again occurs according to:

$$\phi = \frac{\psi}{\psi + \dfrac{\rho_F}{\rho_M}(1-\psi)} \tag{5.42a}$$

$$\psi = \frac{\phi}{\phi + \dfrac{\rho_M}{\rho_F}(1-\phi)} \tag{5.42b}$$

where φ and ψ are the volume and weight proportions, and ρ_F and ρ_M are the densities of the fibre and the matrix.

Two methods are usual for the calculation of the bearable forces of a laminate. The so-called network theory is suitable for the quick preliminary design, and the 'classical laminate theory' (CLT) is used for a more accurate calculation.

In network theory, as the name states, it is assumed that only the fibres do the carrying, and they form a continuous 'network'. The matrix is ignored, as the E-modulus and strength are much lower than that of the fibres. The properties (E-modulus and strength in fibre-longitudinal direction) of the individually arranged fibre layers are determined with consideration of the fibre strength, the fibre volume content, and the thickness of the individual layers, and then transformed into a common laminate coordinate system. In this way the carrying capacity of the laminate is obtained.

For the CLT the weave of the non-woven fabric is separated into individual fibre layers (e.g. the multi-axial non-woven fabric according to Figure 5.19 into four layers) and their mechanical values are calculated according to, for instance, the Puck rule. The expansion stiffness matrices for the individual layers can be calculated according to Hooke's elasticity rule Equation (5.13). The matrices must all be transformed into a unified laminate coordinate system and added together taking the layer thicknesses into account. The flexibility matrix is obtained by the inversion of the matrix obtained in this way by means of Equation (5.14). The results of this are the E- and shear moduli as well as Poisson's ratio for the laminate.

The calculation method according to network theory and the CLT are described in greater detail in [12]. The calculation of the limiting carrying capacity or the failure of components of fibre-bound materials can be carried out by various methods. The most used are the so-called failure hypotheses according to Puck, Hashin, Tsai-Hill or Tsai-Wu. These will not be further discussed within the scope of this introduction; further literature is found in [12] or [15].

The long-term or operational strength of rotor blades of FRP occurs most often according to the Palmgren-Miner method. In this the time-varying loads of the rotor blades are divided into so-called load collectives. A load collective (i) is described by means of an average stress σ_{m_i} a stress vibration width $\Delta\sigma_i$, and the number of load cycles n_i with these stresses. Each load collective (i) causes a part-damage d_i. The sum of all part-damages due to the load collectives must not exceed the value of 1. As the Palmgren-Miner rule, especially for fibre-bonded materials, is not uncontroversial because it has certain weaknesses, high factors of safety should be set when using this limit. Detailed information on the calculation of the damage amount can be found, for instance, in the GL Guidelines [5].

5.6   Production of Rotor Blades

5.6.1   Structural Parts of the Rotor Blades

Rotor blades for large wind turbines reach dimensions of more than 80 m length, 5 m wide and 4 m thickness. The manufacture of such large parts of fibre-bonded materials presents a big challenge to production technology. With a weight of approximately 20 t and a volume content of 55% and a fibre weight proportion of approximately 73%, it means that 14.5 t of glass fibre and 5.5 t of resin must be used for a rotor blade. Added to this, the fibres must be soaked with this amount of resin in a few hours. Although it is possible to interrupt the soaking process, it

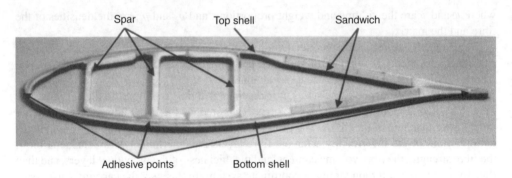

Figure 5.23   Typical cross-section of a rotor blade with three spars

mostly means a reduction of the strength of the laminate in the interruption region and additional production effort in the continuation of the soaking.

The structure of large rotor blades usually consists of the following parts (see Figure 5.23):

Spar or spar-chord combinations: The spars are mainly there to absorb the lateral bending forces. They are either made of solid laminate or of sandwich construction for increasing the shear-buckling strength required by the high lateral forces. In order to absorb the shear stresses, the main section of the fibres is arranged in the ±45° direction. In a spar-chord combination, parts of the overall thickness of the top and bottom shells are produced with the spar because in the region of the spar in the two shells the bending stresses are also very high, and the laminate thicknesses must be great there. Also in this way the later combination of the spars with the top and bottom shells is simpler. The spars are produced separately in one piece. According to their strength requirements they reach over a large part of the blade length with the exception of the blade tip and the connection region to the hub. There the cross-section is circular and has sufficient strength due to the large diameter and great thickness.

Connection with the hub: The connection serves to link the rotor blade with the hub. Because of the large forces to be transferred, the connection is made with a corresponding number of large bolts or threaded rods. The introduction of forces in components of FRP must be very carefully carried out and distributed over a greater area than for metals, as local stress increases are only bearable to a small amount due to the low elongation at fracture of the FRP. For small rotor blades, ring-shaped inserts of aluminium for accepting the connection bolts can be glued onto the blade. The ends of such inserts at the transition metal/FRP should be very soft and thin (see Figure 5.24a).

In this way the differences in the expansion of the metal and FRP are slight and there will be no large stress peaks. For large rotor blades the forces are either transferred by means of laminated-in lateral or T-rods ('IKEA' principle) or glued-in threaded rods. The number, sizes and lengths of the rods are adapted to the forces to be transferred and the thickness of the blade shells in the connection region. Here, too, care must be taken that great stress peaks in the laminate are avoided, due to the large forces to be transferred. The lateral rod connections with the laminate can be reinforced with roving strings in that one winds the rods in a sling and allows them to run out flat in the laminate. If threaded rods are foreseen, then the accepting holes are drilled into the laminate afterwards and the rods are glued in.

(a)   (b)

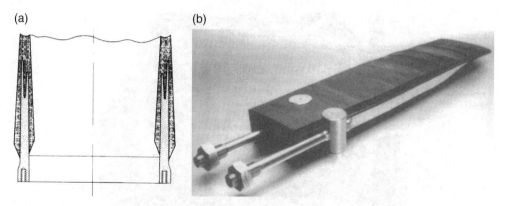

Figure 5.24   Bolt connections for the rotor blade with the hub: (a) glued; (b) T-bolts

Figure 5.25   Portal for automatic design of the fibre non-woven fabric (CAD model). Source: MAG
Renewable Energy

Stringers for buckling reinforcement: Stringers are an arrangement mainly for increasing
the buckling strength in the longitudinal direction of the blade. The lateral arrangement as ribs
would increase buckling strength only a little, since long, small fields have to be reinforced. If
the length/width relationship of a buckling field is greater than ≈ 3.5, then a larger relationship
has no further influence on the buckling strength.

    If, on the other hand, one halves the width of a buckling field b, then the critical buckling
stress is increased four-fold, this increases the buckling strength by $1/b^2$. The contribution of
the stringers to the bending strength is relatively small because of their position. The stringers
can be prefabricated as columns in GFK or with a sandwich core and laminated in the top shell
and/or bottom shell (depending on which is endangered). Alternatively they can also be glued
in after completion of the shell.

Upper shell: The upper shell forms the suction side of the profile. It is normally the largest
element of the rotor blade. For manufacture, first a negative mould is produced and the positive
form taken from it. The glass fibre non-woven fabric or the weave is inserted into it and soaked

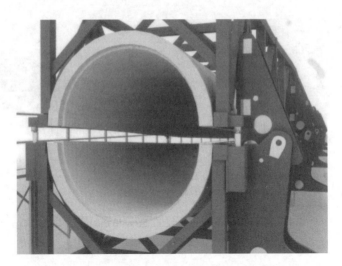

Figure 5.26 Lockable mould for producing one-part rotor blades (CAD model). Source: MAG Renewable Energy

in resin. If carbon fibres are foreseen due to strength considerations, they are arranged in the regions of the spars. The carbon fibres should be placed to the outer side of the shell as they have a greater E-modulus than the glass fibres. An arrangement of the carbon fibres further inside for production reasons makes no sense.

Bottom shell: The bottom shell forms the pressure side of the profile. Its manufacture is in the same manner as the top shell.

Alternatively the top and bottom shells can be manufactured in one piece. For this purpose, hinged moulds are used that are folded together after inserting the fibres (see Figures 5.25 and 5.26). The fibres are pressed against the forms by means of one or more inflatable balloons and the resin is injected. The balloons can subsequently be removed. The advantage of this production method is that the later gluing of the two shells is done away with. In this way the weak spots in rotor blade are eliminated (see below). The disadvantage is that the building-in of spars for reinforcement is very difficult or impossible. However, stringers can be provided.

5.6.2 Composite Manufacturing Methods

Two methods are used for the laminating process (soaking the fibres with resin): hand lamination and injection technology in various methods of application.

Hand lamination: In hand lamination, the cleaned and polished mould is first provided with a separation substance so that the component does not adhere to it and can be easily removed from the mould. Then a so-called gel-coat layer of pure resin as a surface protection for the component is applied. A first layer of fibres (non-woven fabric, weave) is then placed on the not-fully hardened and thus still sticky gel coat and soaked with resin. The fibre layer is compressed by hand with large rollers in order to achieve the greatest possible fibre volume content and to remove air bubbles as far as possible. While this layer of resin is not fully hardened, the next fibre layer is placed, soaked and compressed, and so on, until the required laminate thickness has been reached. Depending on the speed of hardening, this process can take several hours for larger rotor blades. Then the laminate is allowed to harden completely. Only then may the component be removed from the mould as the not-fully hardened resin may deform. It is not possible to achieve high volume contents with hand lamination. For this the thickness of the component for a given fibre quantity must be increased, meaning that more resin is used and the blade becomes heavier and more expensive.

Simple injection method: The simple injection method offers advantages compared with hand lamination only with the use of UP and VE resins, as the release of sterol is drastically reduced. After the preparation of the mould and the application of the gel coat, all the fibre layers are placed dry into the mould. Then the mould is closed from the top with a cover or elastic film. Then the resin is injected at a low pressure into the interstices of mould and film. The air displaced by the resin is allowed to escape through a hole in the opposite end. Depending on the size and shape of the component, several injection and vent positions should be provided in order to ensure complete venting and soaking.

Vacuum injection method: The preparations are similar to those of the simple injection method, but instead of the resin displacing the air, a vacuum is induced at the opposite end. This achieves the following advantages. (1) Because of the external pressure the fibres are pressed together, and less resin is used for the same fibre quantity, making the fibre volume content greater. (2) The thickness of the laminate and flow velocity can be influenced by the regulation of injection pressure and vacuum. (3) The number and size of air bubbles are greatly reduced and therefore the quality is improved.

With this technique the resin must have a lower viscosity than with the above methods, as the fibres are pressed together by the external pressure and are less permeable to resin. In order to improve the soaking of the fibres, use is made of flow-aids and films that are permeable to air but impermeable to resin. Several injection and vacuum connections are provided in order to ensure that such a large component as a rotor blade is completely soaked with resin within the pot time.

Soaking behaviour can be simulated using the computer by means of special programs; experimental investigations on a large rotor blade are too expensive. Generally the method is demanding in its application but delivers good laminate qualities.

Directly after hardening the laminate should be tempered in order to achieve a higher strength and quality. The same results cannot be reached with later tempering. In tempering the component should, if possible, be left in the mould and should be slowly and evenly heated (over 8–10 hours), then kept for about the same time at a constant temperature, and then slowly cooled again. The higher the temperature, the better are the results. The temperature is dependent on the resin, possibly the sandwich material and the form. It should be between 60–90°C. Tempering is particularly important for UP and VE laminates as this increases the degree of hardening and thus the strength, and the sterol still in the laminate evaporates faster.

5.6.3   Assembly of the Rotor Blade

If the upper and lower shells as well as spars are manufactured separately, then they must be assembled. The spars are first glued into the upper and lower shells. Then adhesive is applied to the front and rear edges of the two shells as well as the upper surface of the spars, and the

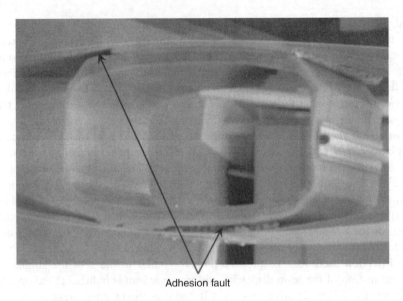

Adhesion fault

Figure 5.27   Adhesion fault at the joint of the spar with the external shell on a rotor blade (photo: author)

second shell is placed on the first one. As the rotor blade is then closed and, depending on the size, no longer accessible, the adhesive surfaces can no longer be checked.

The production tolerances for large FRP components are relatively large. This is because the resins shrink when hardening, epoxy laminates up to 3% depending on the arrangement of the fibres, UP and VE laminates up to 5%. Furthermore, the laminates heat up during the hardening process to more than 100°C and expand correspondingly. The moulds do not expand to the same extent as they are mostly not made of material with the same coefficient of expansion and do not reach the same temperatures as the laminate. Because of the resulting large tolerances, a large surplus of adhesives is used in the assembly in order to bridge the tolerances (see Figure 5.27). A thick adhesive joint, however, has a much lower strength than a thin one. Furthermore, the strength of the adhesive joints can be increased when the components are pressed together, which is hardly possible with rotor blades. For this reason the adhesive positions are weak points within rotor blades and sometimes lead to their failure.

References

[1] Dannenberg, L. 2008. Final report CEWind-Forschungsvorhaben 'Festigkeit von Rotorblättern', FH Kiel.
[2] Szabó, I. 1966. *Einführung in die Technische Mechanik*. Springer-Verlag, Berlin.
[3] Szabó, I. 1964. *Höhere Technische Mechanik*. Springer-Verlag, Berlin.
[4] Holzmann, G. 2002. *Technische Mechanik 3: Festigkeitslehre*. B.G. Teubner, Wiesbaden.
[5] GL (Germanischer Lloyd). 2004. *Richtlinie für die Zertifizierung von Windenergieanlagen*. GL, Hamburg.
[6] Göldner, H. 1991. *Lehrbuch Höhere Festigkeitslehre, Bd. 1*. Fachbuchverlag Leipzig.
[7] Dannenberg, L. 2008. *Vorlesungsskript zu Festigkeit der Schiffe II*, FH Kiel.
[8] Hapel, K.-H. 1990. *Festigkeitsanalyse dynamisch beanspruchter Offshore-Konstruktionen*. Vieweg Verlag, Stuttgart.
[9] Clauss, G. 1988. *Meerestechnische Konstruktionen*. Springer-Verlag, Berlin.

[10] Hauger, W. 1999. *Technische Mechanik, Bd. 3, Kinetik*. Springer-Verlag, Berlin.
[11] Wiedemann, J. 1996. *Leichtbau 1: Elemente*. Springer-Verlag, Berlin.
[12] Michaeli, W. 1994. *Dimensionieren mit Faserverbundwerkstoffen*. Carl Hanser Verlag, Munich.
[13] Fleming, M. 1996. *Faserverbundbauweisen: Halbzeuge und Bauweisen*. Springer-Verlag, Berlin.
[14] Ehrenstein, G.W. 1999. *Polymer-Werkstoffe*. Carl Hanser Verlag, München.
[15] Fleming, M. 2003. *Faserverbundbauweisen: Eigenschaften*. Springer-Verlag, Berlin.

6

The Drive Train

Sönke Siegfriedsen

6.1 Introduction

The machines called wind turbines in this chapter are used for the generation of electrical energy. The drive for these machines is a rotor that draws kinetic energy from the wind as repeller, and uses it to introduce mechanical force into the power plant. As in every conventional power plant, this mechanical force is made up of a rotational movement and the force of this movement results in the torque.

$$P_{mech} = M\omega = M \times 2\pi n$$

If we transfer the above-mentioned term 'power plant' to wind energy then one could replace it by the nacelle. This gondola, also called the machine house, contains the technology necessary for the generation of electrical energy. The rotating parts, thus the torque transferring ones, are called drive train components. If one views these components as a whole, then one talks of a drive train. It starts at the rotor and mostly ends at the generator. As there are various arrangements and variations of the components possible, there are many different drive train concepts and the potential for innovation is very high. The concept selected has a great influence on the head mass and the costs (approximately 30% portion of the overall costs) of a wind turbine, and directly influences the development in the dimensioning of components. Depending on the concept, a drive train can make use of more or fewer standard machine components. However, in the design of these parts, attention must be paid to the special characteristics of the drive component and resulting demands.

Understanding Wind Power Technology: Theory, Deployment and Optimisation, First Edition.
Edited by Alois Schaffarczyk. Translated by Gunther Roth.
© 2014 John Wiley & Sons, Ltd. Published 2014 by John Wiley & Sons, Ltd.

The rotor not only transfers torque to the gondola, but also pitch and yaw moments (see Section 6.7). These loads must be taken into account in selection and design. Due to the poor accessibility caused by the position on a tower, damage due to poor design is an enormous economic burden.

Because of its non-constant direction and availability, the wind as a special driving medium is a particular challenge for technology. Even if it cannot be directly allocated to the subject of drive train, the wind direction and blade angle adjustment will be discussed briefly in the following sections.

As already described in the previous chapters, in principle various rotor types for the generation of torque are possible, and at this point I would like to point out that only wind turbines with a horizontal axis will be discussed.

6.2 Blade Angle Adjustment Systems

The pitch principle is mostly used for rotor blade adjustments in wind turbines with horizontal rotors of the newer generation, especially the multi-megawatt class. In comparison to fixed joining of the rotor blade and the capacity limitations via the stall effect, the technically more complex pitch system offers various advantages that, with increasing sizes of plants, have an overall cost-reducing effect. Despite a plant operation adapted to different wind conditions with each generator type and grid requirement of optimised capacity, and speed of rotation regulation, it is possible with suitable control algorithms to keep load peaks substantially lower, especially in extreme wind situations. This again influences the dimensioning of different cost-relevant components such as rotor mountings, main drives and foundation structures.

In grid operation, the pitch angle for capacity regulation is varied between 0° and approximately 30° (Figure 6.1, left). Further, with a pitch angle of about 70°, a slow rotation of the grid-couple drive train can be achieved that is advantageous for support and lubrication (Figure 6.1, right). A rapid stopping of the rotor including drive train by means of adjustment to a pitch angle of 90° in the so-called feathering setting (aerodynamic brake) is possible.

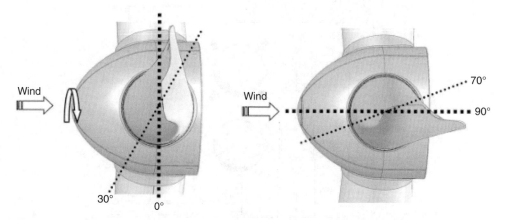

Figure 6.1 Pitch angle from 0° to 30° for grid operation, 70° for grid-coupled trundling, and 90° as feathering setting for complete stop

The following section deals exclusively with the basic structure of the pitch system for adjusting the whole rotor blade about its pitch axis and the different technical features of the required drive. In this, neither the design of the pitch system nor its drive and control will be discussed further.

A requirement of any pitch system, independent of its structure and type of drive, is the rotatable supported connection to the rotor hub. The blade support is normally a dual-row four-point ball bearing (Figure 6.2) that, because of its structure, can transmit high radial and axial load component introduced to a high degree. Whether the inner or outer ring is fixedly connected with the rotor hub whilst the other ring takes up the rotation of the rotor blade is always dependent on the preferred model and positioning of the drive (Figure 6.5).

More important in the decision is the selection of a blade connection size that is suitable for the rotor blades (performance class) and wind conditions (type class) that are already established on the market. Thus, either a newly developed, In-house rotor blade can also be sold for other plants, or use can be made of a larger selection of outside-developed rotor blades for one's own plant development. Dependent on the loads, the result is the design determination with the required bearing diameter and the lightest possible cast body design of the rotor hub.

Historically, three successive concepts for three- and two blade rotors have appeared for the adjustment or targeted setting of a pitch angle. In the simplest concept an actuator, synchronised by means of a mechanical coupling drive, adjusts all the blades at the same time. In order to stop the blades in case of failure of this actuator, a separate mechanical rotor brake is necessary that will bring the drive train to a stop against the maximum rotor moment from maximum revolutions. The resulting loads occurring in the drive train and the short-term high heat loss to the brake would require an economically unfeasible dimensioning of the affected components for plants in the multi-megawatt class.

For this reason, a concept has been established in which each rotor blade is equipped with its own actuator including all system components, for a once-off adjustment of the blade in feathering condition. In capacity-regulated operation, all rotor blades, only synchronised by the controls, are set to the same angle. If one actuator fails, then the remaining ones with their linked rotor blades can independently bring them to feathering. In this way the load on the

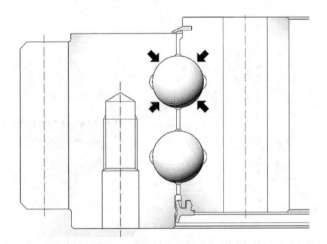

Figure 6.2   Section of a dual-row four-point bearing

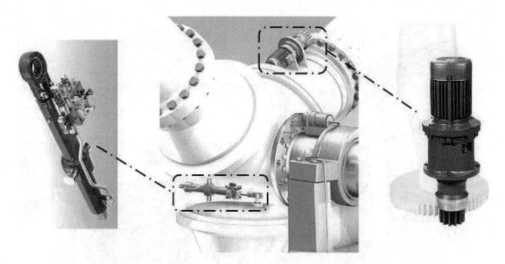

Figure 6.3 Pitch system with hydraulic (left) or electric (right) actuators (Bosch Rexroth)

drive train and the mechanical rotor brake can be sufficiently limited and an economically sensible dimensioning can be achieved. Also the rotor revolutions can be held below the critical in any conceivable case without recourse to the mechanical brake.

For some years, attempts have been made to develop this concept further. The wind conditions on the circular surface formed by the rotating rotor are not homogeneous. With increasing plant sizes and rotor diameters, the differences of the wind velocity and turbulences in the region of the rotor surface become ever greater. This means that at a defined time every rotor blade experiences a different load. The idea is simple: the pitch angle of each rotor blade is individually controlled (individual pitch control – IPC) such that the load on each rotor blade is the same and in total the rotor is kept to the nominal capacity or the maximum part capacity. The currently still unsolved problem consists, on the one hand, of the instrumental acquisition of the loads and, on the other hand, the development of control algorithms that permit a sufficiently fast pitch angle change. The rapid rotation of the rotor blade over its whole length of currently up to approximately 70 m without appreciable twisting and the resulting lack of the aerodynamic effect, with the simultaneous entry of parasitic vibrations, is a challenge to the drive and instrumentation.

Of primary importance for with systems with only one actuator is the arrangement of a hydraulic cylinder in the stationary gondola. The stroke of the piston is transferred via a rod through the rotor and possibly the hollow drive shaft into the rotating rotor hub, and converted there by means of coupling rods to all the connected rotor blades to give the same swing movement.

In the systems used in present multi-megawatt plants, a distinction is made between electrical and hydraulic actuators (Figure 6.3). The transmission of the supply service and the signals between the stationary gondola and the rotating rotor hub is by means of a slide ring unit, or a combination with a hydraulic rotation hole (Figure 6.4). Depending on the structure of the drive train, this interface sits at the front or rear end of the hollow rotor shaft or the free end of the drive entry shaft.

In order to ensure, in the case of failure, the simply designed interface of the above-mentioned aerodynamic brake and to stop the plant safely, the pitch system must maintain a swing

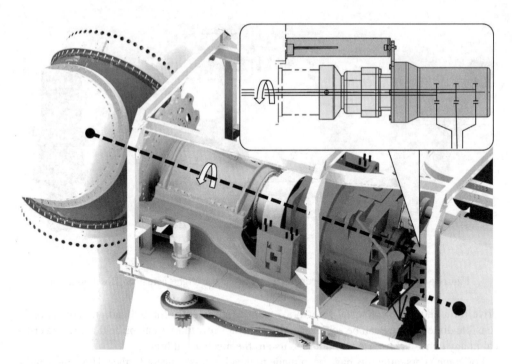

Figure 6.4 Transmission of supply service and communication between stationary gondola and the rotating rotor

movement of at least 90° in the rotor hub. Besides the control technology for the actuator, this also includes a corresponding energy store.

Electrochemical lead-gel accumulators, and in newer concepts also increasingly ultra-capacitors, are used as energy stores of electrical systems. These latter are comparable with bubble or piston-storage hydraulic systems:

- loading capacity for a safe full-swing movement;
- short-term loading and operation readiness;
- sufficient length of life also with frequent discharging.

In their most common basic forms, electrical systems comprise a motor drive combination with mostly three-step epicyclic gears.

The driven-side drive pinion gear meshes either directly into the teeth of the inner or outer ring of the blade mount (Figure 6.5), or is coupled to it by means of a toothed belt. The drive is mostly fixedly mounted to the rotor hub and rotates with the blade support ring connected to the rotor blade. There are also designs in which the drive is fixed to the rotating blade mount ring and rotates about the toothed fixed blade mount ring.

Despite disadvantageously increased effort during erection, maintenance and energy requirements, this arrangement can substantially improve the stiffness against ovalising in the blade connection region of the carrier plate required for the drive.

Compared with this, an innovative step is the quite new development of a ring-shaped direct drive connected to the blade mount. The homogeneously generated electromagnetic torque

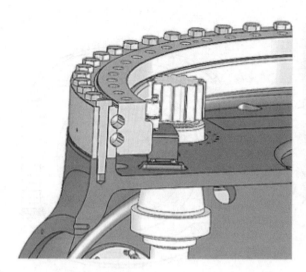

Figure 6.5 Motor drive combination fixed to the rotor hub turns the inner or outer ring of the blade mount (REpower Systems)

along the circumference of the blade connection is converted directly into the blade rotating movement.

Without the tooth play of intermediate drives, this design allows a substantially increased dynamic, which is an advantage with individually adjustable blades. However, this expensive technology has not been used on any plant.

The hydraulic systems usually include double-acting cylinders that are linked to the rotor hub and the rotatable blade mount ring. The stroke of the piston is converted into a swing movement of the blade mount ring with connected rotor blade. For application with a single cylinder, the stroke for a 90° swing movement requires a cylinder length that cannot be accommodated within the limits of the rotor blade connection. Therefore, in most known designs the cylinder protrudes forwards out of the rotor hub (Figure 6.6).

In an alternative design, the swing movement is obtained via two small cylinders and via an intermediate swivel lever arranged one behind the other. This arrangement is situated completely within the blade connection diameter.

Both designs have in common that the basic supply for the operation of the cylinder must be from hydraulic aggregates in the stationary gondola and must therefore be carried out by means of hydraulic rotating connections. The technically common aggregates with tank are not suitable for installation in the rotating rotor hub.

An innovative step could be the electro-hydrostatic actuator (EHA) that has been in use in the aviation industry for the last two decades. This is a hybrid drive that combines electric servomotor technology with hydraulic drive and control technology. The actuator has only electrical interfaces to the outside and the hydraulics are integrated completely autonomously.

The best pitch system concept ...
... does not exist. The questions of whether, for instance, electrical or hydraulic actuators, chemical or static energy storage, blade mounts with toothed inner or outer rings are to be installed, must be answered individually for the whole concept every time. What matters for

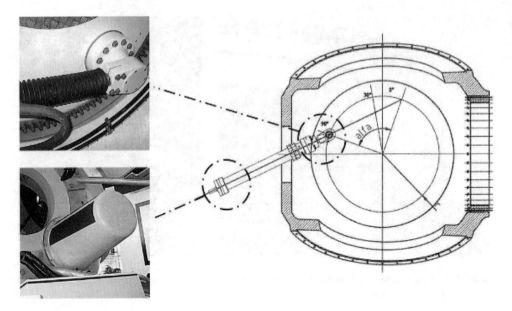

Figure 6.6   Arrangement of a single cylinder in hydraulic pitch systems (Vestas)

the operator of a plant is the return on his investment. Therefore it is of primary importance that the input operation on the part of the plant functions without interruption as long as the wind at the site blows advantageously.

Energy stores are in extensive use in site areas with a high degree of grid fluctuations and stoppages. Here, static storage methods such as ultra-capacitors or compressed oil storage have an advantage over electrochemical lead-gel accumulators. There is presently increasing attractiveness of ultra-capacitors in electrical systems, but in the past use was made of mostly hydraulic systems.

On erection sites with frequent very low external temperatures, the necessary heating of the system components with purely electrical systems is much easier to carry out. Oil lines, valves and compressed oil storages would have problems here.

On erection sites with often occurring very high temperatures, cooling for hydraulic systems by means of oil-return cooling in the stationary gondola is much easier. The cooling of converters and motors of electrical systems in the rotating rotor hub has up to now only been achieved by means of outside air drafts. This latter, insofar as the shielding of the internal space of the rotor hub from contaminated outside air (offshore, desert, steppes, etc.) is attempted, can lead to additional design-dependent effort. The same conditions are to be taken into account for an alternative fluid cooling circuit as for providing a hydraulic pitch system.

For various reasons, a hydraulic installation without leakage in a wind turbine is very hard to achieve. In the industry there are many instances of rotor hubs with their insides completely oil-soaked, as well as smeared rotor blades and spinners of plants with hydraulic pitch systems. Besides the increased danger of fires and service failures, there is a seeming contradiction with the concept of 'green energy'.

Finally the (further) development of the blade angle adjustment systems will be dependent upon the preparedness of cooperation between the apparatus and component manufacturers.

6.3   Wind Direction Tracking

6.3.1   General

Wind direction tracking, known in the trade as the azimuth drive or yaw system, ensures that the gondola is kept pointing in the primary wind direction. For this purpose the gondola is rotatable on the tower head. Large modern plants possess an electrical or hydraulic drive that takes care of the tracking. Small wind turbines have a weather vane that automatically turns the gondola into the wind (Figure 6.7).

A weather vane can be dispensed with when the plant is a small idling plant (rotor in direction of wind behind the tower). Old designs of smaller or medium-sized plants and historical windmills possess a side wheel for automatic weather tracking (Figure 6.8). This section describes the various possibilities for arranging wind direction tracking components and their function.

6.3.2   Description of the Function

The main task is to keep the gondola facing the wind. The drives do not trigger immediately for every change of wind direction. The controls trigger when, after a certain time, an average fixed-angled direct flow has stabilised. Otherwise the wind direction tracking system would carry out many small movements too often. The deviation of the wind direction to the wind turbine longitudinal axis is measured by means of a weather vane on the gondola. The drive works usually via several gears (pinions) that work on a gearing on the tower head bearing.

The number and size of the drives depends on the size of the plant. During tracking the process must be damped by brake pads or prestressed glide bearings, as otherwise it could produce a to-and-fro impact on the gondola within the gearing and in this way substantially shorten the life of the components. The yaw brakes ensure that the gondola is held in the wind when the drive is stopped. The drives themselves usually have brakes that support the stoppage of the gondola. A further function of the drives is cable unwinding. A cable loop is

Figure 6.7   Wind turbine with weather vane (Leading Edge Turbines Ltd)

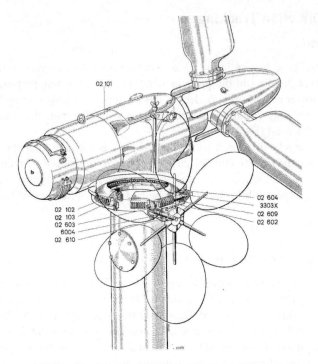

02 101

02 604
3303X
02 609
02 602

02 102
02 103
02 603
6004
02 610

Figure 6.8   Wind turbine with side wheel (Allgaier)

situated in the upper part of the tower and this must not have more than two or three windings. By means of the controls and corresponding sensors that monitor the number of windings, an unwinding of the cable is carried out at a suitable time by the yaw drives.

6.3.3   Components

6.3.3.1   Tower Head Bearing

The tower head bearing in most cases consists of a large ball bearing with external or internal gears designed as a four-point bearing (Figure 6.9), either single or double row. A special form of the tower head bearing is the sliding bearing (Figure 6.10), into which the damping function is integrated. The disadvantage here is the larger drive capacity that is required due to the high friction.

6.3.3.2   Drives

A drive usually consists of a drive motor with a drive-side pinion, several coaxial epicyclic gear steps and a flanged AC motor (Figure 6.11). The AC motor usually has an integrated stop brake. Other types of drive, such as with worm gears or hydrostatic motors, are seen much less. The characteristic of the AC motor with a defined tipping torque is used as overload protection for the gears.

Figure 6.9 Single-row four-point bearing (Roth Erde)

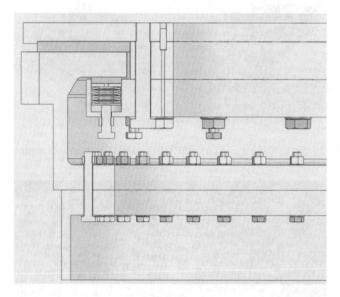

Figure 6.10 Glide bearing (aerodyn)

6.3.3.3 Brakes

The azimuth brakes normally consist of a brake disc and several hydraulic fixed calliper disc brakes that are controlled by means of a corresponding aggregate (Figure 6.12, right). Normally two pressure levels are used: a smaller pressure for damping and a greater one for braking.

Other systems work with electrical brake systems (Figure 6.12, left) or can dispense completely with the disc brakes if, for instance, a glide bearing is installed and the brakes are integrated into the drive.

Figure 6.11 Drives for wind direction tracking (Bonfiglioli)

Figure 6.12 Brake callipers: electrical (EMB Systems), hydraulic (Svendborg Brakes)

Figure 6.13 Encoder (aerodyn)

6.3.3.4 Sensors

Various sensors are used in order to acquire the position of the gondola and to initiate a cable unwinding in good time. In most cases a cam-operated switch is used that detects the position for cable unwinding. In addition, the absolute position of the gondola is acquired with an encoder (see Figure 6.13). The sensors are mostly set by means of a pinion on the bearing gears.

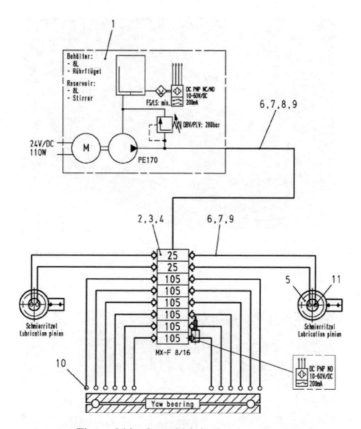

Figure 6.14 Central lubrication (aerodyn)

6.3.3.5 Central Lubrication

The yaw system usually also has a built-in lubrication installation in order to provide the bearing races and gears with grease. Here special systems for the wind industry have been developed by the suppliers (Figure 6.14).

6.3.4 Variations in Wind Direction Tracking Arrangements

There are four different arrangements for wind direction tracking:

1. Internal gears of the azimuth bearing with internal azimuth brake.
2. Internal gears of the azimuth bearing with external azimuth brake.
3. External gears of the azimuth bearing with internal azimuth brake.
4. External gears of the azimuth bearing with external azimuth brake.

In the first variation, the gondola is turned by means of an azimuth bearing with internal teeth (Figure 6.15). The azimuth disc brake and the azimuth drive are situated inside the tower. The advantage of the arrangement is that this variant can be well protected from outside influences

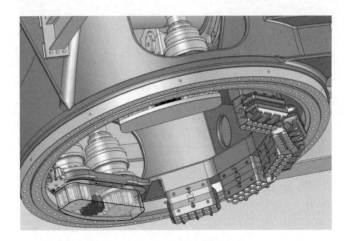

Figure 6.15 Internal gearing of the azimuth bearing with internal azimuth brake (brake disc not shown) (aerodyn)

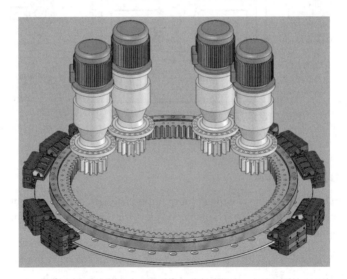

Figure 6.16 Internal gearing of the azimuth bearing with external azimuth brake (aerodyn)

(soiling, moisture, transport damage) and also the flow of force through the bearing is very advantageous. The disadvantages are that there is little space for all the components, that the bolts for the azimuth bearing ring are outside the tower, and that grease can be deposited on the brake disc.

The second variation also has an azimuth bearing with internal gears, but here the azimuth disc brake is outside the tower (Figure 6.16). The azimuth drive, however, is still inside the tower. The flow of force in this case is very advantageous. A further advantage is the large diameter of the brake disc, whereby fewer brake discs are needed. In addition, the grease of

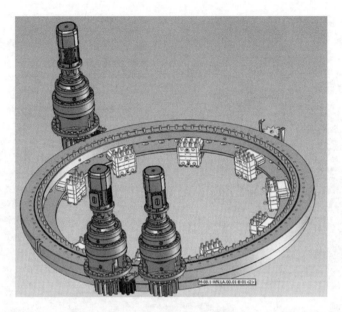

Figure 6.17   External gearing of the azimuth bearing with internal azimuth brake (aerodyn)

the gears can no longer be deposited on the azimuth brake disc. In this variation, the bolts of the azimuth bearing rings are also outside the tower and are therefore a disadvantage. The overall width of the main frame, however, is larger than for an internal brake.

The third variation consists of external gears of the azimuth bearing with internal azimuth brake. Here the azimuth drive is situated outside and the azimuth disc brake on the inside of the tower (Figure 6.17). This arrangement has many advantages:

- better gondola access;
- better possibilities for placing the azimuth drive;
- better access for servicing the azimuth drive;
- grease for the teeth is not deposited on the azimuth brake discs;
- greater number of teeth on the gears leading to less torque on the drives.

The overall width of the main frame is very large and thus a disadvantage of this variation. In addition, the flow of force through the bearing is somewhat disadvantageous, meaning that the tower flange must be thicker.

The fourth variation with external gears and external brake has not yet been put into practice, and so there are no examples.

6.4   Drive Train Components

The various components of the drive train will be described in the following, and discussed independently of the various drive train concepts.

Figure 6.18 GE-2.5-MW drive train (GE)

6.4.1 Rotor Locking and Rotor Rotating Arrangements

In the operation of a wind turbine, the rotor lock has no direct functions, and is only used for servicing or repairs at the drive train. In order to work on the rotating components the plant must first be stopped. For this, on the one hand, the rotor can be stopped with the aid of an aerodynamic brake procedure (see Section 6.2) or, on the other hand, a mechanical brake (see Section 6.4.4) can be used. However, the latter is conceived rather for an emergency and as a holding brake.

In the case of servicing work at the drive train, or in the hub of the rotor, the service personnel must be protected against any rotating movement of the drive train. With the mechanical brake, due of its method of functioning (traction), a certain amount of slip cannot be excluded and the rotor blade, even in the 'feathering condition', can generate torque, so the so-called rotor lock is integrated into the drive train. This presents an interlocking connection between a rotating component and a non-rotating component, mostly the machine mounting. The simplest and most often used principle is for a pin to be driven hydraulically or electrically into a holed plate integrated in the drive train (Figure 6.18). For smaller plants the insertion of the pin is often carried out by hand or with the aid of a thread.

The holed plate is often positioned in the neighbourhood of the rotor hub. This maximises the number of fixed components and provides the advantage of also permitting a safe exchange of drives with corresponding support in case of emergency. However, in this case the arrangement is subject to extreme loads by the rotor and must be dimensioned accordingly.

Besides the lock immediately at the rotor, there are also thoughts of placing the lock behind the transfer gears. An example here is the intervention into a toothed brake disc in combination with a rotor turning device that is placed there. This has the advantage that the loading from the torques due to the conversions is correspondingly reduced. Yet it must be remembered that all loads also act on the gears. In addition, in the case of damage to the drive, no locking would be possible.

Because manual movement of the locking position via a target brake is difficult and the rotation by hand (only possible with transmissions of the smaller capacity class) bears the risk of injury, a possibility for rotating angle adjustment, a rotor rotating device or a tower drive should be incorporated. There have been several attempts to produce this drive. Besides additional drives that mesh via a pinion into the brake disc (Figure 16.8), driver-integrated solutions from suppliers or hydraulic linear positioners can also be used. The use of the generator as motor, or automatic target braking with the aid of the aerodynamic and mechanical brake, is also used in practice.

Rotor rotating devices, however, are not used exclusively for adjusting the rotation angle setting for locking, but also in the erection of the individual blades. However, here the devices are designed for much greater capacities due to the large static loads resulting from the inertial eccentricity of the rotor, and usually only come into play as mobile units during erection.

6.4.2   Rotor Shaft and Mountings

The connection of the turning rotor with the rest of the drive train, as well as the standing basic structure of the wind turbine, is the task of the rotor shaft and its mountings. The rotor shaft, to which the hub is usually flanged, transmits the generated torque to the downstream drive train. The rotor loads are led into the grounding structure of the gondola by means of the bearings, which make the connection between rotating and stationary components.

The flow of force is a decisive criterion in the selection of a suitable solution, but compactness is also a deciding factor for the structural design, and depends on the philosophy of the developer. The mounting concept selected has an influence on the whole drive train and length of the gondola. Added to this is that the ground structure, also called the machine carrier, is substantially determined by the mounting concept. These points have a direct influence on the tower head mass and the associated costs.

The rotor bearings are highly loaded machine elements. In order to prolong the life of these bearings and to ensure a 'friction-free' operation, a central lubrication installation is provided for the rotor bearings, the blade bearings, the azimuth bearing and the bearings of the generator. The type of lubrication depends on the bearing concept and the sealing technology used.

Whereas oil lubrication of the drive is often used in a drive-integrated rotor mounting, grease lubrication is often used for housings of the rotor bearings. Today use is made of automatically controlled grease lubrication pumps that supply the bearings at regular intervals with fresh grease. The old grease is pressed out of the bearing through the seals, often labyrinth seals, and into a receiving container. The lubrication with oil in the described compact design is basically structured as a circuit. However, there is extra effort for the sealing system of the bearing. Besides an oil tank and an oil filter, a cooler is also required for cooling the heated oil.

6.4.2.1 Double Mounting

Double bearings are a concept in which the rotor shaft is arranged on two roller bearings. These transmit all the rotor loads through the machine carrier in the tower. Behind the rotor shaft the torque necessary for acquisition of energy acts on the next drive train component. Depending on the drive train concept, this can be a gear (see Figure 6.19) or a generator. The gears, or the generator in the gearless plants, with double shaft mountings are not loaded with forces of bending moments from the rotor, but only with the torque. This applies also for the later-described torque support solution. An exchange of drives is possible without dismounting the rotor, which is not possible with the three-point bearings.

The result of the two bearings and the distance between them is that the rotor shaft in this solution is very long and thus very heavy. In many developments it is produced as a forged shaft because of its strength advantages. In order to accommodate the energy supply lines of the pitch system, the rotor shafts are provided with a large central hole. For large plants, use is also made of hollow cast shafts.

In this mounting variation, a distinction is made between double bearings with separate housings (Figure 6.19) and double bearings with a common housing (Figure 6.20). Here the latter solution is a much more compact structure. Swing bearings (Figure 6.21, right) are usually used with two individual bearing housings in order to equalise the incorrect alignment of the housings to each other and the deformation of the shaft. The same type of bearings is also used in the three-point bearing concept described below. Also other types of bearings can be

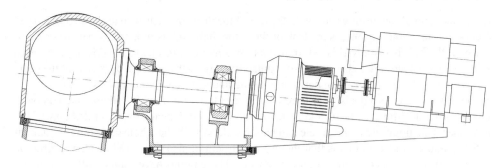

Figure 6.19 Rotor shaft with two individual mounts (aerodyn)

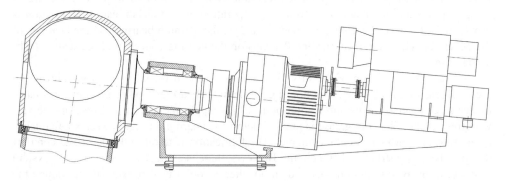

Figure 6.20 Bearings in a closed housing (aerodyn)

used that have less play than swivel joint bearings for the double shaft bearings and single-part housings, and this is seen as an advantage in terms of reliability.

A special solution for the double bearing concept is the support of the rotor on a stub axle projecting in front. Here the two bearings are arranged directly on the stub in the hub and transfer the loads directly from the rotor into the stub axle of the machine carrier (Figure 6.46). The rotor shaft in this solution only has the task of transferring the torques to the drive or the generator.

6.4.2.2 Three-point Support

In the three-point support the rear rotor shaft bearing is integrated into the gears (Figure 6.22). This means that the rear bearing housing is unnecessary and the shaft is much shorter. The three-point support gets its name from the connection points to the machine carrier. Besides the front rotor shaft bearing that is directly connected to the machine carrier, one also adds the side drive supports. The total component group is therefore very compact, but still quite accessible. However, this solution also has disadvantages: besides the effects of the load on the gears, in the case of damage these cannot be replaced without much effort. Yet the three-point support is a widely-used concept with many implementations.

Figure 6.21 Types of bearings: double-row tapered roller bearing (right); tapered roller bearing (left) (FAG)

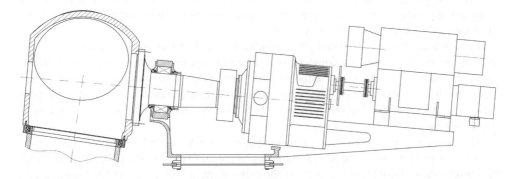

Figure 6.22 Drive train with three-point support (aerodyn)

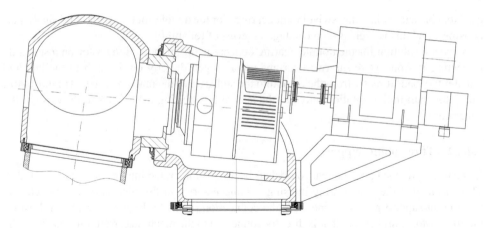

Figure 6.23   Drive train concept with torque support (aerodyn)

6.4.2.3   Torque Support

The torque support, today mostly a double-row tapered roller bearing (Figure 6.21, right), is used as the sole support of the rotor shaft. Besides the effective forces, the support must especially bear the tilt and yaw moments of the rotor. This property lent its name to this type of support and is currently the most compact support concept. This is due primarily to the very short rotor shaft. In addition, this concept is suited to further function integration in the drive train structure (see Section 6.5.4).

The solution with the torque support (see figure 6.23) makes great demands on the machine carrier. This base structure, today mainly produced as a casting, must be structured in a support-surrounding manner in order to be able to transfer the high loads into the tower.

6.4.3   Gears

Gears are the heart of wind turbines with conventional drive trains. Because the drives have components whose damage causes very high downtimes for the plant, and are financially a large part of the overall costs, there has been much innovation and development in this field.

The primary aim has been to improve the reliability and servicing-friendliness of the gears. This has led to many concepts that have had variable success in the market. This part of the chapter will provide an overview of the most common geared drives, and will mention some fundamental thoughts about gearing variants.

Gears in wind turbines are usually multi-stepped, and their purpose is to convert the large torque provided by the rotor into a high number of revolutions with which the generator is driven.

Whereas smaller wind turbines in the region of 100 kW are equipped with pure spur geared drives, for drives with higher capacities the first drive level is carried out by means of epicyclic gears. Above a capacity of approximately 2.5 MW, also the second step is used as an epicyclic step (Figure 6.24).

Whereas spur gear steps only have one point over which the complete torque is transferred, epicyclic steps have more points corresponding to the number of epicyclic gears over which the torque can be distributed. This relieves the components and leads to a much more compact method of construction.

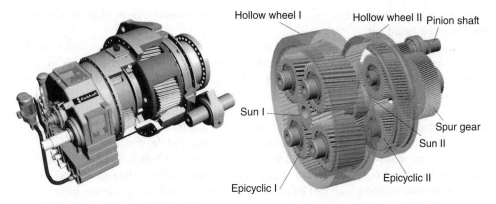

Figure 6.24   3.6 MW drive with two epicyclic gear steps and one spur gear step (Eickhoff)

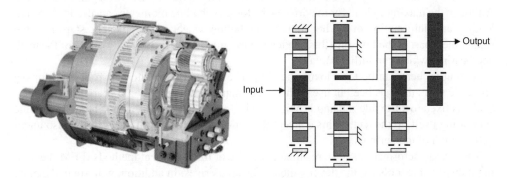

Figure 6.25   Drive with capacity branching (Bosch Rexroth)

The number of epicyclic gears used varies in common geared drives between three and five. In the epicyclic gear steps in conventional drives, the epicyclic gears are driven and the sun gear represents the drive, whilst the hollow wheel is stationary. The individual epicyclic gears therefore, besides the contact to the sun, always have contact with the hollow wheel which leads to support of the torque.

In addition, many manufacturers follow the concept of the capacity branching with gearing variants that distribute the incoming torque evenly over several steps, and thus branch out the capacity string (Figure 6.25). This concept has the advantage, among others, that the individual steps are subject to a smaller and more even loading and a more compact construction can be achieved. In this, methods of construction are achieved in which the epicyclic gears remain stationary and drive the hollow wheel.

The last step is a spur wheel in order to achieve an axial displacement between incoming and outgoing shafts. This displacement is necessary because the gear is provided with a hollow shaft which is used for passing the cable through to the rotor.

A so-called slip ring bushing is attached to the drive, and this transfers the power supply and electrical signals by means of sliding contacts. From there they are then fed through the hollow shaft in the gondola. With hydraulic pitch systems, a hydraulic rotating bush is attached to the hollow shaft. This presents a problem with drives that operate exclusively via epicyclic steps (so-called coaxial geared drives). As the generator is situated immediately behind the drive output and there is therefore no space for attaching a corresponding bush, other complex solutions must be found here.

Coaxial geared drives are used, for instance, with so-called hybrid drives in which the drive has two epicyclic steps and is directly connected to the generator. This operates somewhat more slowly and is thus larger (see Sections 6.5.3 and 6.5.4). Here, as an example, the bush can also be led through the generator, which would lead to a complex generator concept.

The teeth for spur gear and epicyclic steps are arranged straight or slanted. The advantage of the slanted teeth is in the even force transfer and quieter running. These advantages are offset by the increased effort in manufacture and erection, and the resulting costs.

Modern geared drives are developed with the aim of high numbers and series production, which is why it pays to use many cast parts. For this reason the epicyclic carriers and housings of the drives are nowadays almost exclusively of cast components. These are highly loaded because, depending on the drive train concept, additional loads from the deformation of the drive train or the rotor loads affect them. An example of this is the three-point mounting. Although the maximum load for the drive is the torque, bending moments can also lead to high loading and deformation of the housing. As the hollow wheel of the epicyclic steps is at the same time a part of the housing, this can lead to irregular intervention and loading of the teeth and bearing, which can inflict damage.

A solution here can be the Flexpin bearing. The Flexpin is an alternative bearing for the epicyclic gears that has been in use with various manufacturers for some years. Here the epicyclic is fixed to a flexible rod and is able to swing on its axis slightly whilst at the same time supporting itself on the teeth. This concept thus ensures correct tooth intervention also under deformation of the periphery.

These large deformations have resulted in the use of finite element methods (FEM) becoming indispensible tools for the development of geared drives. In addition, with time, dynamic aspects have also been taken into account. With the multitude of gears and cycling frequencies of the bearings, the geared drive has the greatest potential for resonant excitations in the drive train. Special software is used for the determination of these excitations, and comprehensive investigations are a feature of the drive verification.

The linking of the drive to the wind turbine structure normally takes place by means of two side supports of the housing which transfer the forces via elastic elements into the plant structure. These have the property of only taking up the torque; the inherent weight of the drive itself is taken up by the connection to the rotor shaft.

Besides the conventional and already-mentioned drive variations, there are also various concepts whose developments are being furthered by plant manufacturers, and in part describe completely different ideas. For instance, there are drive variations with variable transmissions that provide a constant speed of revolution to the generator. The Voith WinDrive presents such a solution (Figure 6.26). This is a combination of epicyclic gears and hydrodynamic converter which are positioned between the main drive and the generator.

The epicyclic geared drive of the Voith WinDrive is a superposition drive in which only a part of the input capacity is fed to the generator and the rest of the energy acts on the rotatably supported hollow wheel of the first level, with variable speed of revolution via the converter and via the epicyclic steps. The control of the conversion is carried out by adjustable guide vanes in the hydrodynamic converter.

The advantage of this concept is that load peaks are damped by the dynamic coupling and the frequency converter can be dispensed with.

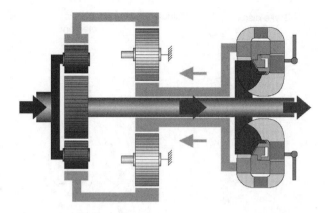

Figure 6.26 WinDrive (Voith)

The concept of variable translation has also been taken up by other manufacturers. Besides the hydrodynamic variant by Voith, there are also attempts to work with electromagnetic converters, hydrostatic drives or CVT [continuously variable transmission] tension media drives.

6.4.4 Brake and Coupling

The coupling is the connection element between drive and generator and, besides the pure transmission, fulfils several functions. The linking between drives and generators usually occurs with clamp rings.

Because of the flexible supports of generator and geared drives, the coupling (see Figure 6.27) must be able to compensate for the movement of the two components. Besides this, it must be able to equalise certain incorrect erection alignments between gear shaft and generator shaft. Mostly double-element rigid rotating couplings are used.

In addition, couplings are equipped with overload protection which are meant to protect the gears from torques that can occur on the generator side. These are caused primarily in the case of generator short-circuits in which the generator is blocked whilst the torque continues to be input by the rotor. Without the overload protection, very high loads would act on the bearings and teeth in the gears which would lead to failure of the components. The overload coupling is mostly in the form of a slip coupling which is set to a maximum torque and slips for excess torque. The transmission of the torques in normal operation occurs by means of friction contact.

A further property of couplings is their electrical insulation. In the case of a lightning strike, the lightning passing along the blade would strike throughout the whole drive train, whereby especially the generator would be damaged. For this reason the couplings are equipped with an intermediate shaft which is normally made of glass fibre-reinforced plastic. Alternatively, variations are available that posses a middle tube of carbon fibre or steel and in which, in the steel variant, the flexible element has the task of electrical insulation.

The coupling also has the function of transmitting the brake force via a brake disc to the drive train. Normally the brake is situated directly at the drive outlet and is bolted to the drive housing. The brake disc is normally of steel.

For large plants with three independently-working pitch actuators, the brake is not normally used to stop the plant in full operation. It is only used in order to achieve complete stoppage

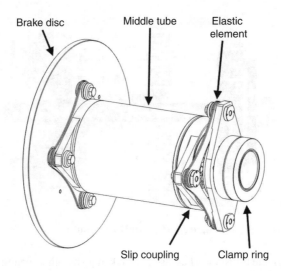

Brake disc   Middle tube   Elastic element

Slip coupling   Clamp ring

Figure 6.27   Coupling (KTR)

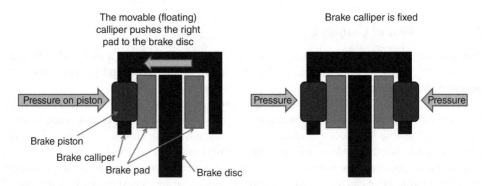

The movable (floating) calliper pushes the right pad to the brake disc

Brake calliper is fixed

Pressure on piston

Pressure

Pressure

Brake piston
Brake calliper
Brake pad   Brake disc

Figure 6.28   Principle of the floating calliper brake (left) and fixed calliper brake (right) (aerodyn)

of the drive train after the plant has been stopped largely by the pitch of the blades. In order to generate a redundancy of the brake system with stall plants or plants with central pitch cylinders, the mechanical rotor brake must be able to stop the rotor (emergency stop). Thus, the brakes are larger or more brakes are required.

Because of the lack of transmission, rotor brakes for directly driven plants are also larger. Because of the construction of directly driven plants, which have greater gondola diameter due to the generator dimensions, the brakes are generally of a larger size but basically more brake callipers are used. Two brake callipers are mostly used for plants with gears and a capacity of >2 MW, whereas smaller plants still manage with a single brake.

There are two types of brakes: fixed calliper and floating calliper brakes. In the case of floating calliper brakes, the pressure is applied to one of the two friction pads. The other pad is connected directly to the calliper, which is floating and can move freely on a guide pin. In the fixed calliper brake the two halves of the calliper are fixed together and possess two hydraulic pistons (see Figure 6.28).

The advantage of the floating calliper brake is that it can align itself and thus transmits no axial forces on the coupling and via this route into the gears and the generator. For this reason it places less demand on the erection accuracy. Fixed calliper brakes must be installed much more accurately in order to ensure that the brake disc is exactly centred between the friction pads.

The advantage of the fixed calliper brake is that it can absorb much higher forces, which means that in some cases fewer brakes are needed. The fixed calliper disc brake is cheaper, but cannot be used if an axial movement of the gear shaft is to be expected.

Rotor brakes are operated either electrically or hydraulically. Electrical brakes have only been on the market for a few years and are not yet widespread. This is also due to the much higher price compared with the hydraulic brake calliper. However, it is sensible to install them if one wishes to do without a hydraulic aggregate altogether and thus avoid the problem of oil leaks. Besides this, lower servicing costs are to be expected with the use of electrically operated brakes.

The high temperatures that are generated in plants with drives during braking present special challenges to the brake pads. Organic brake pads or sintered metal pads are normally used here. The organic brake pads consist of a binder and various materials such as metal, glass, rubber, resin and various strengthening fibres. They function better than sintered material pads as stop brakes (so-called static brakes) and thus have a higher coefficient of friction. The sintered material pads have a higher temperature resistance. The decision for the pads must be made on the basis of the plant concept, the load cases and the plant controls. Attention must be paid here to the defined servicing wind velocity, which determines the maximum stopping moment of the brakes, and the maximum brake speed, which is the result of the defined maximum speed of rotation for brake failure. For directly driven plants it is better to select organic brake pads as the thermal load is small.

6.4.5 Generator

The generator is also an important part of the drive train. It is described in detail in Chapter 8. There are numerous models for direct-driven plants with a high number of poles, or for geared plants with a low number of poles. For the direct-driven drive train concept, the generator is supported by the already-existing rotor mounts, whereby separately mounted systems also exist. A conventional machine with two mounts is mostly used in the fast-running systems. The mechanical connections of the various generators with the respective drive trains can be found in Section 6.5. The doubly-fed asynchronous machines are most often used for fast-running generators. In the case of current designs, permanently excited (see Figures 6.29 and 6.30) synchronous generators or asynchronous squirrel cage machines, which have the advantage that they do not use slip rings and can adhere more simply to the grid connection conditions via the full power converter, are often used. A disadvantage, however, is that the costs are substantially higher, as with the doubly-fed machines only a third of the current needs to pass through the converter and thus the converter is cheaper. Fast-running generators are mostly designed with four poles. The nominal speed of revolution is mostly 1500 rpm (synchronous) or approximately 1650 rpm (asynchronous). For large generators (>3 MW) or for 60 Hz grid frequency, mostly six-pole machines (1000 rpm synchronous at 50 Hz) are used. The bearings for fast-running machines are a weak point. Although the roller bearings are designed for a 20-year life, practice has shown that the actual life is much less.

Figure 6.29 Permanently excited generator with water-jacket cooling (left) and generator with blower at top (right) (ABB)

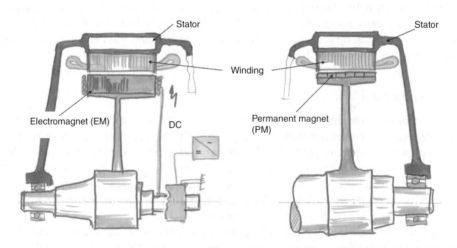

Figure 6.30 Excitation of direct-driven generators (F. Klinger)

The causes of this have not been clearly identified. Meanwhile, the manufacturers have excluded current flows, since the cause as this is countered by means of corresponding earth-ing brushes. Lubrication is mostly ensured by means of an automatic add-on relubrication system. Yet lubrication is often mentioned as the cause of damage. Many machines have recently been re-equipped with ceramic bearings in order to ensure a longer life for the bearings. These possess better properties with poor lubrication and better insulation, but are very expensive. Cooling of fast-running systems is often achieved by means of air–air heat

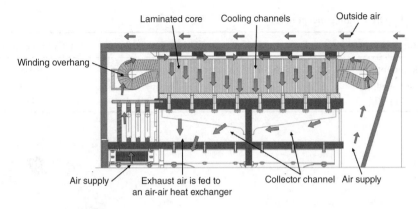

Figure 6.31   Cooling system of a Vensys plant (Vensys)

exchangers. In this, an internal flow of air is generated by blowers that again exchange the heat in a heat exchanger to an external airflow. In addition, with a closed and sealed gondola, a machine with an air–water heat exchanger can also be used.

Whilst with water-jacket cooling the inside of the machine is often not evenly cooled, the disadvantage with air–water heat exchangers is the weak temperature drop compared with air–air exchangers. This is because of the additional heat transfer that this concept brings with it.

Slow-running or medium-speed machines with a greater number of poles have fewer problems with the stability of the bearings. The direct-driven turbines often make use of the rotor mountings at the same time as generator mountings. With these machines, use is made of permanently excited or externally excited machines. In the case of externally excited machines, the slip rings can be dispensed with when an accompanying exciter machine is used. Because of the high price of magnets, externally-excited systems are becoming interesting again, although these cannot be built so compactly. The rotor of the machine can also be designed as an outside runner, which is advantageous from the outside diameter point of view.

The cooling and the air-gap stability is the design challenge for slow-running machines. The influence of the drive train and the support of the air gap is described in Section 6.5. There are numerous possibilities for the cooling medium flow which are also often protected by patents (water cooling, air cooling, and so on). The efficiency of the cooling, besides the force flow density in the air gap through the exciter system, determines the compactness of the machine. Because for reasons of costs, the direct-driven machines have a poorer efficiency than the fast-running ones, the cooling system must be designed to be correspondingly larger. As an example, Figure 6.31 depicts the cooling system of a Vensys plant.

Chapter 8 contains a comprehensive discussion of the electrical design of the generator and converter.

6.5   Drive Train Concepts

In this section, various drive train concepts and their dependence on their rotor mountings will be described, and the drive/generator concepts will be presented and discussed. As the possibilities of combinations are large, this book will limit itself to schemes that have actually been carried out. A basic summary of the possibilities and the sections where they are discussed is shown in Table 6.1.

Table 6.1 Drive train concept matrix

Mounting concept	Drive type: direct	Drive type: 1–2 steps	Drive type: 3–4 steps
Double mounting	6.5.1	6.5.3	6.5.5
Three-point support			6.5.6
Torque support	6.5.2	6.5.4	6.5.7

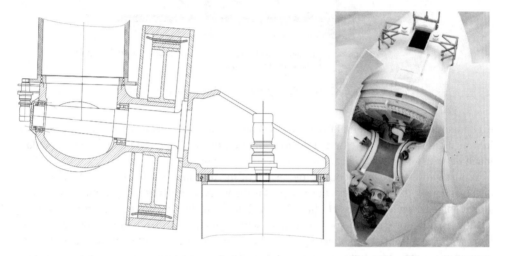

Figure 6.32 Drive train concept (left, aerodyn), and the hub of an Enercon wind turbine (right, Enercon)

It can be seen that two possible combinations with the three-point support (direct-driven concept, and drive train with 1–2 steps) have been left out. Although these variations are possible in theory, there are no practical implementations of them that are known to the author at this time. To be added, though, is that a direct-driven drive train based on a three-point support cannot be implemented in an economically competitive manner because of the loading of the generator housing.

6.5.1 Direct-Driven – Double Mounting

Gearless wind turbines are distinguished by their high reliability and low servicing needs. Disadvantageous are their large dimensions and the comparatively high costs for their construction. The best known manufacturer with a comparatively large market share is the German Enercon Company. Enercon uses externally excited generators that are very heavy and require a large construction volume. Modern variants, for instance Vensys or Siemens, work with permanent magnets and are no heavier than the geared plants. However, the costs are also very high for these manufacturers as the prices for permanent magnets have recently risen steeply.

The mounting concept of the rotor for the Enercon wind turbines is based on two bearings that are arranged on a protruding shaft stub (see Figure 6.32). The hub is directly connected

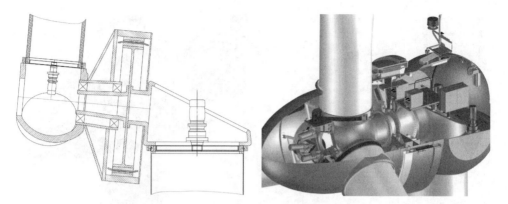

Figure 6.33 Drive train concept (left, aerodyn) and gondola (right, Vensys) of the Vensys 1.5 MW wind turbine

with the internal rotor of the generator. The stator is arranged in the housing, which again is flanged to the machine carrier. The new offshore wind turbine of the ALSTOM Company, a 6 MW turbine with a rotor diameter of 150 m, also makes use of this concept.

However, the advantages described in Section 6.4.2 for the double bearings must also be compared with the disadvantages. On the one hand, the bending of the axle stub influences the air gap of the generator, so that the machine mounting and the axle are correspondingly stiff and must be designed to be heavy. On the other hand, access to the hub is not possible, and also the generator is hard to reach.

The structure shown in Figure 6.33 can be considered to be similar to the Enercon concept. However, this drive train by the Vensys Company has an external generator. The stator is situated inside and is connected to the machine mount. As the axle with the two bearings does not extend through the whole hub, as with the Enercon concept, the hub is accessible in this structure. In addition, the rotating generator housing is not loaded by the rotor loads. However, with this concept the air gap is also influenced by the bending deformations of the axle stub. The resulting heavy design is also obvious here.

A further example of the mounting concept with two separate bearings in connection with a direct-driven wind turbine is shown in Figure 6.34. The concept of the mTorres TWT 1500 and TWT 1.65 is based on a longer, strongly integrated generator that is arranged directly over the tower. This variation has the advantage of a smaller generator diameter and carrying machine mount, but has the disadvantage that up to 90% of the rotating load acts on the generator housing. The generator air gap is also affected.

The solution of the Scan Wind Company (now GE) is based on a very long rotor shaft mounted with two separate bearings with small diameters. Besides the rotor, the generator with stator and rotor are also mounted on the shaft.

This bearing concept has the advantage that the generator is only loaded by the torque. The result is low deformation of the generator and its housing, whereby the air gap of the generator is also not impaired.

As the rotor shaft and the machine mounting for the rotor and generator shown in the left of Figure 6.35 are dependent on the diameter at the top of the tower, the carrier structure can be very large and heavy. Besides this, the solution has the disadvantage that the generator requires its own mounting.

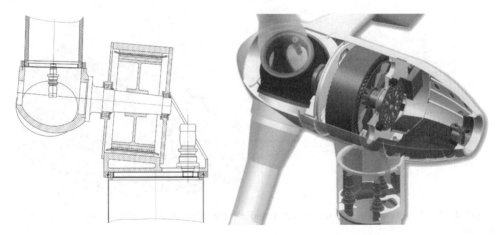

Figure 6.34 Drive train concept (left, aerodyn) and gondola of the mTorres TWT 1500 (right, mTorres)

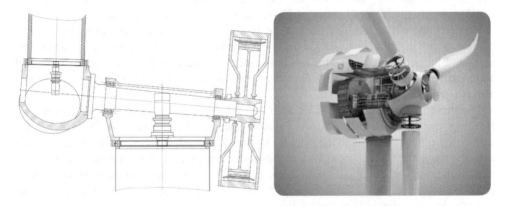

Figure 6.35 Drive train concept (left, aerodyn) and gondola of the GE 4.1-113 (right, GE)

A variation of this concept can be seen in Figure 6.36. The double mountings of the extremely long rotor shaft remain the same. However, the generator housing with the stator of the Heidelberg motor HM 600 is not supported on the rotor shaft, but flanged directly to the machine carrier. This model has the advantage that the generator housing is subjected to very low loads. Also, the generator housing needs no mountings.

However, one must realise that the air gap of the generator is influenced by the bending of the rotor shaft. A corresponding design of the rotor shaft is therefore necessary; in connection with the length of the shaft, this is a challenge that must not be underestimated.

6.5.2   Direct-Driven – Torque Support

As mentioned in Section 6.5.1, here, too, a gearless drive train concept is being considered whereby the advantages and disadvantages will not be mentioned again.

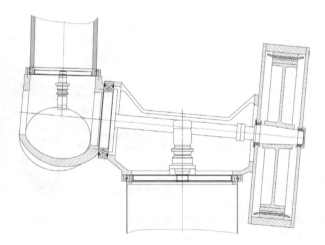

Figure 6.36   Drive train concept of Heidelberg Motor HM 600 (aerodyn)

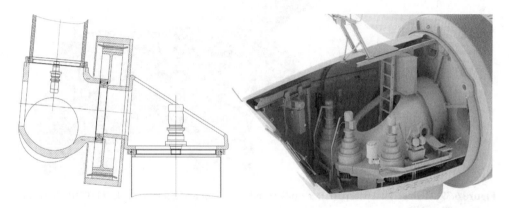

Figure 6.37   Drive train concept (left, aerodyn) and gondola (right, Lagerwey) of the Lagerwey L93 2.5 MW wind turbine

In the following we will concentrate on the direct-driven drive train concepts with torque support. We will begin with a drive train concept that has been used by the Lagerwey Company in various wind turbines. The rotor hub is equipped with a torque mounting which is mostly designed as a double-row tapered roller bearing and is mounted on the machine carrier. It transfers the torque to the internally positioned rotor of the generator. The stator is fixed in the generator housing at the machine carrier (see Figure 6.37).

Although the structure of this variant permits direct access to the hub, it has disadvantages because of the large bearing diameter and the effects of the deformation on the air gap of the generator with reference to the dimensions and the weight of the carrier design.

Other than with the 1.5 MW plant, the Vensys Company's 2.5 MW plant uses a drive train with only one main bearing. However, Figure 6.38 shows that the generator concept has not changed. Also, with this wind turbine the generator is designed as an outside runner with the stator inside and connected with the machine carrier.

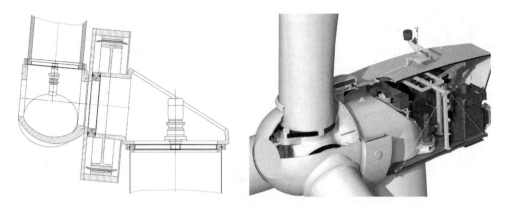

Figure 6.38 Drive train concept (left, aerodyn) and gondola (right, Vensys) of the Vensys 2.5 MW wind turbine

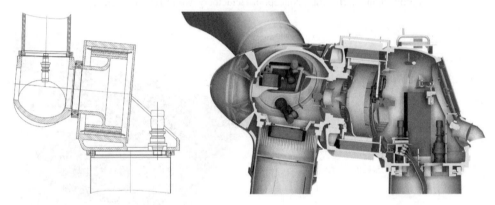

Figure 6.39 Drive train concept (left, aerodyn) and section of a gondola of the LEITWIND Company (right, LEITWIND)

As the structure only differs in the rotor–stator arrangement from the Lagerwey concept, the same advantages and disadvantages apply.

In its products, the LEITWIND Company follows the drive train concept shown in Figure 6.39. Plants with a nominal capacity of 1–3 MW make use of a torque support and a generator with a relatively small diameter. The resulting machine mounting is therefore rather smaller than the previously discussed solutions. However, this advantage is countered by the large bearing diameter of the hub which is correspondingly larger. Also the other disadvantages of 'Direct drives' with torque support must be taken into account here.

6.5.3 One–Two Step Geared Drives – Double Bearings

The drive train concept presented in this and the following sections are based on a 1–2 step geared drive and a slowly rotating generator. This variation, also known as hybrid, attempts to combine the advantages of direct drives (see Sections 6.5.1–6.5.2) and the classic drive train

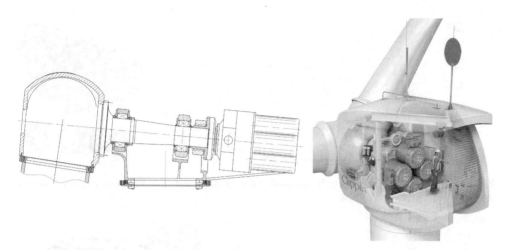

Figure 6.40 Drive train concept (left, aerodyn) and gondola (right, Clipper Windpower) of the Clipper Liberty 2.5 MW

concept with multi-step transmission drives (see Section 6.5.6–6.5.7). Because of the compact drive and generator, advantageous tower head weights are meant to be achieved.

Because of the complete integration of the drive train concept in the machine carrier (see Multibrid M5000 or aerodyn's SCD), the successful implementation achieves its goal. However, it must be remembered that repair possibilities without disassembly of the tower head are very limited.

We will again discuss this solution in the following. It must, however, be remarked that this relatively new method of drive train construction is not yet as widespread as the two solved drive train variations, and that various configuration possibilities are not discussed in this book.

In this section we will first look at the hybrid method of construction with a double mounting. Mention is made here of the Clipper Liberty 2.5 MW wind turbine as an implemented solution (Figure 6.40). In this plant the capacity generated is transmitted to the first gear step and then split up.

The downstream components in the drive train are carried out four-fold as the second drive step. This can be clearly seen in Figure 6.40 (right) which shows the four generators in the gondola.

An advantage can be seen in the good encapsulation of the components from the rotor loads and the resulting simple design. Also, simple generator exchange in the case of failure can be counted as an advantage, thanks to the small size of each generator. With a view of the mounting concept (applies to all double-bearing drive train concepts), the supplier situation is good.

But this solution has its disadvantages. Besides the long and heavy rotor shaft, the design implementation of the mechanical rotor brake is also subject to problems. Added to this is the special development of the drive, with a resulting dependency on a particular supplier.

Gamesa also works with double bearings in connection with a two-step geared drive in its own 4.5 MW plant. A feature here is the chaining of the components (see Figure 6.41). The housing of the rotor mounting is directly connected at the front with the machine carrier. The drive is flanged at the rear, where the generator is again fixed. Here also a tube-type spacer is

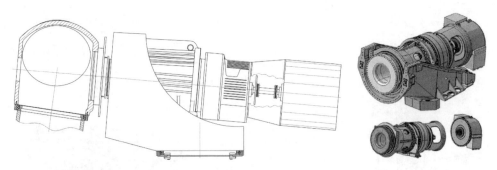

Figure 6.41   Drive train concept (left, aerodyn) and support concept with drive train (right, Gamesa) of the Gamesa 4.5 MW wind turbine

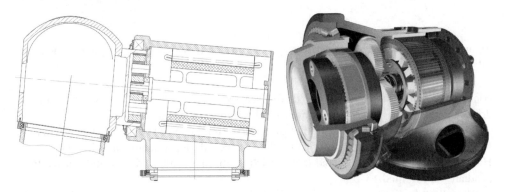

Figure 6.42   Drive train concept (left) and head carrier (right) of the Multibrid M5000 (aerodyn)

used, which includes the coupling. With this solution, achieving the signal and power connections of the components is complex.

6.5.4   One–Two Step Geared Drives – Torque Support

The AREVA M5000, formerly Multibrid, shows a hybrid drive train. It is similar to the drive train concept in Figure 6.42 that shows a geared drive integrated in the main bearing and a generator with a small diameter.

The mounting of the extremely compact drive train of the AREVA M5000 consists of a dual-row ball bearing into which a special form of an epicyclic gear is integrated (see Figure 6.42, right). Besides the bearing and the geared drive, the generator is also integrated into the machine carrier. Because of its absolute function integration and the short rotor shaft, the drive train of this wind turbine has very small dimensions and is therefore very light.

However, this form of construction makes high demands regarding design and implementation. Because of this function integration, the drive as well as the generator is stressed by rotor loads. Whereas, in the case of the generator, the air gap is primarily affected, also the tooth meshing of the drive can be impaired. In addition, a structural solution must be found for access to the gondola. As most of the components are special designs, this can also present problems with suppliers.

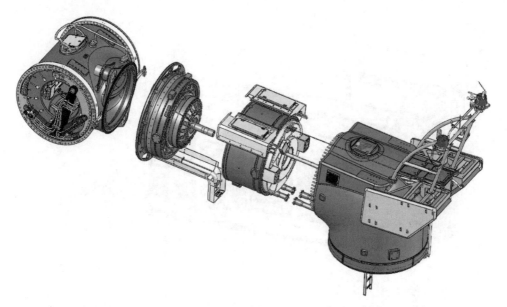

Figure 6.43   Gondola of the SCD 3.0 MW (aerodyn)

Besides AREVA, which builds the M5000, and the SCD concept of the aerodyn company, there are similar concepts with a part-integration of the drive into the machine mounting (e.g. from the WinWind Company).

The SCD concept (Super Compact Drive) is a design of the aerodyn company with a two-blade rotor that is currently being built as a 3–6 MW plant by a Chinese manufacturer. The concept has a modular structure in which the drive and the generator are bolted together in tube form onto the head carrier (see Figure 6.43). All the components have almost the same diameter and are bolted to the outer ring. Thus, the flow of force passes through the housing of the drive train component via the head carrier to the tower.

Basically the head carrier has a similar appearance to a pipe elbow, except that at the one side the drive train is fixed with a horizontal axis and on the other side the head of the wind turbine is mounted on the tower with a vertical axis. Therefore a gondola with covering in the normal sense no longer exists. All peripherals are arranged on two levels in the head carrier and tower. Such full integration only makes sense when the manufacturers have a large production facility and a large number are built.

6.5.5   Three–Four Step Geared Drives – Double Mountings

The double mounting with multi-step drives is the classical model of a drive train. The basic structure is described in the rotor shaft and mounting section.

If we view the drive train concept in connection with a double rotor mount, then it must be said that various designs are possible. The best-known solutions and the ones most often implemented in practice are shown in Figures 6.19 and 6.20 and are described in Section 6.4.2.

Both variants pass the largest part of the rotor loads through the bearings directly to the machine carrier, so that the drives and the generator are only subjected to torque. Because of

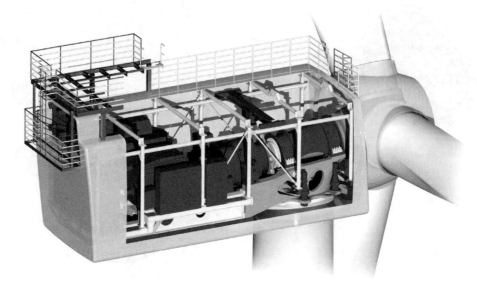

Figure 6.44 Gondola of the aeroMaster 5 MW of aerodyn Energiesysteme GmbH (aerodyn)

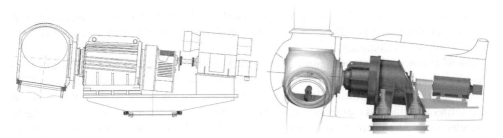

Figure 6.45 Drive train concept (right, aerodyn) and implementation of the FL 1500 by Fuhrländer (left, Fuhrländer)

the many developed plants based on this concept, the supplier situation is correspondingly good. In order to present only a few examples of the types of wind turbine, the variants with separate mounts are represented by Vestas V80, Siemens SWT 3.6-107 or the REpower 5M.

The variants with a closed bearing housing are mainly the wind turbines from GE (e.g. GE 2.5xl), Vestas 2 MW Grid Streamer or the aeroMaster-Family (1.5 MW (see Figure 6.44), 2.5 MW, 3.0 MW, 5.0 MW) of the aerodyn Energiesysteme GmbH. As already mentioned in Section 6.4.2, the advantage of this arrangement is the use of zero backlash gears. This leads to an increase in their reliability. Of primary importance is that axial movement of the drive train is prevented, which is also to the good of the gears.

The following concepts are also based on double mountings, yet are somewhat different. The concept in Figure 6.45 (left) is dominated by a bearing housing, which, besides the two roller bearings, also possesses a connection for the gears. The torque supports on the machine mounts, which are normal in the previous solutions, thus fall away. Figure 6.45 (right) shows the implementation by the Fuhrländer Company.

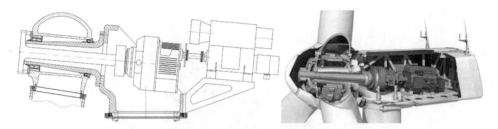

Figure 6.46   Drive train concept (left, aerodyn) and gondola (right, ALSTOM) of the ALSTOM ECO 100 with Pure-Torque technology

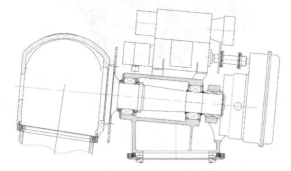

Figure 6.47   Drive train concept of the Komal KWT 300, the HSW 1000 and the Dan-Win27

As already mentioned in Section 6.5.1, the ALSTOM Company uses a rotor mount in which the roller bearings are arranged on a stub axle. The aim of directing the bending moments strongly into the tower and only to transmit torque to the drive train (Figure 6.46) is the same as in the cases of the direct-driven offshore plant. The only special feature of the otherwise conventional drive train is the rotor shaft running in the stub axle. It emerges from the front end of the stub axle and is flanged there to the hub.

Two further drive train configurations based on double mountings are shown in Figures 6.47 and 6.48 and are mainly found in older wind turbines. Figure 6.47 shows a concept in which the generator is arranged on the housing of the rotor mounting and the geared drive is used for providing the axle offset. As the drive train runs in a U-type form, it is short and high, so that an additional generator carrier can be dispensed with.

Figure 6.48 shows a partly function-integrated drive train. In this design the rotor mount is implemented directly in the gears. The result is a relatively short and light drive train. However, the gears are subject to strong loads in which mostly the tooth insertion is impaired. In the past, mainly smaller plants such as the Tacke TW600 or the DeWind D4 were built with this type of drive train.

6.5.6   Three–Four Step Geared Drives – Three-Point Mountings

The concept of the three-point mounting has already been described in detail in Section 6.4.2. Because of the many types of implementations of the drive train with this type of mounting, only a few will be mentioned. Besides REpower (MD 70/77, MM 82/92, 3.3M) mention

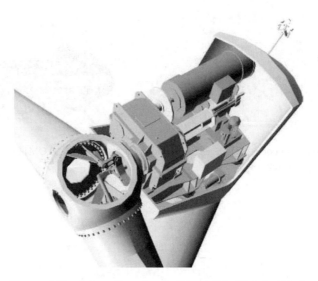

Figure 6.48   Drive train concept of the DeWind D4 (DeWind)

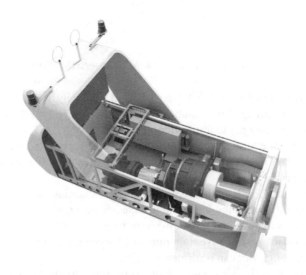

Figure 6.49   Gondola of the Vestas V112-wind turbine (Vestas)

should be made of Vestas (V82, V112, see Figure 6.49), Fuhrländer (MD 70/77), Nordex (S70/77), GE (1.5, 3.6) and Siemens (previously AN Bonus 1.3 MW, 2.3 MW). As with every standard design of this variation, advantageous supplier situations with regards to components can be expected here as well.

A drive train concept with a very similar design is shown in Figure 6.50 (left). Here, only a variable torque converter is inserted in the drive train between the gears and the generator. This has the task of keeping the speed of revolution on the output shaft to the generator constant and thus to simplify the grid connection of the wind turbine. The aim is to save the frequency converter that is necessary for plants with fixed transmission conditions and variable

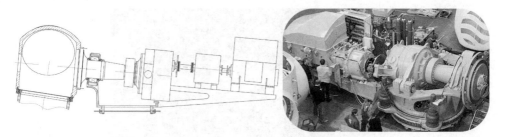

Figure 6.50 Drive train concept (left, aerodyn) and gondola (right, DeWind) of the DeWind D8.2

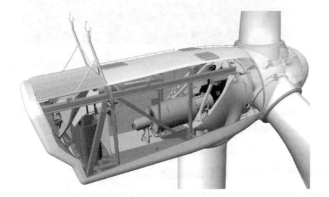

Figure 6.51 Drive train with torque support and integrated multi-step gearing of the Vestas V90 (Vestas)

generator revolutions. However, as the design of the actual drive train is unchanged, thus also a gearing with higher transmission is required; the advantages and disadvantage of the previous design can be expected.

In the example of DeWind (Figure 6.50, right), a hydrodynamic converter is used as is described in Section 6.4.3.

6.5.7 Three–Four Step Geared Drives – Torque Support

In the last category of the geared drive train matrix, mention is made of drive trains that use gears with high transmissions and fast-running generators. The mounting concept used here is the torque mounting mentioned in Section 6.4.2 (Figure 6.23).

Next we will discuss a solution in which the separated drive train is combined with the said torque mount. The result of this is primarily a short drive train. However, basically standard components can still be used. Only the torque mounting can lead to availability problems due to its special development.

The most prominent implementation of this drive train is probably the 5 MW offshore wind turbine of the Bard Company, but also Mitsubishi (MWT 92/95), Fuhrländer (FL 2500/3000) and Unison (U 88/93) can be mentioned here.

A torque support is also used by Vestas for the V90-3.0 MW (Figure 6.51). In order to minimise the drive train still further, the multibrid concept (see Section 6.5.4) makes use of

Figure 6.52 Destroyed epicyclic gear step (Gothaer Versicherung)

function integration, and the rotor mounting is arranged in the gearing. However, in contrast to the multibrid, the generator is arranged separately and the plant uses a three-step gearing which is only partly enclosed by the machine carrier.

6.6 Damage and Causes of Damage

Damage to wind turbines is always an important subject at congresses and discussion forums (e.g. Figure 6.52). The damage and associated repair costs can make up a substantial portion of the power maintenance costs. The insurance companies no longer cover all damage that can occur, so the operator often carries a fairly large risk.

In the early years of the wind industry at the end of the 1980s and the start of the 1990s, design methods were still inadequate, so damage occurred due to design errors. This has improved in the past few years, but there is still room for improvements in the reliability of the plants. The short development times of ever-larger plants with very short testing times are the main problems. The series production of some new types of plants already begins before all technical malfunctions have become known and removed by means of redesign.

The main causes of damage are:

- design errors;
- lack of quality in production;
- wind events at the site, for which the plant is not designed;
- inadequate servicing.

Unfortunately there are still no reliable statistics in the publications of the various institutes and companies on damage. For instance, some statistics blame the geared drives as the main source of failure (see Figure 6.53), while others mention the electrical systems such as converters and pitch systems as the main cause for stoppages (see Figures 6.54 and 6.55). It seems to emerge that geared drive damage still takes up a substantial portion of the repair costs, such that a wave of new developments for direct-driven plants has taken place in the recent years.

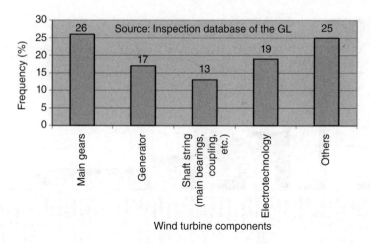

Figure 6.53   Defined malfunctions of wind turbines according to statistical data (Gothaer Versicherung)

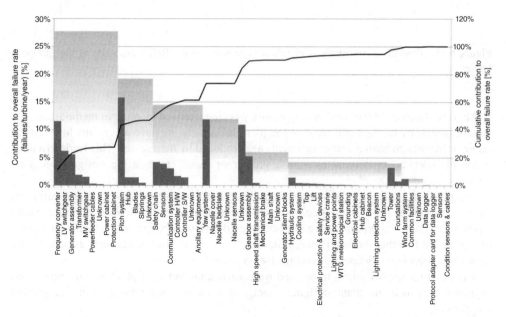

Figure 6.54   Failure rates according to the Reliawind study of the European Union

6.7   Design of Drive Train Components

The design of drive train components in wind turbines is distinguished by the fact that the wind energy industry is a young industry in which the procedure for furnishing proof of design is very modern. Tools such as the finite element method (FEM) and simulations of dynamic processes have long been a part of the standard repertoire of the developer of wind turbines.

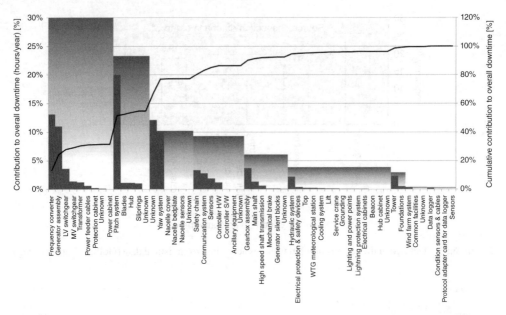

Figure 6.55 Downtime contributions of failures according to the Reliawind study of the European Union

Also the structure of the wind energy industry is characterised by modern methods, shown, among others, by the fact that most of the plant manufacturers have a very shallow vertical range of manufacture, thus few components are produced by themselves. This leads to a multitude of independent companies with a greater or lesser stake in the design of a wind turbine.

The plant manufacturers themselves coordinate and monitor the development, carry out their own proof of the structural components and undertake the subsequent erection. These types of structures require intensive links between the various companies, which is in contradiction to the fact that the wind energy industry is a very restrictive industry, as patents and know-how represent a decisive advantage (see Section 6.8).

It is usual for specifications to be used for coordination between plant manufacturers and suppliers in which the plant designer exactly defines the requirement of the respective components.

Guidelines play a large role in the design of wind turbines and their components, and certification is based on them. They define what proofs are required and with what safety features the developments are to be provided. The *Richtlinie für die Zertifizierung von Windkraftanlegen [Guidelines for the certification of wind turbines]* of the Germanischen Lloyd and the IEC 61400 are internationally recognised guidelines according to which wind turbines are often designed. The guidelines themselves make special demands on the components as regards safety factors, the residual safety, and the procedure for furnishing proof, in which strong reference is made to the existing specific component standards from DIN, EN, ISO and IEC.

A load calculation based on statistic evaluation of wind data is made at the start of development of every plant. The data are used to simulate the loads that act on the plant. Decisive

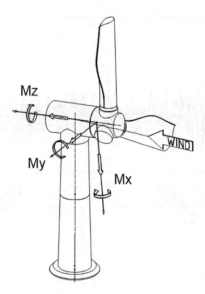

Figure 6.56 Coordinate system hub–rotor shaft (aerodyn)

here is the design of the blades. Besides the length and area of the blade, they can also have many other characteristics that decisively influence the occurring loads (stiffness, inertia). Thus, the development of the blade and the load calculation are two processes that take place at the start of the development of the plant, usually in parallel, and have a strong influence on each other.

In the load simulation, the rough dimensions for the plant are determined and several sections are made to which coordinate points are fixed and for which the resulting loads are determined. Each component design refers to a particular coordinate system. A distinction is made between coordinate systems such as blade root, tower head or rotor shaft. The hub–rotor shaft coordinate system is the prime system for the design of the drive train.

A possible coordinate system is shown in Figure 6.56. Refer to this coordinate system for further understanding, all the named loads in this section.

Wind turbines are designed for a life of 20 years, whereby the load calculation refers to static and stochastic evaluation in order to determine the loads for this period of time. In the above-mentioned guidelines, exact load cases are defined from which time series are generated which, in turn, contain information on duration and features of the external plant conditions for defined operating conditions. In addition, the information is given on how often the time series occur in the overall plant lifetime.

In the evaluation of the loads, a distinction must be made between operating loads that occur during the normal course of the plant operation, and extreme loads under particular conditions such as the failure of components, and which do not enter into the operating strength verification.

For evaluation and utilisation in the verification, the operating load time series are summarised into load collectives with the rain-flow-count process to LDDs (load duration distribution).

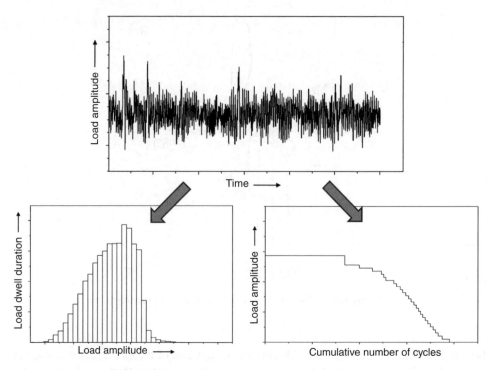

Figure 6.57   Time series in LDD and RFC depiction

6.7.1   LDD

In the load duration distribution (LDD) the loads are divided into several steps and evaluated, however long the respective load step occurs during the time series (load dwell time). Multiplied by the frequency of occurrence of the respective time series during the life of the plant, the result is a certain hour number for the respective load step, on the basis of which damage for a component can occur. The frequency of occurrence results from the statistic distribution of the individual load cases that are allocated to particular wind velocities.

In the drive train design, the LDDs of the torque are used for the design of the gear teeth. Figure 6.57 shows an example of how the LDD of a load component is derived from a time series.

6.7.2   RFC

In the load collective is a count of the load change play that a component experiences in its lifetime. As this is not as easily added up as the LDDs, a special method of counting must be used: the so-called rain-flow-count (RFC).

The load cases are investigated over the course of the load cycles and evaluated as to what load changes occur per load case. Multiplied by the frequency with which the load cases occur in the overall lifetime of the plant, it can then also be determined how frequently components must tolerate the load changes. The RFCs contain no information on how high the loads are, but only how often the change of the load occurs. If, in the verification, the average stress influence is also important, then so-called Markov matrixes are used in which the collective is additionally classified according to an average stress effect. The selected number of load change levels influences the accuracy of the calculation results. The use of load collectives in the procedure of furnishing proof is actually only a simplification for shortening the calculation processes. More accurate results are obtained with the direct use of the time series.

Nowadays this is used in the operational strength calculation of large structural components. This involves the use of special software. Figure 6.57 shows an example of how a RFC of a load component is derived from a time series.

Besides the pure drive torque (Mz), forces such as bending moments also act on the drive train, whereby the drive train is mainly configured for the bending moments. A distinction is made between tilting moments (My) and yaw moments (Mz) that swing about the horizontal or vertical axes (see Figure 6.56). The biggest deformations of the drive train derive from these moments and are determined by means of FEM. In the design of the drive train components, both the extreme loads as well as the operating loads are taken into account. Wind turbines have comparatively high load change numbers. They are designed for a life of 20 years according to the requirements of the guidelines. The operating strength calculations are carried out according to modern calculation methods in which damage calculations (time strength) with different Wöhler curves are used. This chapter will not deal with the details of this calculation.

The loads for the detailed design of the functional components, such as the pitch system or the yaw system, are evaluated by means of special programs. The results allow the dimensioning of the drives.

As the development of wind turbines shows strong tendencies to light construction, deformations in the development and design caused by the materials used play a large part. The components in the drive train are often provided with soft mounts in order to minimise this deformation and the occurrence of large restoring forces. The result of this is that the drive train has a higher potential to oscillation excitations. In this the components can stimulate each other which can again result in increasing damage effects, up to and including component failure. In order to prevent such effects, increased investigations have been carried out in the past few years on the drive trains, which meanwhile have become a part of the certification. Here the drive train of the plant is simulated by means of special software. This software contains all relevant components and information on the potential for excitation, such as stiffness and inertia, speed of rotation, cycling frequencies of bearings and frequencies of tooth meshing. In the actual simulation the complete operational range of the plant is gone through and evaluated. The software is able to determine the frequencies of the whole drive train and in which operating conditions energy is generated due to change effects. These can be allocated to the individual components. Thus, it is possible to view the behaviour of every component in the context of the plant and to prevent potential excitations due to variation in the stiffness of the design.

The first software solutions for the complete dynamic investigation of drive trains were only able to depict the torsional behaviour of drive trains. The best known software here is the DRESP rotation vibration simulation program, which is still widely used. Modern simulation software, meanwhile, is able to take several degrees of freedom of the components into account; one speaks of multibody simulation software. Figure 6.58 shows a model of a drive train derived from such software.

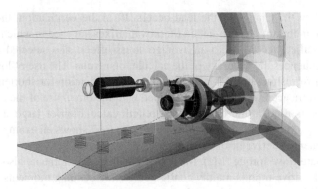

Figure 6.58   MKS model with SIMPACK software

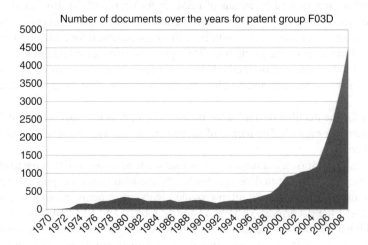

Figure 6.59   Annual new applications

6.8   Intellectual Property in the Wind Industry

The importance of protective rights in wind energy technology has increased substantially in recent years. The application numbers in the patent group 'wind energy plants F03D' has increased approximately ten-fold in the past ten years. At present, approximately 25,000 inventions have been recorded in the whole group.

Anybody wishing to develop, build, sell or operate wind energy plants must master this multitude of granted or applied-for patents. In this, the protective rights and possible conflicts must be analysed and evaluated with reference to own design solutions or processes. This is especially difficult for not-yet-granted protective rights, as at present it is unclear how large the protective field will be in the case of a granting of the application. The whole is made more difficult in that a new application is only published after 18 months. In the case that protective rights exist that particularly interfere with one's own requirements, or there is even an infringement, one must investigate how far these protective rights can be bypassed by means of other technical solutions, or even destroyed. For this it is necessary to determine the state of the technology that affects the patent claims and that was not available to the examiner at the patent department.

An overview of all published protective rights can be found in various Internet databases, some of which have good search possibilities. Patent databases are also an excellent tool for generating one's own ideas for further developments. Comprehensive research on the degree of newness and degree of inventiveness must also be carried out in order to check good ideas for their protective possibilities.

Approximately 90% of the technical knowledge of protective rights is only published in patents and not in other publications. Patents can be downloaded in full from the Internet portal of the European Patent Office (www.espacenet.com).

Figure 6.59 shows the course of worldwide patent applications over the past four decades. Before the start of the 1970s there was almost no activity in the application of wind energy patents. Thinking only turned to the use of wind energy with the first oil crisis in 1972. This was also the case with the first patent applications. In the following 25 years around 200 applications per year were added. At the turn of the century the worldwide total of patents was approximately 5000. After that the number of patent applications rose drastically so that today approximately 5000 patents are added each year.

Figure 6.60 shows the distribution of all the patents over the individual countries, whereby possibly foreign nationals in other counties are included. It can be seen that the greatest activities in patent applications are in the countries with great importance for the use of wind energy. Among the leading countries are China, the US and Germany. Despite their low usage of wind energy, countries such as Japan, Korea and Australia have high application numbers.

Figure 6.61 shows the distribution of the protective rights applications according to the applicants. This graph is sorted according to the WO [World Intellectual Property Organization (WIPO)] applications and not according to national first applications because WO applications have greater importance due to the numerous possible nationalities. The General Electric Company leads in the national first applications with approximately 660, but less than 10% were transferred to world applications. With Vestas, the world market leader, there were 450 first applications and approximately 370 were transferred in the international process. Among the other top applicants are companies such as Mitsubishi, Enercon, Siemens, LM Glasfiber and Gamesa.

6.8.1 Example Patents of Drive Trains

Three patent applications for different designs of drive trains are presented in the following. The selection of a drive train with a high transmission, a direct-driven plant without gears, and an intermediate solution with only two gear steps were taken at random. Worldwide there are more than 1000 patents for drive trains and all the solutions have their advantages and disadvantages.

Electrical Energy Generating Installation Driven at Variable Rotational Speeds, with a Constant Output Frequency, US 2012/0038156A1

The invention relates to an energy-generating installation, especially a wind power installation, with a drive shaft connected to a rotor, a generator, and with a differential transmission

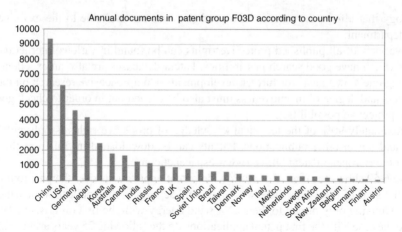

Figure 6.60   Distribution of patent documents according to countries

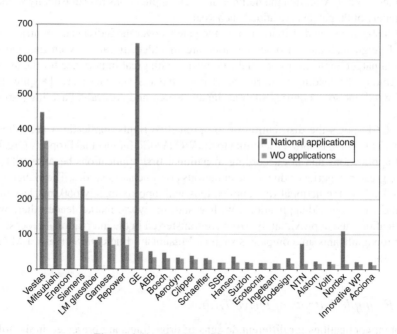

Figure 6.61   Top 25 applicants (up to 2012)

with three drives and outputs, a first drive being connected to the drive shaft, output to a generator, a second drive to an electrical differential drive, and the differential drive being connected to a network via a frequency converter (see Figure 6.62). This objective is achieved according to the invention in that the frequency converter can be controlled for active filtering of harmonics of the energy-generating installation, especially of the generator. For this reason, a reduction of harmonics need not be considered or need be considered only to a lesser extent in the design of the generator.

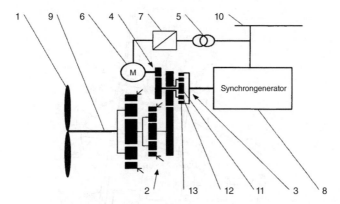

Figure 6.62   Rotation variable gear drive

Patent Claim 1

Energy-generating installation, especially a wind power installation, with a drive shaft connected to a rotor (**1**), a generator (**8**), and with a differential transmission (**11 to 13**) with three drives and outputs, a first drive being connected to the drive shaft, one output to a generator (**8**), and a second drive to an electrical differential drive (**6, 14**), and the differential drive (**6, 14**) being connected to a network (**10**) via a frequency converter (**7, 15**), characterized in that the frequency converter (**7, 15**) can be controlled for active filtering of harmonics of the energy-generating installation, especially of the generator (**8**).

Wind Turbine DE 102 39 366

The invention relates to a new type of wind turbine in which the structural simplification and reduction of the mass permits a reduction in the structural length of the rotor/generator unit projecting from the side of the carrier tower. According to the invention the wind turbine that solves this task is characterised in that the blades of the wind rotor are arranged on the rotor ring and/or an axial extension of the rotor ring.

Advantageously, the invention permits the rotor to be set back in the direction of the carrier tower so that a central, middle section of the blade, in an extreme case in which the rotor ring alone forms a hub of the wind rotor, does not protrude further than the generator from the carrying tower in the direction of the rotor rotating axis. In a particularly preferred manifestation of the rotor or the rotor ring, possibly with the extension of a tube-shaped carrier structure, is mounted on a carrier structure containing the carrier windings, whereby, in contrast to other wind turbines, a special large reduction of material and weight is achieved when the radius of the carrier structure or the mounting approaches the air gap radius of the generator (Figure 6.63). In this case even for weak material-saving design of the generator parts, the air gap geometry of the generator is sufficiently constant. The possibly extended rotor ring can be rotatably arranged on several mountings axially arranged to each other or a single mounting preferably at the transition to the extension. In the case that the rotor ring alone forms the rotor or a part thereof, then it is necessarily mounted at two points, whereby the carrier winding is arranged between the two mounts.

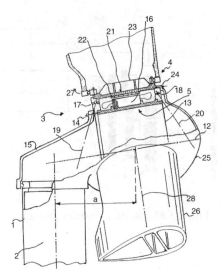

Figure 6.63   Directly driven wind turbine

Patent Claim 1

Wind turbine, especially for a nominal capacity in the megawatt range, whose wind rotor (**4**) is directly linked with a magnet (**22**) carrying rotor ring (**21**) of a generator (**5**) characterised in that the blades (**26**) of the wind rotor (**4**) are arranged on the rotor ring (**21**) and/or an axial extension (**31**) of the rotor ring (**21**).

Wind Turbine with Load-Transmitting Component DE 10 2007 012 408

The object of the invention is to create a drive train that enables a very compact, light-weight, and thus cost-effective total construction and ties the main components such as the rotor bearing, gearbox, generator, and wind direction tracking unit into the force transmission from the rotor into the tower (Figure 6.64). It is to be ensured at the same time that the individual components, in particular gearbox and generator, can be mounted separately and also handled individually for repair work. In the case of the invention, the components gearbox, generator, and wind direction tracking unit are arranged in separate casings that are bolted together. The respective casings are designed as supporting structures for transmitting the maximum static and dynamic rotor loads. Also the rotor bearing is bolted on to the gearbox casing and transmits the rotor loads into the gearbox casing. The gearbox casing transmits the loads into the generator casing. The generator casing in turn transmits the loads into the head support that in turn introduces the loads via the azimuth bearing into the tower. As a result of this design, the casings of the components assume the twin function as load-transmitting element and as mounting element for the individual parts of the components.

This design makes it possible for the machine to be very light-weight and thus cost-effective and also a nacelle shroud to be dispensed with, since all components are of such a design that they can be exposed to weathering. For reasons of assembly it makes sense to design the gearbox casing and the generator casing as two separate casings; however, it can also be designed as one piece.

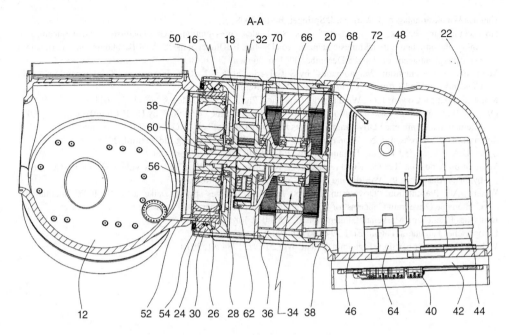

Figure 6.64   Wind turbine with load-transmitting components

Patent Claim 1

Wind turbines with at least one rotor blade **(10)**, a hub **(12)**, a gearbox casing **(18)** receiving a gearbox **(32)**, a generator casing **(20)** receiving a generator **(34)**, a head support **(32)**, a tower **(14)**, and an azimuth bearing **(42)** supporting the head support rotatably on the tower **(14)**, characterized in that the rotor bearing **(16)**, the gearbox casing **(18)**, and the generator casing **(20)** are arranged between the hub **(12)** and the head support **(22)** and designed as load-transmitting components and are joined to each other via screw connections **(54, 38)**.

Further reading

aerodyn Energiesysteme GmbH, Präsentation: "Up-to-date and innovative turbine concepts" von Dipl.-Ing T. Weßel und Dipl.-Ing P. Krämer auf der "1. IQPC International Conference Drive train Concepts for Wind Turbines" in October 2010.

Allgaier-Werke GmbH, Betriebsanleitung für die Allgaier Windkraftanalage System Dr. Hütter Type WE 10/G6, Uhingen 1954

Burton, T.; Sharpe, D.; Jenkins, N.; Bossanyi, E.: Wind Energy Handbook, 2. edition, John Wiley & Sons, 2011

ChapDrive AS, Präsentation: "ChapDrive hydraulic transmission for lightweight multi-Megawatt wind turbines" von K.-E. Thomsen & P. Chang at the "2. IQPC International Conference Drive train Concepts for Wind Turbines" on 17 October 2011 in Bremen

ESS Gear, Presentation: "Gearbox Concepts" by H. P. Dinner at the "2. IQPC International Conference Drive train Concepts for Wind Turbines" on 17 October 2011 in Bremen

Gasch, R.; Twele, J. (Hrsg.): Windkraftanlagen, 5. Auflage, Vieweg + Teubner, Stuttgart, 2007

GL Garrad Hassan, Presentation: "Failure rates and the impact of failures on cost of energy" by B. Hendriks at the "2. IQPC International Conference Drive train Concepts for Wind Turbines" on 17 October 2011 in Bremen

Hau, E.: Windkraftanlagen, 4. Auflage, J. Springer, Berlin 2008

Hochschule für Technik und Wirtschaft des Saarlandes, presentation: "Getriebelo-se Windenergieanlagen: Anlagentechnik und Entwicklungstendenzen" von Dr.-Ing. F. Klinger at the "2. VDI-Fachkonferenz: Getriebelose Windenergieanlagen" on 7 and 8 December 2011 in Bremen

Magnomatics, Presentation: "Magnomatics" by S. Calverley at the "2. IQPC International Conference Drive train Concepts for Wind Turbines" on 17 October 2011 in Bremen

Manwell, J. F.; McGowan, J. G.: Wind Energy Explained, 2nd Ed., John Wiley & Sons Ltd., 2011

Orbital 2, Presentation: "Variable Ratio Drivetrains for Direct Online Genration" by Dr. F. Cunliffe at the "2. IQPC International Conference Drive train Concepts for Wind Turbines" on 17 October 2011 in Bremen

Sørensen, J. D. ; Sørensen, J. N.: Wind energy systems, Woodhead Publishing Ltd, Oxford, 2011

Spera, D.A. (Ed.): Wind Turbine Technology, 2nd Ed., ASME Press, New York, 2009

Voith Turbo Wind GmbH & Co. KG, Presentation: "WinDrive – Large Wind Turbines without Frequency Converter" by Dr. A. Basteck at the World Wind Energy Conference 2009 in Korea

Winergy AG, Presentation: "Drive train Concepts for Wind Turbines" by M. Deike at the "2. IQPC International Conference Drive train Concepts for Wind Turbines" on 17 October 2011 in Bremen

Deutsches Patent- und Markenamt, Patent: DE 10 2007 012 408 A1, 2008

Europäisches Patentamt, Patent: EP 1 394 406 A2, 2004

Weltorganisation für geistiges Eigentum, Patent: WO 2010/121784 A1, 2010

7

Tower and Foundation

Torsten Faber

7.1 Introduction

Today's wind turbines soar to dizzying heights. A wind turbine (WT) with a 100 m tower and an overall height including the rotor blades of 150 m competes with the tower of Cologne Cathedral. The current largest wind turbines with more than 5 MW capacity can even reach hub heights of 140 m and soar to an overall height of 200 m, far above the Cologne tower.

These impressive large structures are designed for a life of 20 years. The plant manufacturers guarantee an availability of 98% including servicing. The full-load hours expected onshore is around 2000 hours per year, with a load change of 10^9. Offshore, on the high seas, almost twice as many full-load hours are expected.

The standard tower today is a steel tube and consists of individual tower sections that are fixed to each other by means of ring flanges with prestressed bolts (see Figure 7.1). The tower is connected to the foundation by up to 160 bolts.

For instance, with a 2 MW plant (tower height 80 m), the wind that acts on the rotor creates a horizontal force on the tower head of 1500 kN. This corresponds to the weight force of almost 100 cars, each of 1.5 tons ($100 \times 1500 \text{ kg} \times 9.81 \text{ m/s}^2 = 1472 \text{ kN}$). This force pulls the tower head in the direction of the wind, and with the tower as a lever at the base (foundation) this generates a moment to be resisted of 110,000 kNm.

The fatigue load that the tower is required to absorb is up to a billion load changes with bending amplitude of 15,000 kNm; in a simplified calculation this is called a damage equivalent of 1 level collective. This moment at the tower base is caused by a load compared with the weight force of 11 cars that pull the tower head backwards and forwards approximately 500 million times over 20 years lifetime of the plant (see Figure 7.2).

Understanding Wind Power Technology: Theory, Deployment and Optimisation, First Edition.
Edited by Alois Schaffarczyk. Translated by Gunther Roth.
© 2014 John Wiley & Sons, Ltd. Published 2014 by John Wiley & Sons, Ltd.

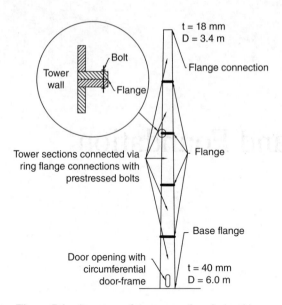

Figure 7.1 Structure of the tower of a wind turbine

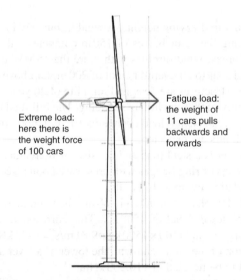

Figure 7.2 Loads on the tower of a wind turbine

In the production of a standard wind turbine, the tower represents a third of the overall costs. This is followed by the rotor blades with a proportion of approximately 20%. With regards to the costs, the foundation and bolts play a subsidiary role with only 2.5–5% and 1% respectively.

The WT is defined as a structure without bolts from the tower head flange downwards, according to the guidelines of the Deutschen Instituts für Bautechnik [1] [German Institute of Construction Technology]. The tower is thus the upper part to the base flange without anchor bolts. The tower head bolts belong to the machine mountings and are thus verified according to machine construction standards. The base anchor bolts are normally verified according to foundation statics.

7.2   Guidelines and Standards

The relevant standards for certification include the *Richtlinie für Windenergieanlagen* [*Guidelines for wind turbines*] of the Deutschen Institut für Bautechnik (DIBt) [1]. This guideline was created by the 'Wind energy plants' project group of engineers from research and industry. It is concerned with the proof of stability for the WT structure, the tower and the foundations, and the special effects on the same.

A further important standard is the *Richtlinie für die Zertifizierung von Windenergieanlagen* [*Guideline for the certification of wind turbine plants*] by the Germanischen Lloyd (GL) [2]. This guideline deals with the whole WT and is understood to be the common theme that recommends the Eurocodes (EC) and the DIN standards. International standards can be used when they are equivalent to the named standards. An important standard for the steel construction of a wind turbine is Eurocode 3 [3]. This covers subjects such as fatigue, material loading ability and crack propagation of steel structures.

A further important guideline to be mentioned here is the *Wind Turbine Generator Systems – Part 1: Safety Requirements* of the International Electrotechnical Commission (IEC) [4]. Calculations that are not described here can, for instance, be carried out according to the guidelines of the Germanischen Lloyd.

The guidelines of Det Norske Veritas (DNV) are an alternative to the GL guidelines. The DNV has been developing standards for carrier systems and components of wind turbines since 2001. Besides specialising in type certification of large megawatt turbines, the DNV, as well as the GL, each year offers numerous publications for offshore certification and practical recommendations.

7.3   Tower Loading

7.3.1   Fatigue Loads

Unlike normal structures, with a WT the own-weight portion in the calculation of stress is relatively small. In contrast, the bending moment part of a WT caused by the wind is the decisive stress portion. As a rule, the WT must withstand these dynamic fatigue loads for 20 years. In order to be certain that a WT will withstand these loads, exact dynamic simulation of the extreme and fatigue loads on the whole system is necessary. A fundamental condition for availability is that the plants and their components withstand the external loading that they are subjected to during the planned period of usage.

The section loads for calculation of individual components of a wind turbine, in practice, are carried out using computer-aided simulation in which both the stochastic load events of variable intensity as well as wind velocity and direction act on an idealised wind turbine structure (see Figure 7.3).

In a stochastic load result, the section sizes are determined for turbulent wind and possibly offshore irregular waves. In a deterministic load result, the sections are determined resulting from discrete gusts and discrete regular waves.

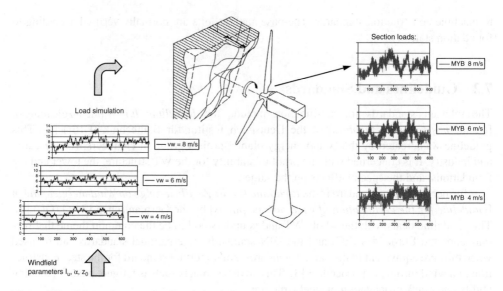

Figure 7.3   Schematic example of load simulation for an onshore WT. With kind permission of Springer Science+Business Media

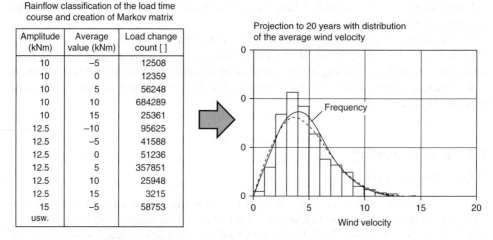

Figure 7.4   Classification and projection of the course of the loads over 20 years. Reproduced by permission of GL Renewables Certification

Because of the enormous amount of calculation, it is not possible to simulate the whole desired usage duration of 20 years. Instead, small time periods are considered that mirror the respective individual events and load cases that can occur (see Figure 7.4). These include start and stop processes, grid failure, gusts, slanted flow, and normal operation, as well as a combination of them. The load times series acquired from these time windows are summarised by means of static methods taking the probabilities of their individual occurrence into account, so that there is a time series for each viewed section size component for the desired usage duration. This is carried out for various (up to several hundred) dimension sections. The extreme and operational loads that are important for the static dimensioning of the various load cases

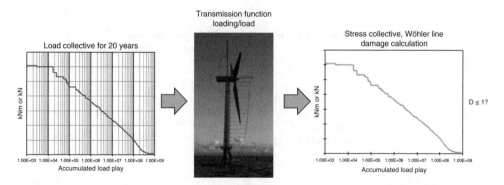

Figure 7.5 Length of life estimate. Reproduced by permission of GL Renewables Certification

can also be extracted from the generated data. For both the dimensioning as well as for the residual life, it must be remembered that the actual occurring loads can deviate substantially from the simulated loads, depending on the site.

As the further calculation of time series is only carried out in exceptional cases because of the great effort, the load fluctuation widths for the individual dimensioning sections are usually classified by means of a suitable counting method (rainflow count), which results in the so-called Markov matrices. Usually these can be dealt with more easily with sufficient information content.

For instance, for the fatigue calculation of steel structures with linear relations between external loads and internal section sizes and stresses, the Markov matrices can be reduced in a further step directly to fluctuation with collectives, with fluctuation play count allocated to each fluctuation width (see Figure 7.5).

Otherwise, a FE calculation must be made for structures such as the tower head flange in order to determine the non-linear transmission function, and must be taken into account in the evaluation. A further difficulty is when the material resistance is dependent on the average value, as is the case with reinforced concrete.

7.3.2 Extreme Loads

Besides the fatigue loads, a wind turbine must also resist extreme loads for short periods of time. In this the bending moment usually increases with the wind velocity. From a certain wind velocity, however, it can be reduced again as the pitch regulation rotates the rotor blades ever further from the wind. If the wind velocity becomes too great, the wind turbine switches off, the rotor blades are turned out of the wind and the bending moment sinks almost to zero.

The event of a 50-year wind gust is seldom relevant to the measurements. Much more critical is a combination of different extreme wind events. In this, extreme wind events are combined as regards their probability and the resulting loads are calculated. This calculation is very complex and can vary in result due to the wind fields simulated at the start.

The art of load calculations is primarily in the simulation of the actual loads to correspond to the real loads of a wind turbine. Much experience and understanding of the whole WT system is necessary for this.

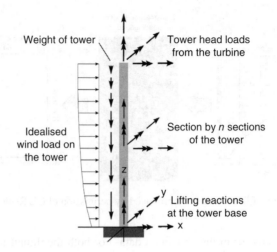

Figure 7.6   Loads on the tower of a wind turbine

7.4   Verification of the Structure

The proofs of concept of the construction technique mostly make use of the beam or shell theory. These theories do not consider the local design details or welds.

As described previously, the calculations of loads in the past ten years have led to increasingly more accurate methods. If ten years ago, in the calculation of internal forces at the tower, the calculation of loads on the tower head was state of the art, a few years later this was first interpolated linearly between the tower head and the base and today the internal forces are calculated individually at all relevant positions. As shown simplified in Figure 7.6, the loads acting on the tower are those from the machine, the inherent weight of the tower and the wind load on the tower.

7.4.1   Proof of Load Capacity

For the proof of the load capacity against material failure, the maximum extreme stress is statically determined. In this the greatest stress covers all loads. This side is called load side σ_{Fd}. Situated on the other side is the resistance side σ_{Rd}. Here the proof of the resistance capability of the material with consideration of the geometry (structure and section) is defined. The following must apply: $\sigma_{Fd} \ll \sigma_{Rd}$, meaning that, to exclude damage, the resistance side must be at least equal or greater than the loaded side.

In construction, the concept of proof with part factors of safety on the loading side and on the resistance side applies. On the resistance side the part factors of safety are dependent on the material γ_m. For fatigue loading the part factors of safety are further dependent on inspectability and damage consequences.

For the fatigue load, γ_m is in the region of 0.9–1.25, depending on good or bad accessibility and the consequences in case of damage. If the damage only leads to an interruption of the operation, then γ_m is in the range of 0.9–1.0; if it leads to damage to the wind turbine, then γ_m is higher. In the case of total damage to the wind turbine or danger to people, then γ_m is at its highest.

Figure 7.7 Stability failure of the tower in the region of the door opening. *Source*: AXA versicherung AG

Besides the material and cross-section, thickness also plays a large role in the proof of stability. The stress verification is only based on the law: stress = load and cross-sectional resistance. Thus, depending on the loading, the material is either compressed or pulled apart. Stability failure is understood to be buckling or kinking, such that the system deviates from the system level or axis.

For the determination of the tower in the border condition of load capacity, all the proofs of stress (fracture), the proof of stability against buckling and kinking and the proof of operational strength for fatigue must be carried out. In addition, places that need special investigation are, for instance, the flange of the tower and the door opening (see Figure 7.7).

The eccentric load input and the associated torsion resistance moment must be verified for the tower head flanges, usually by means of detailed FEM calculations, and local stress peaks must be investigated at the door opening. A special feature to be watched with lattice towers is the torsion softness and the means of connection of the individual lattices that are often produced as slide-resistant prestressed bolted connections. Very short clamping lengths must be prevented.

7.4.2 Proof of Fitness for Use

Besides the border condition of the load capacity, proof is also required of the fitness for use. Here the deformation is investigated to see whether the space between rotor blade and tower is sufficient, taking the imperfections and deformations into account. The partial factor of safety for the resistance side is then $\gamma_m = 1.0$.

7.4.3 Proof of Foundation

For the determination of a foundation, a distinction is made between the external and internal load capacity. In the border condition of the internal load capacity, the relevant stress verifications and proof against fatigue must be obtained. If the foundation is normal reinforced

Figure 7.8 Tipped-over foundation. *Source*: Beton Kontrollsysteme GmbH

concrete, then one quickly reaches the limits of the normal proofing method. In the calculation of the fatigue strength, the force flow determination with an analogue skeleton model in which the concrete can tear out, in the case of a tension and compression loading, is no longer permissible. With cyclic dynamically loaded piles, the tensile stresses are to be critically investigated as there is little experience in this field.

Besides the limiting condition of the internal load capacity, proof of the limiting condition of the external load capacity must also be brought. Included in this is the proof of tipping over (see Figure 7.8), ground breaking, sliding, lift and settling.

Applicable for all structural changes is that the effect on the stiffness of the WT must be checked because the plant conditions and possibly the controls and also the loading change.

7.4.4   Vibration Calculations (Eigen-Frequencies)

The eigen-frequencies for the whole structure as well as for each component must be determined. For instance, the Campbell diagram must be used to investigate whether the eigen-frequencies are outside the operating region of the excitation frequencies. Possibly dampers could be used here.

Eigen-frequencies play a large role in the tower of a wind turbine, because if a system is excited at its eigen-frequency it could lead to resonance. The tower shakes and this can lead to damage, up to and including complete failure of the tower.

The first and second eigen-frequencies of the tower are normally relevant, and with lattice masts also the torsion eigen-frequency. The exciting frequencies are caused by the revolutions of the blades of the WT, with their values depending on the number of rotor blades, normally 3P.

In order to prevent resonance, the excitation frequencies must be at least ±5% different from the eigen-frequencies of the tower. Operation in the neighbourhood of the resonance region is only permitted in exceptional cases with the use of operating vibration monitoring.

Figure 7.9 shows the Campbell diagram. This diagram is a method of preventing the overlapping of the stimulating frequency with the eigen-frequencies of the individual components. The x axis shows the speed of revolutions of the operating region in revolutions per minute. The y axis shows the eigen-frequency of the components in hertz. Here the eigen-frequency of

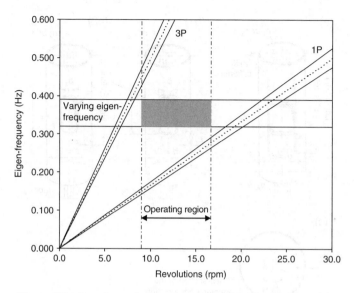

Figure 7.9   Rotation and eigen-frequency in the Campbell diagram

the component is shown with a deviation of ±5%. In order to compare the revolutions with the eigen-frequency, the frequency must be converted.

This is carried out by the equation: $y = 1/60 \times (1P)$. This graph is drawn into the diagram including two graphs with ±5% deviation in order to fulfil the requirements. As a total of three rotor blades stimulate the system, the same is carried out for three times the value, so that a graph with the following equation is given: $y = 3/60 \times (3P)$.

Now the region highlighted in grey and the two linear graphs (1P and 3P) must not intercept, as is the case here.

For towers with little stiffness, the first eigen-frequency lies below the 1P lines. These are mainly long slim towers. Towers with medium stiffness are the most common, so that the first eigen-frequency lies between the 1P and 3P lines. Very stiff towers are unusual; in the diagram they lie above the 3P lines.

In most cases the excitation frequency of the rotating rotor blades (3P) crosses the first eigen-frequency of the WT. The second eigen-frequency is usually higher than the 3P frequency.

Dynamic dampers are relevant for the dynamic behaviour and thus for the safety of the wind turbine. Their influence must be taken into account in the design of the tower. In contrast to the verification of the stress, there is no *safe side* for the dynamic proof.

7.5   Design Details

Relevant design details must be calculated separately by means of simplified methods, such as the finite element method (FEM). Relevant design details of a wind turbine are possibly the door frame, flange connections with eccentric prestressed bolts, as well as welded connections.

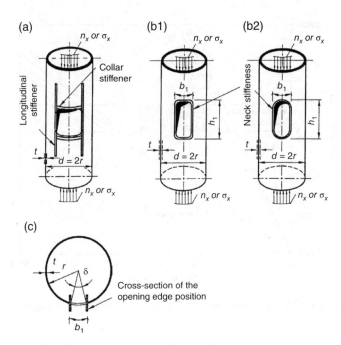

Figure 7.10   Examples of door openings [1]. Reproduced by permission of GL Renewables Certification

7.5.1   Door Openings in Steel Tube Towers

Openings in the tower shell lead to local load concentrations so that complex geometries of openings can only be calculated using FEM. The load concentration factor for nominal loading and an undisturbed tower shell has a value of 1, and for a round unstiffened opening a value of 3.

Figure 7.10 shows the various openings in the walls of steel tubular towers.

7.5.2   Ring Flange Connections

The standard flange is connected by means of eccentric prestressed bolts. The non-linear flow of force of the bolts requires special calculation methods. The proof against extreme loads can be carried out without consideration of the prestressing. The proof against fatigue of the bolts, however, requires complex calculations because of this non-linear flow of force.

Deviations of the ring flange lead to a significant increase in the bolt force and can therefore lead to damage. A perfect flange has no deviations; the maximum deviation may not exceed 3 mm. Deviations must be included in the FEM calculations. Examples of various points of error are shown in Figure 7.11.

7.5.3   Welded Connections

In welded connections the permissible stresses of the basic material are reduced. For fatigue loads, the nominal load is calculated in a simplified manner via detailed categories. The behaviour of welded seams under fatigue loads is poor due to the groove that always occurs in a welded seam. This means that welded connections are relegated to substandard groove case classifications.

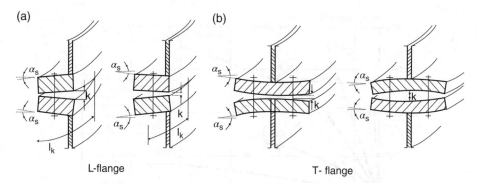

(a)   (b)

L-flange   T- flange

Figure 7.11   Poor flange connections [1]. Reproduced by permission of GL Renewables Certification

In the calculation of structural loads, the geometry of the connection but not the welding geometry must be considered (e.g. FE calculation).

7.6   Materials for Towers

Towers for WTs can be manufactured from various materials. The materials must be especially able to withstand fatigue loadings. In this they possess very different properties (see Figure 7.12 and Table 7.1). The most usual materials are mentioned below.

7.6.1   Steel

The steel tube tower is the standard tower structure and belongs to the most commonly built towers in the wind energy industry. Among others, this is because the fatigue behaviour of steel is homogeneous, isotropic and has been well researched. Steel has the highest degree of automation and prefabrication and is relatively economical in production. Erection on site is simple and quick. Fatigue capability is weakened by the welding of steel segments and flanges.

The tube tower consists of several tower sections that are connected with flanges by means of prestressed bolts at the site. The individual tower sections are welded together from individual flat steel plates. Then the steel plates are rolled into a tubular shape through three rollers (see Figure 7.13) and a certain eigen-stress is induced. The resulting radius of the tube walls is dependent on the pressure applied.

The flanges are normally formed out of a single steel block. The quality of the steel used depends upon the thickness and the requirements of use such as construction component category, steel strength classification, as well as onshore or offshore usage.

7.6.2   Concrete

Concrete towers, in contrast to steel towers, have no transport limitations in diameter and have fewer corrosion and stability problems. They are connected by means of site-prefabricated pieces normally with prestressing. Unstressed reinforced concrete towers are seldom erected with site concrete. The crack formation and fatigue behaviour of this type of structure is unfavorable.

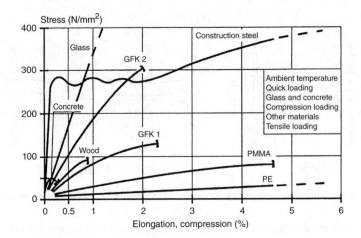

Figure 7.12 Stress-elongation lines of various materials to Wesche: construction materials for carrier components, Bauverlag 1996. Reproduced by permission of Bauverlag BV GmbH

Table 7.1 Modulus of elasticity and temperature elongation for various construction materials

Material	Modulus of elasticity E (MPa)	Temperature elongation coefficient α_T (1/K)
Glass	70,000	9×10^{-6}
Steel	21,0000	12×10^{-6}
Stainless steel	170,000	10–16×10^{-6}
Aluminium	70,000	24×10^{-6}
Concrete	20,000–40,000	10×10^{-6}
Timber parallel to fibres	10,000–17,000	4–6×10^{-6}
Timber vertical to fibres	200–1200	25–60×10^{-6}

7.6.3 Timber

Timber was the primary material in the history of wind energy. Timber was increasingly replaced by steel only at the start of the twentieth century, and now it hardly has a role to play as a material for wind energy. However, the Timber Tower Company has specialised in the construction of timber towers and makes use of the natural advantages of this material.

A timber tower uses a connection system of planked plywood plates. The individual components are assembled at the plant site from a timber body with a diameter of 7 m at the base and 2.4 m at the head. The tower is hexagonal, octagonal or dodecagonal.

Timber as a pure natural product is CO_2-neutral, and after the end of its life the wind turbine can be recycled without problems.

Among the technical advantages is transport in 40 ft containers without a heavy transporter being necessary. Up to now, hub heights of up to 200 m are planned.

The use of timber for the tower of a wind turbine has not yet been researched extensively. For this reason there is a lack of experience, and special attention is being paid to the means of connection.

Figure 7.13 Manufacture of a tower section: steel plates are rolled through three rollers to the inside. *Source*: Vestas

7.6.4 Glass Fibre-Reinforced Plastic

Glass fibre-reinforced plastic (GFK) is used as standard for rotor blades of WTs. There is no experience as yet in the use of GFK for the tower of a wind turbine. The material is very versatile with good carrying capacity properties. It possesses high strength with relatively low weight. The high costs and difficult production methods are a disadvantage.

7.7 Model Types

There are numerous tower types for WTs. They differ in cross-section, material and carrying concept. The most common forms of towers are discussed below.

7.7.1 Tubular Towers

The standard tower for wind turbines and thus the most widespread is the tubular tower. The greatest advantage is with the centro-symmetrical cross-section. Here the resistance to loading is the same in all directions. Furthermore, a circular cross-section is the stiffest cross-section for torsion.

Welding tower sections

Welding the flange to
the tower section

Production of a flange

Welding technique:
UV arc welding

Figure 7.14   Production of the flange and welding the tower section. *Source*: Vestas

The tubular tower lends itself to a high degree of prefabrication. Normally the tower consists of several steel tower sections that are connected with flanges by means of prestressed bolts at the site (Figure 7.14). The erection on site is simple and quick. A disadvantage is the limitation of the diameter due to transport under bridges (max. $D = 4.20$ m).

7.7.2   Lattice Masts

Lattice masts are not very widespread, yet their form of design offers a number of advantages. Here, too, a high degree of prefabrication is possible. There are no transport limitations due to size because of the light construction method and the final structure that is assembled from small parts. A disadvantage is the relatively large amount of work in erecting the structure on site because of the many components that must be assembled here. Added to this is the high degree of servicing work because of the large number of components and bolt connections. Also, the resistance of the tower as regards torsion moments can be critical.

7.7.3   Guyed Towers

The tower of the first large wind turbine (3 MW) in Germany at the start of the 1980s, termed GroWian (**Große Wi**ndenergie**an**lage) [large wind turbine], was a guyed tower. Unfortunately, because of numerous material problems and the lack of experience at the time, the plant could only be operated for few hours. However, important research results were acquired from it.

As for masts of sailing boats, the greatest advantage of a guyed tower is in the force relief through the shrouds that permit a smaller diameter of the tower cross-section in the lower region. With this, material can be saved and transport is made easier. However, a problem is the anchoring of the guy ropes in the ground, among others, because the exact analysis of the ground is often expensive and not always clear. In addition, at present the fatigue behaviour of many soils remains mostly unresearched.

7.8   Foundations for Onshore WTs

This chapter describes the three most common forms of foundations for wind turbines.

7.8.1   Force of Gravity

The gravity foundation is the standard solution for the foundation of a wind turbine. The concept is to absorb the moment of the wind turbine by means of a heavy, large-area foundation. There are round, hexagonal or octagonal foundations, but any centro-symmetrical plan area is possible.

7.8.2   Piles

A pure gravity foundation is only possible with good ground conditions. If the ground is too soft, then the foundation will sink. For this reason the gravity foundation in soft ground will be anchored with additional piles. The variant mostly used is the star form.

7.8.3   Cables

In a guyed system the loads are brought to the ground by means of prestressed cables. This occurs with the use of ground anchors that are fastened to the ground via piles. In this way the tower is relieved and manages with a smaller diameter as well as a thinner wall thickness. This saves material and thus costs. However, the soil conditions between foundation and ground anchors must be precisely investigated as the play of forces in the soil is very complex (see Figure 7.15). These soil investigations are very difficult. However, if they are not carried out, then it can lead to serious complications, including toppling of the WT in soft ground.

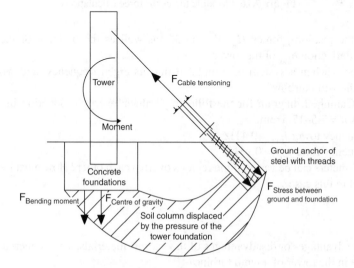

Figure 7.15   Flow of force: guyed tower

7.9   Exercises

Section 7.1

- How high are the highest wind turbines, including the upwards-pointing rotor blades?
- How great is the load to which the wind subjects the tower head?
- What percentage of the costs of a wind turbine is represented by the tower?

Section 7.2

- Name three important guidelines for wind energy plants

Section 7.4

The following wind turbine plant is given (Figure 7.16):

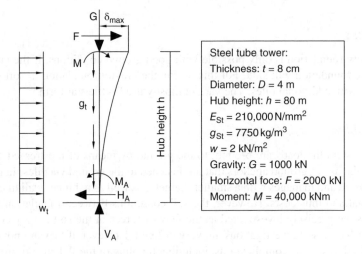

Figure 7.16   Module tower for force calculations

- Calculate the resulting forces H_A, V_A and M_A as well as the moment of inertia and the maximum deflection δ_{max} of the tower.
- What happens when a system is stimulated into its eigen-frequency, and what does this imply for the wind turbine?
- Draw the Campbell diagram for the following values: energy production in the range of revolutions $n = 7.5$–17.5 rpm.
- Eigen frequency tower $f_{ET1} = 0.4$ Hz.
- Eigen frequency rotor $f_{ET2} = 0.9$ Hz.
- Which parameters can be varied if the regions overlap each other? How must the parameters be changed in this case?

Section 7.6

What are the advantages or disadvantages of the various materials, steel, concrete, timber and GFK, for use in the tower of a wind turbine?

Section 7.7

How many structural models do you know? What are the advantages and disadvantages of the various models?

7.10   Solutions

Section 7.1: 200 m, 140 million Nm (same as the weight of 100 cars), 33%.

Section 7.2: Deutsches Institut für Bautechnik (DIBt)

- Germanischer Lloyd (GL)
- International Electrotechnical Commission (IEC)
- Det Norske Veritas (DNV) Abschnitt 7.4.

Section 7.4

$$A = \pi \frac{D^2 - (D - 2t)^2}{4} = 0.9852 \, \text{m}^2$$

$$g_{St} = \frac{7750 \, \text{kg/m}^3 \times 9.81 \, \text{m/s}^2}{1000} = 76.03 \, \text{kN/m}^3$$

$$g_t = g_{St} A = 74.90 \, \text{kN/m}$$

$$w_T = w \, D = 8 \, \text{kN/m}$$

$$H_A = w_T \, h + F = 2640 \, \text{kN}$$

$$V_A = G + g_t \, h = 6992 \, \text{kN}$$

$$M_A = w_T \frac{h^2}{2} + Fh + M = 225{,}600 \, \text{kNm}$$

$$y_1 = \frac{w_T h^4}{8EI} = 0.103 \, \text{m}$$

$$y_2 = \frac{Fh^3}{3EI} = 0.859 \, \text{m}$$

$$y_3 = \frac{Mh^2}{2EI} = 0.322 \, \text{m}$$

$$\delta_{max} = 1.28 \, \text{m}$$

$$I = \frac{\pi}{64} (D^4 - (D - 2t)^4)$$

$$M'' = G \, \delta_{max} = 1284 \, \text{kNm}$$

The system reacts with strong vibrations, i.e. it swings, and this leads to resonance. The excitation due to the rotation of the rotor blade must not overlap the region of the eigenfrequencies of the rotor blades and the tower.

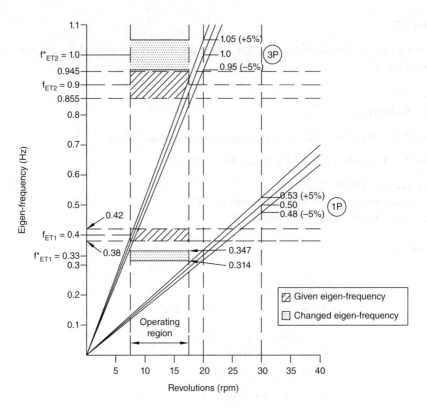

Figure 7.17 Campbell diagram solution

The eigen-frequencies of the components or the revolution region can be varied. The eigen-frequency of the rotor blade must be increased by means of harder materials, or stiffer cross-sections, or the speed of revolution must be reduced. The eigen-frequency of the tower must be reduced by means of softer materials, a stiffer cross-section, or the speed of revolution must be increased (see also Table 7.1).

Section 7.6:

Steel

Advantages:

- Higher grade of prefabrication
- Relatively cheap
- Erection at site simple and quick

Disadvantages:

- Limitation of diameter because of transport
- Welding of raw tower structure at construction site not possible

Concrete

Advantages:

- No limitation of size
- No corrosion or stability problems
- Smaller diameter possible

Disadvantages:

- Costs for site arrangements for site fabrication very high
- Otherwise, again, limitation of size of diameter

Timber

Advantages:

- Timber as a purely natural product is CO_2-neutral
- At the end of the life of a wind turbine it can be 100% recycled
- Transport in 40 ft containers without heavy transporters possible

Disadvantages:

- Little researched

GFK:

Advantages:

- High strength
- Low weight

Disadvantages:

- High costs
- Difficult production
- No experience

Section 7.7:

Tubular tower

Advantages:

- Centro-symmetrical cross-section
- Resistance to loading in all directions the same
- Circular cross-section the section with greatest torsional stiffness
- High grade of prefabrication
- Erection at site simple and quick

Disadvantages:

- Limitation of diameter due to transport

Lattice mast

Advantages:

- Because of the light construction method, assembled from small parts, there are no transport limitations due to size
- High grade of prefabrication

Disadvantages:

- Greater degree of maintenance work due to great number of bolted joints as well as components
- Great effort of construction at site due to the many components
- Also the elasticity of the tower can be critical with regards to the torque

Guyed tower

Advantages:

- Load relief allowing for smaller diameter of the tower
- Material saving and simplified transport

Disadvantages:

- Problems with anchoring the cables in the ground
- Exact analysis of the soil structure is difficult and expensive

References

[1] DIBt. 2004. *Richtlinie für Windenergieanlagen, Einwirkungen und Standsicherheitsnachweise für Turm und Gründung*, Schriften des Deutschen Instituts für Bautechnik, Reihe B, Heft 8, Edition March 2004. Deutsches Institut für Bautechnik (DIBt), Berlin.
[2] Germanischer Lloyd. 2010. *Richtlinie für die Zertifizierung von Windenergieanlagen.*
[3] Eurocode 3: Bemessung und Konstruktion von Stahlbauten; German edition EN 1993-11: 2005.
[4] International Electrotechnical Commission (IEC). 2005. *Wind Turbine Generator Systems. Part 1: Safety Requirements, 3*. IEC.

8

Power Electronics and Generator Systems for Wind Turbines

Friedrich W. Fuchs

8.1 Introduction

Mechanical power that is acquired from the wind with the aid of the rotor of the wind turbine is transformed into electrical energy by means of the generator and fed into the electric grid. For this purpose, various suitable concepts are available [1,2]. Often a gear drive is interposed for converting the torque in order to be able to select the speed of revolution of the rotor and generator to the optimum, independently of the wind speed. On the electrical side, the generator is to be designed for the rated frequency and voltage of the grid by means of a transformer.

The power acquired from the rotor of the wind turbine and transferred to the shaft of the generator depends upon the velocity of the wind and the speed of rotation of the rotor, as shown in Figure 8.1. The rising curves for slow revolutions and the falling curves for high revolutions show a clear maximum. The curves are valid for constant pitch or blade setting angles.

When operating a three-phase AC generator directly connected to the grid, the rotor feeds directly into the grid with a fixed frequency and with a fixed speed of revolution proportional to this grid frequency. In the case of the synchronous machine, the speed of revolution is fixed exactly; in the case of the asynchronous machine there are small deviations depending upon the load.

An extension of the concept of the generator with a fixed speed of revolution is called the 'Danish concept' and was widespread in the 1980s [3]. In this, an additional second generator or a second set of coils in the first generator is used to achieve a second speed of revolution by means of switchover.

Understanding Wind Power Technology: Theory, Deployment and Optimisation, First Edition.
Edited by Alois Schaffarczyk. Translated by Gunther Roth.
© 2014 John Wiley & Sons, Ltd. Published 2014 by John Wiley & Sons, Ltd.

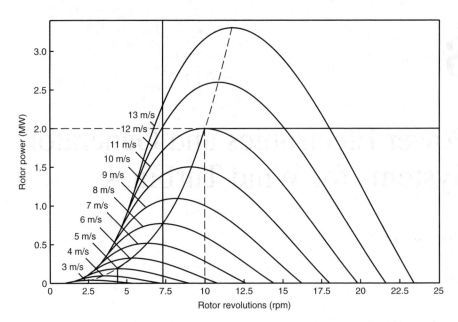

Figure 8.1   Power-revolution diagram of a rotor arrangement of a wind turbine and characteristic curve of a generator at fixed (vertical line at about 7.5 m/s) and at variable optimum frequencies (starting at about 4 m/s); example of a 2 MW plant; parameterising wind velocity, pitch angle constant; from 11 m/s power constant at 2 MW

Because of the non-optimum energy yield of directly coupled asynchronous or synchronous generators, due to the high loads at switchover in the drive train and the difficulty in realising the power requirements in the grid input, this concept is hardly used today [3]. The exception is in plants of the lowest power. This non-optimum energy is shown by the vertical line in the power-revolution diagram (Figure 8.1). With fixed speeds of revolution, the optimum power in the maximum power curve can only be attained by wind velocities of 7–8 m/s.

As a result, today the concept of a variable speed generator and rotor is state of the art. The speed of revolution is controlled depending on the wind so that the optimum power is gathered from the wind. In the power-revolution diagram (Figure 8.1) this is shown by the line that cuts the maximum of the power curves. It is intended to keep the revolutions always at the value where the respective wind velocity gives the maximum power.

Therefore, the rotor and thus the generator revolutions must be varied depending on the wind velocity. In this, the pitch angle (angle of the rotor blades turned out of the rotation area) must be kept constant as long as the rated power is not achieved. This region below the rated power is called the power-optimising region. Only if the wind is so strong that the rated power is exceeded is the rotor blade taken out of the wind by changing the pitch angle (power-limiting region, vertical line in Figure 8.1 at 2 MW).

The ability to control the speed of revolution is achieved when a frequency convertor is placed between the generator and the grid. On the grid side it is operated with fixed amplitude and frequency of the grid and, on the generator side, can generate voltage of controlled amplitude and frequency as is required for operation with variable speed of revolution.

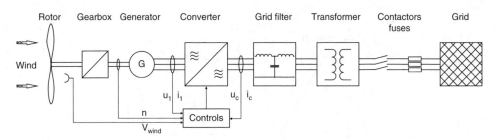

Figure 8.2 Structure and components of the drive train and the grid feed-in of a wind turbine with a fully-rated converter

The structure of the components of the drive train as well as the grid feed-in of a wind turbine with convertor-fed generator is shown in Figure 8.2. This shows a plant equipped with a fully rated converter. The converter converts the full power of the generator.

The rotor of the wind turbine converts the power of the wind into mechanical power. A gear drive is often used so that the low speed of revolution of the rotor can be converted to faster revolutions of the generator. The generator converts the mechanical power into electrical power. A frequency converter converts the frequency of the variable speed generator into the fixed grid frequency and feeds this power via transformer, grid filter, contactors and fuses into the grid. Besides the plants with full power converters that are used for asynchronous or induction generators with short-circuit rotors and synchronous machines, plants with partly rated converters are also in use in which the stator of the generator is directly connected to the grid and the rotor is connected to the grid by means of a converter. The induction machine with slip-ring rotor is used for this, the system being called a doubly-fed induction generator.

The electrical components of the variable-speed wind turbine are discussed in this chapter. These are the generator, the frequency converter, the controls of the generator and converter, the transformer, the grid filter and other electrical components. In addition, various power-electronic and generator concepts will be described for wind turbines according to the state of the art.

8.2   Single-Phase AC Voltage and Three-Phase AC Voltage Systems

As an introduction, some basics of electrical energy engineering will be presented in order to improve understanding of the following sections for those not well versed in the subject.

Stationary conditions and sine- and cosine-shaped courses of the electrical quantities are assumed.

Electrical power is transported by means of AC quantities, converted and distributed. Power is generated by means of electrical current and voltage. These quantities, for instance, the voltage written as $U(t)$ or $u(t)$, are here characterised by a fixed amplitude $\hat{U}$, and change cosinusoidally with time t proportionally to the angular frequency $\omega = 2\pi f$ of the voltage.

$$u(t) = \sqrt{2}\tilde{U} \times \cos(\omega t + \phi_U) = \hat{U} \times \cos(\omega t + \phi_U) \qquad (8.1)$$

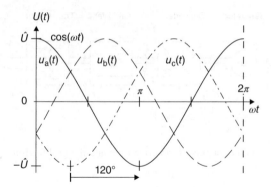

Figure 8.3 Variation in time of the voltages of a cosinusoidally symmetrical three-phase AC system with $\varphi_U = 0°$

where $\hat{U}$ is the peak value, $\tilde{U}$ is the $\hat{U}$, root mean square value, ω is the angular frequency value, and φ_U is the phase angle value, always of the voltage.

For the usual case, a phase displacement ϕ_U of the voltage other than zero is assumed. For higher generated power, in Germany above 3.7 kW (16 A) in 230/400 V grids, a three-phase AC grid is used for energy transportation, whereby in general also a neutral conductor is available. Wind turbines of this higher power are connected to such three-phase grids. This kind of network possesses three-phase voltages and these are sine-shaped and symmetrical. In this case the voltages all possess the same amplitude $\hat{U} = \sqrt{2}\tilde{U}$, and the shape is sinusoidal or cosinusoidal. They have the same angular frequency ω and a phase displacement of 120° between the phases.

The equations for the three phases, called R, S and T, or a, b and c, or L1, L2 and L3, describe these types of voltages here as star voltages. Figure 8.3 shows the variation in time of the voltages.

$$u_a(t) = \sqrt{2}\tilde{U}_{star} \cos(\omega t)$$

$$u_b(t) = \sqrt{2}\tilde{U}_{star} \cos\left(\omega t - \frac{2\pi}{3}\right) \qquad (8.2)$$

$$u_c(t) = \sqrt{2}\tilde{U}_{star} \cos\left(\omega t - \frac{4\pi}{3}\right)$$

If these grids with their generators and loads really are symmetrical, then the calculation in one phase is sufficient, as then the values of the other phases will be known. These are the same as for the first phase with a dedicated phase shift of 120° or 240°.

The performance of these single-phase and three-phase grids can be calculated with the aid of the complex AC calculation, also called the phasor calculation. The complex AC quantities or phasors are denoted by underlining, as $\underline{U}$:

$$\underline{U} = \frac{1}{\sqrt{2}}\hat{U}e^{j\phi_U} = \frac{1}{\sqrt{2}}\hat{U}(\cos\phi_U + j\sin\phi_U) \qquad (8.3)$$

Figure 8.4 Selected positions of the axes of the complex quantities

The complex values can be depicted in diagrams in the complex area. Here the diagrams are arranged such that the real axis is always upwards and the imaginary points to the left (suitable for the energy technology, see Figure 8.4). In addition, in this chapter the load phasor system is used in general, so in the case of consumption of power there is always a positive active power.

The variation in time can be determined with the knowledge of the angular frequency ω from the complex value:

$$
\begin{aligned}
U(t) = U_{star}(t) &= \mathrm{Re}\left\{\sqrt{2}\underline{U}\times e^{j\omega t}\right\} \\
&= \mathrm{Re}\left\{\sqrt{2}\,\frac{1}{\sqrt{2}}\hat{U}e^{j\phi_U}\times e^{j\omega t}\right\} \\
&= \mathrm{Re}\left\{\hat{U}\left(\cos(\omega t+\phi_U)+j\sin(\omega t+\phi_U)\right)\right\} \\
&= \hat{U}\cos(\omega t+\phi_U)
\end{aligned}
\tag{8.4}
$$

The power and its term are important for the characterisation of the wind turbines. The power in such three-phase symmetrical systems is calculated for the active power P that can affect work, the reactive power Q and the virtual power S according to the following equations. In this the star voltage is used and it is assumed that both voltage and current show a phase displacement (φ_U and φ_I). Effective F (root mean square) values of voltage and current are also shown in the following by the characters U and I without further indication.

$$
\underline{U}_{star} = U_{star}\,e^{j\phi_U}
\tag{8.5a}
$$

$$
\underline{I} = Ie^{j\phi_i} = I_{\mathrm{Re}} + jI_{\mathrm{Im}}
\tag{8.5b}
$$

$$
\underline{I}^* = Ie^{-j\phi_i} = I_{\mathrm{Re}} - jI_{\mathrm{Im}}
\tag{8.5c}
$$

$$
P = 3U_{star}I\cos(\phi_U - \phi_I) = 3\,\mathrm{Re}\left\{\underline{U}_{star}\underline{I}^*\right\}
\tag{8.5d}
$$

$$
Q = 3U_{star}I\sin(\phi_U - \phi_I) = 3\,\mathrm{Im}\left\{\underline{U}_{star}\underline{I}^*\right\}
\tag{8.5e}
$$

$$
\underline{S} = P + jQ = 3\underline{U}_{star}\underline{I}^*
\tag{8.5f}
$$

$$
S = \left|\underline{S}\right| = 3U_{star}I
\tag{8.5g}
$$

For calculating the power, the conjugated complex value $\underline{I}^*$ of the current $\underline{I}$ is required.

8.3   Transformer

The purpose of the transformer is to transform voltages to higher or lower amplitudes and/or to carry out a separation of potentials on the primary and secondary side. In the following introduction of the basics, the following is assumed:

- saturation is neglected;
- constant ohmic resistance; and
- symmetrical structure of three-phase transformers.

8.3.1   Principle and Calculations

In its simplest case, the transformer consists of a magnetic conducting iron core, on which two coils are installed, the primary coil 1 and the secondary coil 2, as shown in Figure 8.5. In order to simplify the picture, each coil is shown with only two turns on both sides. Primary and secondary coils are well coupled together by means of the main flux in the iron core (Φ_{12} by means of coil 1 into coil 2, and correspondingly Φ_{21} by means of coil 2 into coil 1). Low parts of the total flux do not reach the opposite coil and they are designated as leakage flux ($\Phi_{1\sigma}$, $\Phi_{2\sigma}$).

Both coils are flown through the same main flux. The numbers of turns of coil windings are w_1 on the primary side and w_2 on the secondary side. This results in primary and secondary interval voltages U_{1i} and U_{2i} generated by the common primary and secondary main flux ($\Phi_{1h} = \Phi'_{2h}$):

$$U_{1i}(t) = w_1\left(\frac{d\phi_{1h}}{dt}\right); \quad U_{2i}(t) = w_2\left(\frac{d\phi_{2h}}{dt}\right); \quad \phi_{1h} = \phi_{2h} \tag{8.6}$$

It is recognised that the voltage transformation of the internal voltage from the secondary side to the primary side can be set by ratio of the number of turns of the windings w_1/w_2:

$$\frac{\tilde{U}_{1i}}{\tilde{U}_{2i}} = \frac{w_1}{w_2} \tag{8.7}$$

The voltages can be written as the product of the inductivity and current:

$$U_1:(t) = w_1\left(\frac{d\phi_{1h}}{dt}\right) = L_{1h}\left(\frac{di_1}{dt}\right) \quad \text{complex:} \quad j\omega L_{1h}\underline{I}_1 = U_{1i} \tag{8.8a}$$

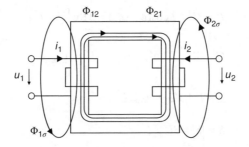

Figure 8.5   Single phase transformer, structure and fluxes

$$U_{2i}(+) = w_2 \left(\frac{d\phi_{2h}}{dt} \right) = L_{2h} \left(\frac{di_2}{dt} \right) \quad \text{complex:} \ j\omega L_{2h} \underline{I}_2 = \underline{U}_{2i} \qquad (8.8b)$$

The voltages are brought into complex notation. In order to complete the occurring effects, the voltage drop from the primary and the secondary ohmic resistances (R_1, R_2) and the voltage drop in the magnetic leakage path, in the associated leakage inductivities ($L_{1\sigma}$ and $L_{2\sigma}$), are additionally included. This results in the voltage equations in the standard notation as complex equations:

$$\underline{U}_1 = R_1 \underline{I}_1 + j\omega L_{1\sigma} \underline{I}_1 + j\omega L_{1h} \left(\underline{I}_1 + \underline{I}'_2 \right) \qquad (8.9a)$$

$$\underline{U}'_2 = R'_2 \underline{I}'_2 + j\omega L'_{2\sigma} \underline{I}'_2 + j\omega L_{1h} \left(\underline{I}_1 + \underline{I}'_2 \right) \qquad (8.9b)$$

The secondary voltage and the secondary current have additionally been multiplied with the turn number ratio $\ddot{u} = w_1/w_2$ or its inverse value on the primary side, which is characterised by a line behind and over the letter ($\underline{I}'$). The result of this is that the last terms of the two equations are the same. In this the following relationships and definitions apply:

$I'_2 = \dfrac{1}{\ddot{u}} I_2$ Secondary current referenced to the primary side

$U'_2 = \ddot{u} \times U_2$ Secondary voltage referenced to the primary side

$\underline{I}_\mu = \underline{I}_1 + \underline{I}'_2$ Magnetising current

$L_{1\sigma}$ Primary leakage inductance

$L_{1h} = L'_{2h} = \ddot{u}^2 L_{2h}$ Primary main inductance equal to secondary inductance referenced to the primary side

$R'_2 = \ddot{u}^2 R_2$ Secondary resistance referenced to the primary side

$L'_{2\sigma} = \ddot{u}^2 L_{2\sigma}$ Secondary leakage inductivity referenced to the primary side

R_1 Primary side resistance

8.3.2   Equivalent Circuit Diagram, Phasor Diagram

The result of the referencing of the secondary values to the primary side is that there are equally large induced voltages in the equations for primary and secondary sides. For this reason it is now possible to make one common equivalent circuit diagram from the primary and secondary sides, both sides connected, for application in the energy field, as shown in Figure 8.6. This permits an analysis of both voltages, which in general have very different dimensions, via one equivalent circuit diagram. This equivalent circuit diagram does not contain the true potential separation between the primary and secondary sides. This can be inserted by means of adding an additional ideal transformer in series that only covers the voltage conversion ($w_1 = w_1$, $R=0$, $L_\sigma=0$).

Besides the ohmic losses in the primary and secondary circuits, there are also losses in the iron, which have to be included for proper characterisation. These can be well represented approximately by means of a parallel resistance R_{fe} to the main inductance.

From the voltage equations it is also possible to draw a common phasor diagram of the primary and secondary sides of the transformer in the complex plane. An example is shown in Figure 8.7. It should be noted that in this diagram the ohmic and leakage voltages are shown larger than in real transformers in order to make them clearer.

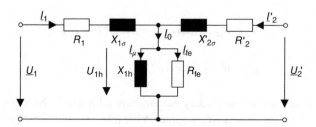

Figure 8.6 Single-phase transformer; complete equivalent circuit diagram for application in the energy field

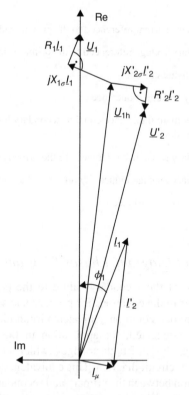

Figure 8.7 Single-phase transformer; phasor diagram, R_{fe} neglected, ohmic and leakage voltage drops greatly enlarged

8.3.3   Simplified Equivalent Circuit Diagram

For simplified applications, especially for calculating the short-circuit behaviour of the transformer or when regarding the transformer as a component in the grid, a simplified equivalent circuit diagram is used that is also called a short-circuit equivalent circuit diagram. In this, use is made of the fact that the shunt impedances are much greater than the longitudinal impedances and the shunt impedances are ignored:

$$R_1, R'_2, X_{1\sigma}, X'_{2\sigma} \ll R_{Fe}, X_{1h} \tag{8.10}$$

The result is the equivalent circuit diagram as shown in Figure 8.8. This is a simple series R-X network. The following applies:

$$\underline{I}_1 = -\underline{I}'_2 = \underline{I} \tag{8.11a}$$

$$\underline{Z}_{sc} = R_{sc} + jX_{sc}; \quad R_{sc} = R_1 + R'_2; \quad X_{sc} = X_{1\sigma} + X'_{2\sigma} \tag{8.11b}$$

The short circuit voltage U_{sc} is used for characterising the transformer. The transformer in a test is short-circuited secondarily. The primary voltage is increased until the primary rated current flows. The simplified equivalent circuit diagram is used for the determination. The resulting rated primary voltage for secondary short-circuit at nominal current is determined according to:

$$\underline{U}_{1SCN} = \underline{Z}_{sc}\underline{I}_{1N} \tag{8.12}$$

This voltage represents the voltage drop at the transformer impedances at rated current. It is much smaller than the primary rated voltage. If one takes the rated short-circuit voltage in comparison to the rated voltage then, independently of the voltage class of the transformer, it

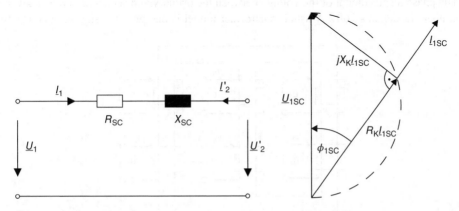

Figure 8.8   Simplified equivalent circuit diagram and associated phasor diagram of the single-phase transformer ($I_1 = I_{1SC}$, $U_1 = U_{1SC}$)

results in a relative dimension that describes the longitudinal impedances. This is called the relative short circuit voltage. The information is usually given in percentages:

$$u_{SC} = \frac{U_{1SCN}}{U_{1N}} \times 100\% \tag{8.13a}$$

$$\underline{u}_{SC} = \frac{U_{1SCN}}{U_{1N}} = \frac{R_{SC}I_{1N}}{U_{1N}} + j\frac{X_{SC}I_{1N}}{U_{1N}} = u1_{SCR} + ju_{1SCX} \tag{8.13b}$$

The value for short-circuit voltage of power transformers is between approximately 4–12%, depending on power and structure. Certain values are power-dependent economically; higher or lower values can be achieved by means of suitable development and production, but generally result in higher costs.

8.3.4 Three-Phase Transformers

Three-phase transformers are used for the transformation of three-phase electrical systems. The use of three single-phase transformers would be possible, but is uneconomical and has some technical disadvantages. One combines the three individual transformers into a single transformer. A typical structure of three-phase transformers as used in wind turbines is shown in Figure 8.9.

The core consists of a rectangular basic shape with two rectangular cut-outs. The vertical parts are called legs, the horizontal ones yokes. The primary and secondary coils for each phase of a three-phase system are wound about each of the three legs. Depending on the model, these coils are both primary and secondary, for instance, switched in the star or delta form (star: connected to an end part, other ends provide phase connection; delta: the start of one coil and end of the next are connected and at the same phase tap-off). Due to the propensity for uneven loading and the need to extinguish certain harmonics, a combination of star and delta connections is normal.

The designation of the coils is shown by the abbreviation D for the delta connection, Y for the star connection, Z for the zig-zag connection, whereby an upper-case letter is the primary and lower-case is the secondary switching.

The phase displacement of the voltage between the primary and secondary coils is given as a number: 12 denotes 360°. A usual transformer model is one with the star-delta connection

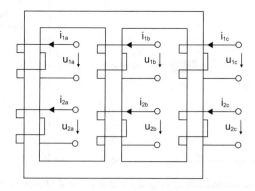

Figure 8.9   Three-phase transformer, core structure form; build-up of coils, exemplary representation

Table 8.1   Switching forms of three-phase transformers

Winding connection	Conversion	Characteristic number (angle)
Yy0	w_1/w_2	0
Dy5	$w_1/\sqrt{3}w_2$	5
Yd5	$\sqrt{3}w_1/w_2$	5
Yz5	$2w_1/\sqrt{3}w_2$	5

with a neutral conductor that shows a phase displacement of primary to secondary of 150°. This has the abbreviation Dyn5. The designation n shows that this is a loadable neutral conductor on the secondary side. Typical switch forms of three-phase tranformers are presented in Table 8.1.

The different types of connections have different features [4] with regard to unsymmetrical loading, which, however, should seldom occur with wind turbines with three-phase circuits. Star-star circuits with secondary-side neutral conductors (Yyn) without equalisation coils in the normal models (three legs, jacket, five legs) are suitable only for small single-phase loading with feeds of up to 10% of the rated power. The other circuit types are fully suitable for single-phase loading or in-feeds. The circuit types Dyn5 and Yzn5 are often used.

Additionally with regards to the transfer or extinguishing of harmonics, there are characteristic differences between the types of circuits. As in this case we are dealing with the extinguishing of low-frequency harmonics (fifth, seventh, etc.) [5], it is not of particular importance for wind turbines with sine wave converters.

Transformers, which for wind turbines of higher power are directly connected to the power converter, are designated as power converter transformers. Because of the operation at the power converter, these transformers are loaded by current and voltage harmonics. At higher frequencies these lead to increased losses in the iron core due to hysteresis and eddy current losses, as well as eddy current losses in the copper. The transformers must be designed for this type of operation.

Three-phase transformers are normally relatively symmetrical in the individual phases. For this reason it is permissible in normal cases to reduce the transformer electrically for calculation to a single-phase equivalent circuit diagram and to view it in the single-phase depiction. The voltage equations are given here again.

$$\underline{U}_1 = R_1\underline{I}_1 + j\omega L_{1\sigma}\underline{I}_1 + j\omega L_{1h}(\underline{I}_1 + \underline{I}_2')$$ (8.14a)

$$\underline{U}_2' = R_2'\underline{I}_2' + j\omega L_{2\sigma}'\underline{I}_2' + j\omega L_{1h}(\underline{I}_1 + \underline{I}_2')$$ (8.14b)

8.4   Generators for Wind Turbines

The task of generators for wind turbines is to convert the mechanical power into electrical power. For the purposes of optimising the energy yield in current installations, they are built as variable-speed systems and therefore feed into the electric grid via frequency converters. As well as the function of revolution speed control, the frequency converter also has the function of enhanced control and regulation for the generator as well as for grid feeding.

Various types of generators are used with current new installations of wind turbines [4–7]:

- doubly-fed induction machine, converter at the rotor;
- permanently excited synchronous machine, converter at the stator;
- separately excited synchronous machine, converter at the stator;
- induction machine with short-circuit rotor, converter at the stator.

The various types of generator are introduced in this section.

The conditions for the following analyses of the generator are:

- stationary operation;
- neglecting saturation;
- symmetrical structure;
- constant ohmic resistances.

8.4.1   Induction Machine with Short-Circuit Rotor

The induction machine with a short-circuit rotor is the most widespread machine in the industry. Its power ranges from a few watts for small drives up to the 10 MW range, for instance boiler feed pumps. It is very robust and is sometimes found also in the wind industry. Three-phase induction machines are usually built for capacities from a few kW to about 30 MW. In wind turbines, a connection voltage of 690 V in the low MW range and several kV in the region of 5 MW and above are normal.

8.4.1.1   Structure

The induction machine as shown in Figure 8.10 consists of a stator with a cylindrical hole in which rotates a rotor with a cylindrical volume. Between the stator and the rotor there is an air gap with a size in the region of millimetres to centimetres. The stator has slots, into which are placed current-conducting coils with their windings. These are machines with three-phase symmetrical feeds, i.e. sine wave voltages of the same amplitude with a phase displacement of 120°. The coil systems are always arranged per phase mechanically about 120° per pair of poles. In the circumference there are supply and return conductors ($U - U'$, $V - V'$, $W - W'$). In real machines there are several slots per phase and in each **groove** generally several coils. Only one slot is shown in the picture for the sake of clarity. This arrangement ($U - U'$, $V - V'$, $W - W'$) can be reproduced several times on the circumference of the machine for machines with higher pole numbers. The magnetic poles, incoming and outgoing magnetic fluxes then occur several times around the circumference. This corresponds then to the pole pair count being greater than 1 (polar pair count 2; same coil arrangement twice at the circumference, each individual in a 180° region).

The rotor, as the stator, can be equipped with a three-phase coil as shown in Figure 8.10. In the induction machine with short-circuit rotor, the rotor coil is produced as a short-circuit coil that is shorted in the rotor. In the case of a short-circuit the coil needs not be produced as a three-phase coil. Equally distributed rods that are shorted at the end by means of short-circuit rings are normally positioned around the rotor circumference. This is called the rotor cage.

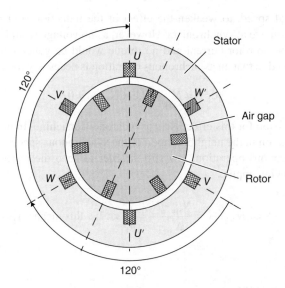

Figure 8.10 Structure of a two-pole induction machine with three-phase coils in stator and rotor; pole pair count $p=1$, number of slots $N=6$ in rotor and stator

8.4.1.2 Basic Function

Assume a machine with three-phase coils in the stator and rotor. The rotor coil is short-circuited. The stator is fed with a sine wave symmetrical three-phase system of frequency f_1. Thus, the machine has a circumferential moving magnetic field. The field rotates with the synchronous speed of revolution, N_{Syn}, per period of stator frequency about a polar pair. This means that for a pole pair 1 the whole circumference per period duration of the stator frequency, and for a polar pair count p only the p-part of the circumference:

$$N_{\text{Syn}} = \frac{f_1}{p} \tag{8.15}$$

The stator has a fixed position; the rotor rotates with a speed of revolution N that is not equal to the synchronous speed of revolution and thus not equal to the circumferential velocity of the stator field. Because of the field that is rotating in regard to the rotor, a voltage is induced in this rotor and fluxes are formed in the short-circuit rotor. This voltage is proportional to the rotor frequency. The rotor frequency f_2 is proportional to the difference between the velocity of the stator field N_{Syn} and the rotor speed of revolution N:

$$f_2 = p(N_{\text{Syn}} - N) = s f_1 \tag{8.16}$$

The rotor frequency is also given as the product of the stator frequency f_1 and slip s. The slip indicates the relationship between the velocities of the rotor and stator field of revolution.

Because of the rotating magnetic field that rotates with the synchronous velocity, and the current induced in the rotor, forces are created on the rotor coils and thus a torque. This drives the rotor in case of a motor operation in the direction of the revolving field. It attempts to lessen the differential rotational speed, to weaken the effect of the induction (Lenz's rule). Thus, the machine attempts to arrive at synchronism. However, as no voltage would be induced into the rotor, there would be no rotor current and no torque would be generated. The behaviour is therefore characterised in that an asynchronous operation is necessary for generating a torque:

$$\text{for } N \neq N_{\text{Syn}} \quad \text{there is} \quad M \neq 0$$

Thus, another expression for this effect is an asynchronous machine. In practical operation, a rotor speed of revolution in the neighbourhood of the synchronous speed of revolution is necessary, thus asynchronous operation, low slip in order to keep the losses small, and which means that the following applies:

$$N_{\text{Syn}} - N \ll N_{\text{Syn}}; \quad s = \frac{N_{\text{Syn}} - N}{N_{\text{Syn}}} = \frac{f_1}{f_2} \ll 1; \quad \text{this means: } f_2 \ll f_1 \tag{8.17}$$

8.4.1.3   Voltage Equations

The equations for the voltage of the rotor quantities machine are based on those for a stationary stator and for a rotating rotor on that representation, as they are measured at the rotor, shown and converted. Then the rotor equation is referenced to the stator (by multiplication with w_1/w_2, shown by a line above/behind the character) and divided by the slip $s = f_2/f_1$. After the conversion the result is the equation for the stator voltage $\underline{U}_1$ and the rotor voltage $\underline{U}'_2$ in the normally used form, and in complex notation as:

$$\underline{U}_1 = R_1 \underline{I}_1 + j\omega_1 L_{1\sigma} \underline{I}_1 + j\omega_1 L_{1h}(\underline{I}_1 + \underline{I}'_2) \tag{8.18a}$$

$$\frac{\underline{U}'_2}{\dfrac{f_2}{f_1}} = \frac{f_1}{f_2} R'_2 \underline{I}'_2 + j\omega_1 L'_{2\sigma} \underline{I}'_2 + j\omega_1 L'_{2h}(\underline{I}_1 + \underline{I}'_2) \tag{8.18b}$$

$$L_{1h} = L'_{2h} = \ddot{u}^2 L_{2h} \tag{8.18c}$$

where R_1 is the stator resistance, R_2 is the rotor resistance, $L_{1\sigma}$ and $L_{2\sigma}$ are the stator and rotor leakage inductivity, and L_{1h} is the stator-referenced main inductivity.

The stator voltage equation corresponds exactly to that for the primary side of the transformer and the relationship correspondingly. The equation of the rotor is very similar to that for the secondary side of the transformer characteristic for the feature of torque generation in the induction machine. In both equations, induced voltage portions appear at main and leakage inductances as well as ohmic voltage drop. Noticeably, it is the frequency-variable rotor-side resistance $(f_1/f_2)R'_2$ in the rotor equation that is one difference in this equation to that of the transformer. The other difference is the special expression of $(f_1/f_2)U'_2$ that is.

The derivation of the equations is explained briefly. The original rotor voltage is converted for the above depiction with the coils relationship w_1/w_2 to the stator as 'referenced to the stator' and shown with an apostrophe. The conversion to the stator occurs with the following equation:

$$\underline{U}_2' = \frac{w_1}{w_2}\underline{U}_2; \quad \underline{I}_2' = \frac{w_2}{w_1}\underline{I}_2; \quad R_2' = \left(\frac{w_1}{w_2}\right)^2 R_2; \quad L_{2\sigma}' = \left(\frac{w_1}{w_2}\right)^2 L_{2\sigma} \tag{8.19}$$

To bring the subsequent resulting internal voltage of the stator and rotor:

$$\underline{U}_{i1} = j\omega_1 L_{1h}(\underline{I}_1 + \underline{I}_2'); \quad \underline{U}_{i2} = j\omega_2 L_{2h}(\underline{I}_1 + \underline{I}_2') \tag{8.20}$$

to the same value, it has additionally been converted with the frequency ratio f_1/f_2.

8.4.1.4 Equivalent Circuit Diagram

The equivalent circuit diagram of the induction machine can be created with the help of these equations (Figure 8.11). It corresponds to the structure of the transformer, whereby it deviates in the value of the variable rotor-side resistance $(f_1/f_2)\,R_2'$ and the rotor voltage. As an example of a short-circuit rotor, here the rotor voltage is set to zero; however, there are also machine types without a short-circuited rotor. In order to take the iron losses into account, its equivalent resistance R_{Fe} could be inserted, as in the transformer, parallel to the main inductance.

8.4.1.5 Phasor Diagram

From the equations of the induction machine in the complex notation, it is possible to create a phasor diagram of the induction machine for the voltage and currents in the complex plane, an example of which is shown in Figure 8.12. A phasor diagram is valid for only one point of operation (torque, speed of revolution).

The stator voltage has been put in the real axis. The behaviour of the induction machine is mainly inductive, so that the stator current in this depiction is positioned in the right half-plane of the coordinate system. From the stator voltage to the induced voltage there is voltage drop at the ohmic stator resistance and at the stator leakage inductance. In the induction machine discussed here with a short-circuit rotor, the rotor-side voltage is generated solely by the voltage drop of the rotor current at the rotor resistance and the rotor-side leakage inductance.

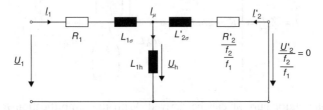

Figure 8.11 Single-phase equivalent circuit diagram of the induction machine with short-circuit rotor

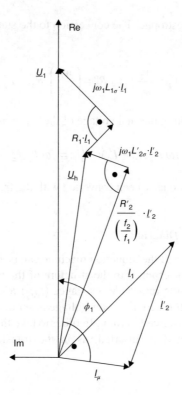

Figure 8.12 Phasor diagram of the induction machine with a short-circuit rotor (Heyland diagram) ($\underline{U}_h = \underline{U}_1$)

8.4.1.6 Heyland Diagram

The locus diagram of the stator current is used for representation and analysis of the operating points of the induction machine with a short-circuit rotor. It is called the Heyland diagram. This diagram can be derived from the voltage equations. The complete voltage equation stator-side can be set up from the created equivalent circuit diagram:

$$\underline{U}_1 = \underline{Z}_1\underline{I}_1 = \left[R_1 + j\omega_1 L_{1\sigma} + \frac{j\omega_1 L_{1h}\left(\dfrac{R_2'}{\left(\dfrac{f_2}{f_1}\right)} + j\omega_1 L_{2\sigma}'\right)}{j\omega_1 L_{1h} + \dfrac{R_2'}{\left(\dfrac{f_2}{f_1}\right)} + j\omega_1 L_{2\sigma}'} \right] \underline{I}_1 \tag{8.21}$$

and from this the stator current can be derived. With the simplification $R_1 = 0$ as well as $L_1 = (L_{1\sigma} + L_{1h})$; $L_2' = (L_{2\sigma}' + L_{1h})$, there is then given:

$$\underline{I}_1 = \frac{\left[R_2' + j\frac{f_2}{f_1}\omega_1 L_2'\right]\underline{U}_1}{jR_2'\omega_1 L_1 + \frac{f_2}{f_1}\left((\omega_1 L_{1h})^2 - \omega_1^2 L_1 L_2'\right)} \tag{8.22}$$

The current locus diagram of the induction machine, as is shown in Figure 8.13, can be developed from this equation. The equation represents a mathematical expression in the form that describes a circle in the complex plane. The design of the current locus diagram is shown in the figure as an example. The Heyland diagram, under the simplifying assumption of $R_1 = 0$, lies with the centre on the negative imaginary axis.

The circle can be drawn with the knowledge of the smallest stator current $\underline{I}_{10}$ at idling (no load, $s=0$, Index 0) as well as the greatest stator current $\underline{I}_\infty$ ($s \rightarrow \infty$, Index ∞, that corresponds to a purely theoretical operating point) or with the short-circuit current $\underline{I}_{1SC}$ ($s=1$, Index SC).

From the equation for the stator current it is easily possible to determine the values for special characteristic currents that are also drawn in the Heyland diagram. For the idling current $\underline{I}_{1.0}$ for $s=0$ there follows:

$$\underline{I}_{1.0} = \frac{1}{j\omega_1 L_1} \times \underline{U}_1; \quad (\angle(\underline{U}_1, \underline{I}_{1.0}) = 90°) \tag{8.23}$$

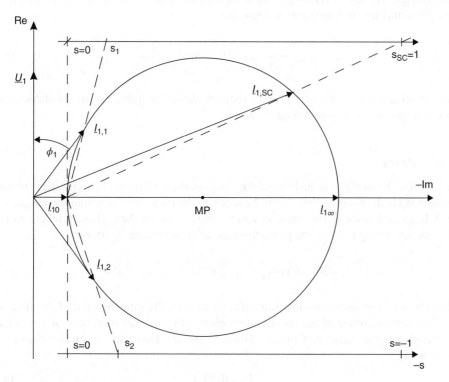

Figure 8.13 Heyland diagram of the induction machine with a short-circuit rotor for $R_1 = 0$

For the ideal short-circuit current, $\underline{I}_{1,\infty}$ applies for $s \to \infty$:

$$\underline{I}_{1,\infty} = -\frac{j}{\sigma\omega_1 L_1} \times \underline{U}_1; \quad (\angle(\underline{U}_1, \underline{I}_{100}) = 90°) \tag{8.24a}$$

$$\sigma = \frac{L_1 L_2' - L_{1h} L_{2h}'}{L_1 L_2'} = 1 - \frac{1}{(1+\sigma_1)(1+\sigma_2)} \left(\text{leakage character}\right) \tag{8.24b}$$

For the starting current or short-circuit current with $n = 0$ and $s = 1$, the following applies:

$$\underline{I}_{1,SC}(s=1) = \frac{R_2' + j\omega_1 L_2'}{R_2' + j\sigma\omega_1 L_2'} \times \frac{1}{j\omega_1 L_1} \times \underline{U}_1 \tag{8.25}$$

Also shown in the Heyland diagram are two further, freely-selected operating points (1, 2) as an example. A parameterising of the operating points with the slip can be carried out linearly as shown on the slip line ($s = 0$, $s = 1$). The intersection point of the line through the end point of the phasor $\underline{I}_{1,0}$, not through the zero point, and through the end point of the respective current phasor (e.g. s_1, dashed line) with the slip line, results in the slip value.

A special point in the operation is the optimum point: the point with the best (i.e. largest) displacement factor $\cos\phi_{opt}$, and the smallest displacement angle of the stator current to the stator voltage. The rated operating point of the machine is often in the neighbourhood of this point. From this the following can be derived:

$$I_{1,opt} = \frac{1}{\omega_1 L_1 \sqrt{\sigma}} \times U_1; \quad \cos\phi_{opt} = \frac{1-\sigma}{1+\sigma} \tag{8.26}$$

This operating point is also very characteristic in the Heyland diagram as the stator current forms a tangent to the Heyland diagram here.

8.4.1.7 Power

The power in the machine is made up of the power of the rotating field P_D, which is the power transmitted via the rotating field, from the stator to the rotor. For the assumption taken here, $R_1 = 0$, it is equal to the stator power P_1 and is calculated for a three-phase machine with the aid of the star voltage $U_{1,St}$ or the phase-to-phase or delta voltage $U_{1,\Delta}$ from:

$$P_D = 3U_{1,St} I_1 \cos\phi_1 = 3U_{1,St} \text{Re}\{\underline{I}_1\} = \sqrt{3} U_{1,\Delta} I_1 \cos\phi_1 \tag{8.27}$$

Here it should be mentioned that name plate data normally state the value of the delta voltage. The vertical portion of the stator current phasor, the real part of the complex stator current, represents the active part of the current in the Heyland diagram (active power) (see Figure 8.14):

$$\underline{I}_{1a} = \text{Re}\{\underline{I}_1\} \tag{8.28}$$

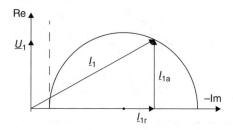

Figure 8.14   Heyland diagram, extracted: active current portion $\underline{I}_{1a}$ and reactive current portion $\underline{I}_{1r}$ of the stator current

The machine always behaves inductively; the stator current has a negative imaginary portion.

The power of the rotating field can be divided into two parts. If the iron losses and the stator resistance are neglected, then the full stator power is transferred to the rotor. It is converted there inside the resistance $(f_1/f_2)R_2'$ on the rotor side. Ohmic losses are created as power losses P_{2L} only in the natural resistance R_2', whilst the remaining power $P_D - P_{2L}$ forms the mechanical power P_{mech}. It can be expressed by the product of torque and speed of revolution.

$$P_D = 3\frac{R_2'}{s}I_2'^2 \tag{8.29a}$$

$$P_{2V} = 3R_2'I_2'^2 = sP_D \tag{8.29b}$$

$$P_{mech} = P_D - P_{2V} \tag{8.29c}$$

$$P_{mech} = (1-s)P_D = 2\pi nM \tag{8.29d}$$

8.4.1.8   Torque

The torque can be calculated from the power of the rotating field:

$$P_D = \frac{\omega_1}{p}M = 2\pi n_{syn}M = 3\frac{R_2'}{s}I_2'^2 \tag{8.30}$$

For this purpose the rotor current is determined dependent of the stator voltage from the equation above. The result is Kloss' formula:

$$\frac{M}{M_{stall}} = \frac{2}{\dfrac{s_{stall}}{s} + \dfrac{s}{s_{stall}}} \tag{8.31a}$$

$$M_{stall} = \frac{3}{\dfrac{\omega_1}{p}}\frac{1-\sigma}{(1+\sigma_1)\sigma 2\omega_1 L_{1h}}U_1^2; \quad s_{stall} = \frac{R_2'}{\sigma X_2'} \tag{8.31b}$$

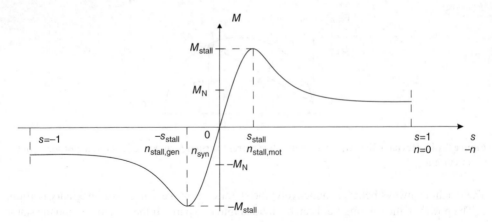

Figure 8.15 Torque–speed diagram of the induction machine with short-circuit rotor at fixed stator frequency

In this, σ_1 and σ_2 are the primary and secondary leakage quantities. They state the size of the leakage inductance, that stands for the non-stator- as well as the non-rotor-penetrating flux linkage, in relationship to the main inductance. The main inductance generates the flux that connects stator and rotor.

$$\sigma_1 = \frac{L_{1\sigma}}{L_{1h}}; \quad \sigma_2 = \frac{L'_{2\sigma}}{L_{1h}} \tag{8.32}$$

The stalling torque M_{stall} provides the highest value of torque and occurs at the stall slip s_{stall}. The course of the torque over the speed of revolution is the typical characteristic curve of a machine. It is shown in Figure 8.15. The right-hand part of the diagram shows motor operation, the left side generator operation.

When idling, the machine is in synchronous operation at the synchronous speed of revolution n_{syn}. In motor operation the speed of revolution is below the synchronous speed of revolution and the slip and the torque are positive. In generator operation the speed of revolution is above the synchronous speed of revolution and the slip and torque are negative. In loaded operation the speed of revolution deviates from the synchronous speed of revolution according to the required torque. Continuous operation is possible with the machine up to the rated slip where, depending on the design, about 30–50% of the stall torque is reached.

The machine is unstable in the region of the speed of revolution above the positive and below the negative stall slip. For motorised operation, for instance, a rising load torque would result in a decline in speed of revolution that would bring the machine to a standstill.

When starting up, the machine runs with slip 1 and continues on the characteristic line in the point with the required torque. For wind turbines according to the 'Danish principle' the characteristic line of Figure 8.15 would be the operating characteristic of the generator.

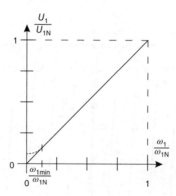

Figure 8.16 Control of the stator voltage of the induction machine with short-circuit rotor for variable speed operation, with constant stator flux linkage and neglecting the stator resistance, valid at $\omega_1 > \omega_{1min}$ (including R_1: dashed line)

8.4.1.9 Control of the Speed of the Induction Machine with Short-Circuit Rotor

When neglecting the stator resistance, the stator voltage is equal to the derivation of the stator flux linkage. With sine wave voltage this results in:

$$U_1(R_1 = 0) = \hat{U}_1 \cos \omega_1 t = -\frac{\mathrm{d}\psi_1}{\mathrm{d}t} = \omega_1 \psi_1 \cos \omega_1 t \tag{8.33}$$

The torque of the machine is formed from the flux in the machine and the electrical loading (current distribution over the air gap) – additionally assuming a correspondingly mutual angle position. The flux is generally kept constant (control of constant flux) and the torque is generated by means of the stator current. For this constant stator flux linkage Ψ_{1N} in the machine, there follows, neglecting the stator resistance for cosine wave quantities:

$$\Psi_1 = \Psi_{1N} = \frac{U_1}{\omega_1} = \frac{U_{1N}}{\omega_{1N}} = \text{const.} \tag{8.34}$$

Therefore a simplified control law applies for variable speed operation. It says that the stator voltage must be adjusted proportional to the stator frequency. This is shown in Figure 8.16. However, in the region of the smaller stator frequency ($\omega_1 < \omega_{1min}$) the ohmic portion increases relatively strongly. For this region it should be included in the analysis, and the result would be the dashed characteristic curve.

The effects of this control on the torque behaviour are now discussed. In the equation of the stalling torque (Kloss' formula) the term $(U_1/\omega_1)^2 = \psi_1^2$ is found, the stator flux linkage, as the sole dependency of the operating point dimension. By controlling for constant stator flux, this term is constantly controlled as described. Thus, the stalling torque remains at its value and also the torque equation for speed control is maintained, independently of the value of the selected stator frequency.

The result of this is that for any desired stator frequency, Kloss' formula has the same validity, written here in dependency of the rotor angular frequency in place of the slip.

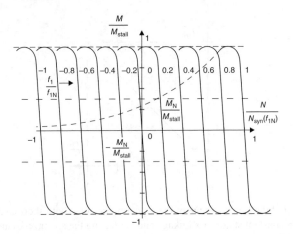

Figure 8.17 Torque–speed curves of the induction machine with short-circuit rotor for control of constant stator flux linkage

$$M = M_{stall} \frac{2}{\dfrac{\omega_2}{\omega_{2\,stall}} + \dfrac{\omega_{2\,stall}}{\omega_2}}; \quad \omega_{2\,stall} = \frac{R_2'}{\sigma L_2'} \tag{8.35}$$

The results of this are the speed–torque characteristic curves in Figure 8.17. Here the characteristic curves are shown over the speed, which is in the reverse direction to the original torque–slip diagram shown above. Therefore, the machine can be operated in the region of speed zero up to the maximum rated speed with a torque to the value of the rated torque, motor ($M>0$) or generator ($M<0$) operation. Continuous speed control is possible from the machine behaviour.

For the power at the rated stator current I_{1N} on the condition $R_1=0$ at this control, here with the use of the delta stator voltages, it can be derived:

$$P_D(I_{1N}) = \sqrt{3}U_{1,\Delta}I_{1N}\cos\phi_{1N} = \sqrt{3}\frac{\omega_1}{\omega_{1N}}U_{1,\Delta N}I_1\cos\phi_{1N} = \frac{\omega_1}{\omega_{1N}}P_{DN} \tag{8.36}$$

Therefore, the power at a given stator current rises with the stator frequency. Corresponding to the results, this operating region is also named as follows: region of constant torques, or region of full flux, or region of active power proportional to frequency.

A principle of the implementation of this type of control in its simplest form is shown in Figure 8.18. The induction machine with short-circuit rotor is fed by means of a frequency converter. This is placed at the entrance to the grid with its assumed fixed voltage amplitude and frequency. The converter is able to adjust the amplitude and frequency of the stator voltage for the machine as required. The converter is operated via the controls in such a manner that the requirements according to the above derivation are adhered to, which is $U_1/\omega_1=$const. This takes place in that the frequency converter is set so that the rated stator frequency is nearly proportional to the rated speed. The stator rated voltage U_1^* must be set with a corresponding factor proportional to the stator frequency. This takes place in a characteristic curve

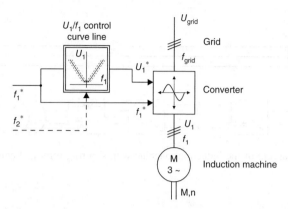

Figure 8.18 Overview diagram for feeding an induction machine with a short-circuit rotor via a frequency converter with characteristic curve control

block (double border). If the stator resistance is to be included, then the rotor frequency f_2 is necessary as a further input quantity and deputises for the torque and the stator current. Depending on this, the characteristic curves are moved slightly up or down.

The structural diagram of the speed control acts to clarify the concept. Today's controls normally operate differently, in general according to the method of field-oriented control.

8.4.2   Induction Machine with Slip-Ring Rotor

Induction machines with slip-ring rotors are primarily used for high power where induction machines with short-circuit rotors cannot be built, for example, for the motor/generator in pumped-storage hydro-power stations with capacities over 100 MW. From the 1990s these have also increasingly been used in variable-speed wind turbines for powers around 1 MW. Today they are probably the most often used generator variant in wind turbines.

The induction machine with a slip-ring rotor has a special advantage when used in the wind turbine. When used as a variable speed generator, in contrast to the other variants with converter at the stator and as shown in the introduction, here the stator is connected directly to the grid and the rotor is connected via a frequency converter as shown in Figure 8.19. For this purpose the designation 'double-fed induction machine' is used.

The frequency converter for this is to be designed only for the required speed range that corresponds to the power region. For wind turbines, the converter is correspondingly to be designed for approximately 30% of the nominal power (part load converter), which becomes noticeable in the lower costs. Converters for all other systems as induction machines with short-circuit rotors that are connected to the stator side must be designed for the full load (full power converter). A disadvantage with the induction machine with slip-ring rotor, however, is that the slip rings require a special servicing and that the machine with the stator is directly connected to the grid. Due to the direct stator coupling with the grid, the behaviour of passing through the grid short circuit is characterised not by an easily mastered stator-side converter of the machine and can be controlled only indirectly by the rotor converter.

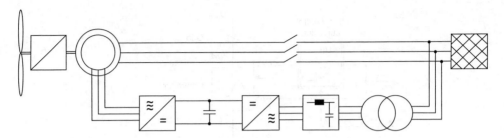

Figure 8.19 Block diagram of the induction machine with slip-ring rotor and converter in the wind turbine

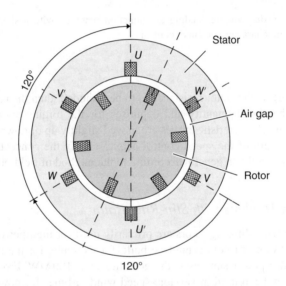

Figure 8.20 Structure of an induction machine with slip-ring rotor (pole pair $p=1$; number of slots $N=6$ in rotor and stator); access to the rotor via the slip rings

8.4.2.1 Structure

As with all three-phase machines discussed here, the stator of the induction machine with a slip-ring rotor is built up as three-phase coils. Three phases (U–U', V–V', W–W') are each displaced about 120° to each other in a pole pair, and each pole pair is arranged at the circumference of the stator. The coils are positioned in the required quantity of slots (see Figure 8.20).

The rotor is equipped with three-phase coils of the same type. For the sake of clarity, the figure only shows one slot per phase and pole. The rotor-side coils are accessible from the stator via slip rings. Slip rings consist of rotating copper rings, at least one per phase, that are electrically insulated from each other, and the rotor iron, and fixed to the rotor. They rotate with the rotor and are connected to the rotor coils. On the stator-side there are carbon brushes for passing on the current. These are fixed to the stator and are pressed on to the slip rings by means of springs.

8.4.2.2 Basic Function

The stator and the rotor of the induction machine with a slip-ring rotor are equipped with a three-phase coil. The stator is connected to a cosine wave symmetrical three-phase voltage system of frequency f_1 and the rotor with a frequency f_2. A rotating field is generated by both coil systems. The result is a magnetic field distributed in a cosine wave over a pole pair that circulates proportionally to the respective feed-in frequency.

The stator field rotates at speed $n_{1,syn}$, thus, during one period of the stator frequency the field rotates over one pole pair:

$$n_{1,syn} = \frac{f_1}{p} \qquad (8.37)$$

The rotor rotates with speed n. The rotor field is generated by the rotor frequency f_2 and thus turns, with reference to the rotor, with a speed of f_2/p. A constant torque is only generated when the fields of the rotor and the stator rotate at the same speed. This is the case when the following applies:

$$n_{1,syn} = \frac{f_1}{p} = n + \frac{f_2}{p} \qquad (8.38a)$$

$$n = n_{1,syn} - \frac{f_2}{p} \qquad (8.38b)$$

The equation therefore represents the operating conditions of the induction machine with a slip-ring rotor. For a given fixed stator frequency f_1, the speed n is the result of subtraction of the controllable rotor frequency f_2/p F prescribed by the rotor-side converter T, from the synchronous speed $n_{1,syn}$. If the rotor frequency is selected as zero, then the rotor speed is equal to the speed of revolution of the stator rotating field, the synchronous speed. This corresponds to a direct current feed-in to the rotor. The behaviour is then that of an externally excited synchronous machine fed in the rotor with DC current. If the rotor frequency is selected with deviations from zero, then the speed can be controlled to other values above or below the synchronous speed. This behaviour can also be called a three-phase excited synchronous machine.

8.4.2.3 Voltage Equations

The voltage equations are those of the induction machine with voltage at the rotor whose derivations were discussed for the induction machine with a short-circuit rotor.

$$\underline{U}_1 = R_1 \underline{I}_1 + j\omega_1 L_{1\sigma} \underline{I}_1 + j\omega_1 L_{1h} (\underline{I}_1 + \underline{I}_2') \qquad (8.39a)$$

$$\frac{\underline{U}_2'}{\frac{f_2}{f_1}} = \frac{R_2'}{\frac{f_2}{f_1}} \underline{I}_2' + j\omega_1 L_{2\sigma}' \underline{I}_2' + j\omega_1 L_{2h}' (\underline{I}_1 + \underline{I}_2') \quad \text{with} \quad L_{2h}' = L_{1h} \qquad (8.39b)$$

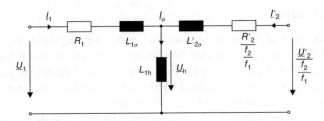

Figure 8.21 Equivalent circuit diagram of the induction machine with slip-ring rotor

where U_1 is the stator voltage, U_2 is the rotor voltage, I_1 is the stator current, I_2 is the rotor current, ω_1 is the stator circuit frequency, R_1 is the stator resistance, R_2 is the rotor resistance, $L_{1\sigma}$ and $L_{2\sigma}$ are the stator and rotor leakage inductances, and L_{1h} is the stator-referenced main inductance.

The conversion of the rotor quantities to the stator, characterised by the apostrophs, occurs, as for an induction machine with short-circuit rotor, with the following equation:

$$U_2' = \frac{w_1}{w_2}U_2; \quad I_2' = \frac{w_2}{w_1}I_2; \quad R_2' = \frac{w_1^2}{w_2^2}R_2; \quad L_{2x}' = \frac{w_1^2}{w_2^2}L_{2x} \tag{8.40}$$

The similarity with the equations of the transformer is obvious. In both equations, induced voltage portions appear at main and leakage inductances and as also ohmic voltage drops. Noticeable is the variable frequency of the rotor-side resistance $(f_1/f_2)R_2'$ and the rotor voltage multiplied by the frequency relationship $(f_1/f_2)U_2'$. For feeding with variable rotor voltage, the latter term becomes the characteristic variable. The equations thus deviate strongly from those of the induction machine with short-circuit rotor.

8.4.2.4 Equivalent Circuit Diagram

With the aid of these equations, the single-phase equivalent circuit diagram of the induction machine with a slip-ring rotor according to Figure 8.21 can be created. It corresponds to that of the induction machine with a short-circuit rotor, except that the rotor voltage here is unequal, zero. In order to take the iron losses into account, the iron resistance R_{Fe} can be inserted as for the transformer, parallel to the main inductance.

8.4.2.5 Phasor Diagram and Current Locus Curve

The behaviour of the induction machine with slip-ring rotor fed by means of a converter is derived here in steps. The derivation here takes place by means of a greatly simplified equivalent circuit diagram without the resistance R_1 and the stator leakage inductance $L_{1\sigma}$ (see Figure 8.22). By means of conversion and neglecting only the stator resistance, this very simplified structure can also be developed from the equivalent circuit diagram with stator-side leakage inductivity [5].

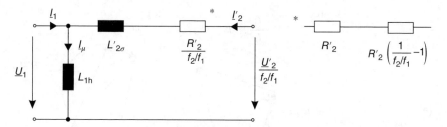

Figure 8.22 Simplified equivalent circuit diagram of the induction machine with slip-ring rotor

In this way an even easier derivation can be achieved, but with reduced accuracy. The rotor resistance is disassembled at the right of the figure in which a portion R'_2 generates the losses and a portion contributes to the mechanical power.

The rotor current is defined from the equivalent circuit diagram of the voltage drop over the longitudinal impedance:

$$\underline{I}'_2 = \frac{\dfrac{\underline{U}'_2}{f_2} - \underline{U}_1}{j\omega L'_{2\sigma} + \dfrac{R'_2}{\dfrac{f_2}{f_1}}} \tag{8.41}$$

If the stator voltage is placed in the real axis and the rotor voltage is disassembled into an active portion in the real and a reactive portion in the imaginary direction, then the following applies:

$$\underline{U}_1 = U_1; \quad \underline{U}'_2 = U'_{2a} + jU'_{2x} \tag{8.42}$$

And it follows for the rotor current:

$$\underline{I}'_2 = \frac{\left(\dfrac{U'_{2a}}{\dfrac{f_2}{f_1}} + \dfrac{jU'_{2r}}{\dfrac{f_2}{f_1}}\right) - U_1}{j\omega_1 L'_{2\sigma} + \dfrac{R'_2}{\dfrac{f_2}{f_1}}} \tag{8.43}$$

If one expands the denominator with complex conjugation in order to obtain a real value, then the numerator is multiplied with the same value and then the numerator is subdivided, then for the rotor current the following is the result:

$$\underline{I}'_2 = \frac{-U_1 \dfrac{R'_2}{\left(f_2/f_1\right)} + U'_{2a}\dfrac{R'_2}{\left(f_2/f_1\right)^2} + \omega_1 L'_{2\sigma}\dfrac{U'_{2r}}{\left(f_2/f_1\right)} + j\omega_1 L'_{2\sigma}U_1 - j\omega_1 L'_{2\sigma}\dfrac{U'_{2a}}{\left(f_2/f_1\right)} + jU'_{2r}\dfrac{R'_2}{\left(f_2/f_1\right)^2}}{R'^2_2/(f_2/f_1)^2 + \omega_1^2 L'^2_{2\sigma}}$$

$$\tag{8.44}$$

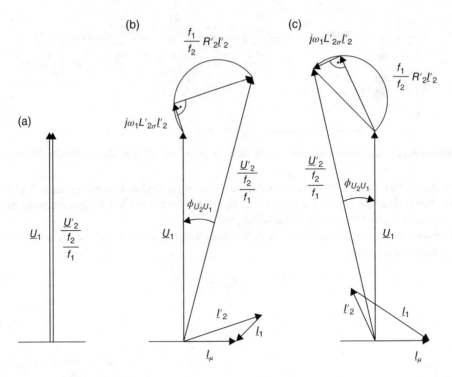

Figure 8.23 Phasor diagram of the induction machine with slip ring rotor and rotor infeed: (a) $U_1 = U_2 \dfrac{f_1}{f_2}$, $P=0$; (b) $U'_2 \dfrac{f_1}{f_2} > U_1$, angle >0; (c) $U'_2 \dfrac{f_1}{f_2} > U_1$, angle <0

The equation contains six terms in the numerator, of which two depend only on the stator voltage and the other four on the rotor voltage. In this way the phasor diagrams can be drawn for various operating points.

The phasor diagram of the induction machine with a slip-ring rotor in Figure 8.23, for various operating points, characterises the machine's behaviour. (a) clarifies the condition of the rotor current zero. From (b) and (c) it can clearly be seen how the rotor current can be set by means of the amplitude and phase angle of the rotor voltage. If the magnetising flux is neglected, the rotor current would be equal to the stator current. When including the magnetising current, the result is the stator current I_1 as shown.

If the stator resistance is neglected, the rotary field power P_{RF} becomes equal to the stator power P_1:

$$P_{RF} = P_1 = 3U_{1star} I_1 \cos \phi_1 \tag{8.45}$$

The actual power losses in the rotor are:

$$P_{2L} = 3R'_2 I'^2_2 = 3R_2 I^2_2 \tag{8.46}$$

and the converter feeds power into the rotor:

$$P_{conv} = 3U'_2 I'_2 \cos \phi_2 \tag{8.47}$$

The result is that the mechanical power is the sum or difference of the rotary field power, the rotor power loss and the power fed via the current converter.

$$P_{mech} = P_{RF} - P_{2L} + P_{conv} = \left(1 - \frac{f_2}{f_1}\right)P_{RF}$$
(8.48a)

$$P_{2L} - P_{conv} = \frac{f_2}{f_1}P_{RF}$$
(8.48b)

Both active and reactive power in the rotary field power and the torque can be calculated with the above equations. Star voltages are always used here.

$$P_{RF} = \text{Re}\{3\underline{U}_1\underline{I}_2^{'*}\} = 3U_1 \frac{-U_1\dfrac{R_2'}{(f_2/f_1)} + U_{2a}'\dfrac{R_2'}{(f_2/f_1)^2} + \omega_1 L_{2\sigma}'U_{2r}'\dfrac{1}{(f_2/f_1)}}{\dfrac{R_2'^2}{(f_2/f_1)^2} + \omega_1^2 L_{2\sigma}'^2}$$
(8.49a)

$$M = \frac{P_{RF}}{\omega_1} = 3\frac{U_1}{\omega_1} \frac{-U_1\dfrac{R_2'}{(f_2/f_1)} + U_{2a}'\dfrac{R_2'}{(f_2/f_1)^2} + \omega_1 L_{2\sigma}'U_{2r}'\dfrac{1}{(f_2/f_1)}}{\dfrac{R_2'^2}{(f_2/f_1)^2} + \omega_1^2 L_{2\sigma}'^2}$$
(8.49b)

$$P_{mech} = 2\pi NM$$
(8.49c)

$$Q_{RF} = \text{Im}\{3\underline{U}_1\underline{I}_2^{'*}\} = 3U_1 \frac{-\omega_1 L_{2\sigma}'U_1 + \omega_1 L_{2\sigma}'\dfrac{U_{2\sigma}'}{(f_2/f_1)} - \dfrac{R_2'U_{2n}'}{(f_2/f_1)^2}}{\dfrac{R_2'^2}{(f_2/f_1)} + \omega_1^2 L_{2\sigma}'^2}$$
(8.49d)

With fixed stator and rotor frequency, thus fixed rotor speed, it can be seen from the equations that both the power and the torque can be controlled because of variations of the components of the rotor voltage (U_{2a}, U_{2r}).

The whole power P_{tot} taken from the grid via rotor and stator that is made up from the stator-side rotary power P_{RF} and the rotor-side terminals power and current converter power P_{conv}, on the assumption that the converter works without losses on the rotor, is:

$$P_{tot} = P_{grid} = P_{RF} + P_{conv} = \begin{cases} \eta P_{mech} & \text{for generator operation} \\ \dfrac{1}{\eta}P_{mech} & \text{for motor operation} \end{cases}$$
(8.50)

The degree of efficiency must be applied differently depending on whether generator operation (η) or motor operation ($1/\eta$) is being used.

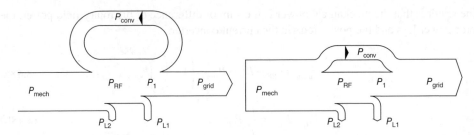

Figure 8.24 Power flow in the induction machine with slip-ring rotor in generator operation (double-fed asynchronous machine; left: sub-synchronous operation, right: super-synchronous operation) (P_{conv}: current converter power; P_{L1}, P_{L2}: stator and rotor power losses)

Because of the many components in the equation, it is not easy to obtain an overview of the flow of power. This can be clarified for various operating points by means of a power flow diagram. In Figure 8.24 the operating regions for wind turbines of sub- and super-synchronous generator operation are shown. In generator operation, the power is taken from the shaft and fed into the grid.

Sub-synchronous, for instance, would be the speed range from $n_{min} = 1050$ rpm to $n_{nom} = 1500$ rpm for a 2-pole pair machine on the 50 Hz grid. In the super-synchronous operation, for instance between 1500 rpm and about 1950 rpm, the power P_{RF} is fed from the stator-side rotary field and P_{conv} from the rotor via the converter into the grid. The grid power is larger by the value of the rotor power than the stator power. With sub-synchronous operation this takes place analogously, but with subtraction of the rotor power from the stator power for grid power. The stator and rotor losses (P_{L1}, P_{L2}) are included additionally in the figure.

8.4.2.6 Speed Control

Speed control of the doubly-fed induction machine is carried out by feeding the rotor by means of the frequency converter. For this purpose the rotor voltage is set to the amplitude U_2 and the rotor frequency to f_2. For operation with torque zero, there is a rotor idling frequency of:

$$\frac{f_{20}}{f_1} = \frac{U'_{2a}/U_1}{1 - \dfrac{U'_{2r}/U_1}{f_{2SC}/\omega_1}}; \quad f_{2SC} = \frac{R'_2}{L'_{2\sigma}} \qquad (8.51)$$

For the case that the rotor reactive voltage equals zero, the rotor idling frequency becomes:

$$\frac{f_{20}}{f_1} = \frac{U'_{20}}{U_1} \qquad (8.52)$$

It can be seen that for the case with a fixed stator frequency that is equal to the grid frequency, the rotor voltage must be regulated dependent on the rotor frequency f_2. The rotor frequency provides the deviation from the synchronous speed of revolution, the speed setting. With synchronism, where the rotor rotates with the speed of the stator rotary field, the rotor frequency and thus the rotor idling voltage are equal to zero. Depending on the operating region when

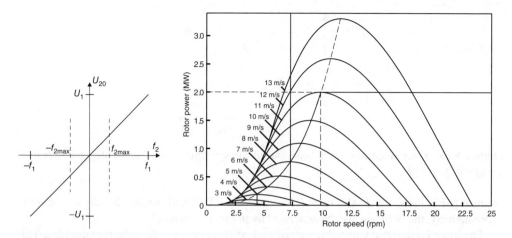

Figure 8.25 Variable speed operation of the induction machine with a slip-ring rotor (doubly-fed induction machine); left: control characteristic curve for the rotor voltage idling with $U_{Rr}=0$ (r: reactive part); right: operating region in the power–speed diagram. Typical region for wind turbines: $-0.3f_1 < f_2 < +0.3f_1$

applied in wind turbines, the typical ranges of rotor frequency are from −30% to +30% around the middle speed. This results in an equal voltage region for the rotor idling voltage (−30 % to +30% of the standstill voltage). For the loaded generator, additional voltage portions are added in order to make the current flow over the rotor possible.

This results in a voltage regulating characteristic curve at the rotor with this operation of the induction machine with a slip-ring rotor (double-fed asynchronous machine), as depicted in Figure 8.25. The course is linear. An operating range of rotor frequency of $-0.3f_1$ to $+0.3f_1$ is typical for wind turbines.

Torque–speed characteristic curves as for synchronous machines are generated. There is a fixed speed for a fixed rotor infeed frequency. A required torque in the permissible power–speed region of the wind turbine can be achieved with this freely selectable speed.

8.5 Synchronous Machines

Synchronous machines are used in various fields. They can be built for the greatest power in the GW range and have the feature of controlled feeding of reactive power. For this reason they are generally used in power stations. On the other hand, this type of machine is also used in the low power region such as for machine tools, devices of the leisure industry, or printers.

The synchronous machine has been in use in wind turbines for a long time, previously only the externally fed variant, but nowadays increasingly also with permanent magnet excitation.

8.5.1 General Function

The variant with external excitation will be used to introduce the machine. The rotor of this variant of synchronous machine is equipped with a DC coil and is fed with direct current. This takes place by means of a power converter that regulates the DC field current. The current is

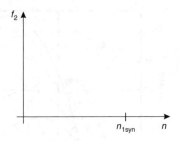

Figure 8.26 Rotor frequency–speed behaviour of the synchronous machine at the grid with fixed frequency f_1 ($f_2 = 0$)

transferred to the rotor via slip rings. This results in a DC field that lies fixedly on the rotor. Alternatively, the field can also be generated by permanent magnets.

The stator is equipped with a three-phase coil, as is also used in the induction machine. This is fed with a cosine wave symmetrical three-phase voltage system. The coil systems of the three phases are displaced by $120°$ in a pole pair to each other. The current in the stator coils generates a sine-wave forming magnetic rotary field that circulates proportionally to the feed frequency. The field moves to a further pole pair for each period of the stator frequency.

$$n_{1,\text{syn}} = \frac{f_1}{p} \tag{8.53}$$

A constant torque is generated when the field of the rotor and the stator rotate at the same speed. The slip is then equal to zero:

$$n = n_{1,\text{syn}} \leftrightarrow s = \frac{\omega_2}{\omega_1} = 0 \quad \rightarrow M \neq 0 \tag{8.54}$$

This is shown in Figure 8.26. There is only one speed operating point for the machine at the grid with fixed frequency.

A further condition for the generation of a torque is that the field directions from rotor and stator have an angular displacement from each other. This is comparable to two rod magnets rotary positioned on an axis which act on each other when displaced. Thus, the machine at the grid generates a torque only with fixed speed, the synchronous speed and with phase displacement of the rotor field relative to the stator field.

8.5.2 Voltage Equations and Equivalent Circuit Diagram

Seen from the feed-in side, the stator with its three-phase coil presents the same relationship, electrically, as the primary side of the transformer. The voltage equation can be taken over from there. The rotor equation is an equation for a DC current circuit that serves only for determining the current in the rotor, the DC I_f. It is given here but will not be given again.

$$\underline{U}_1 = R_1 \underline{I}_1 + jX_{1\sigma} \underline{I}_1 + jX_{1h} \left(\underline{I}_1 + \underline{I}'_f \right) \tag{8.55a}$$

$$U_2 = R_2 I_f \tag{8.55b}$$

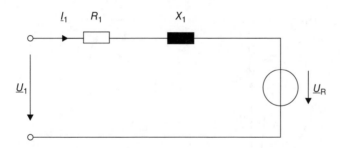

Figure 8.27 Equivalent circuit diagram of the synchronous machine

The important determining equation is therefore the stator voltage equation. The current $\underline{I}'_f$ of the stator voltage equation is the rotor coil current I_f referred to on the stator side and converted. It has the same effect as the direct current in the rotor. Its position or displacement is the result of the position of the rotor relative to the rotary field of the stator.

The portion of the stator voltage that is generated by the rotor current is also designated separately as a synchronous internal voltage or rotor voltage U_R.

$$\underline{U}_R = jX_{1h}\underline{I}'_f \tag{8.56}$$

Thus, the stator voltage in another form is written with $X_1 = X_{1\sigma} + X_{1h}$ as:

$$\underline{U}_1 = R_1\underline{I}_1 + jX_1\underline{I}_1 + \underline{U}_R \tag{8.57}$$

Defined from this is the equivalent circuit diagram of the synchronous machine shown in Figure 8.27. It consists of three elements of which the stator resistance can be ignored with good approximation.

The phasor diagram for the synchronous machine in Figure 8.28 is derived from the voltage equation and from the node rule for the current in the machine. The stator voltage is made up of the synchronous internal voltage and the voltage drops in the inductances and the ohmic resistance. The sum of the stator current $\underline{I}_1$ and the rotor current $\underline{I}'_2$, referenced to the stator, results in the magnetising current $\underline{I}_\mu$. Because of the formation of the voltages at inductivities, several orthogonal angles are formed: between $\underline{I}_\mu$ and the induced voltage $\underline{U}_i$, between $\underline{U}_R$ and $\underline{I}'_f$, between $j\omega L_{1\sigma}/I_1$ and R_1/I_1. The angle between the synchronous internal voltage or rotor voltage and the stator voltage is called the rotor displacement angle θ. The current forms itself by means of the different amplitudes and different phase angles of the stator and synchronous internal voltage.

The operation of the generator is shown in phasor diagrams in Figure 8.28. In this it is presented as an over-excited operation characterised in that the projection of the synchronous internal voltage on the stator voltage is greater than the stator voltage. This leads to a capacitive portion of the stator current. On the other hand, the under-excited operation is depicted. Here, the projection of the synchronous internal voltage on the stator voltage is smaller and the stator current shows an inductive portion. The depiction occurs in the consumer reference phasor system (motorised operation $P > 0$).

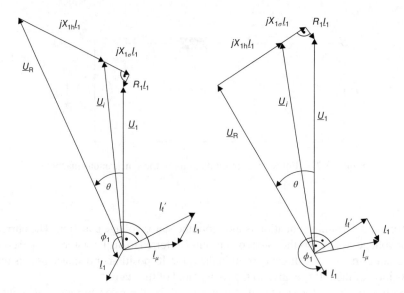

Figure 8.28 Phasor diagram of the externally excited synchronous machine; left: generator operation over-excited ($\theta > 0$, $P < 0$, $\cos \phi < 0$, $Q > 0$, capacitive); right: generator operation under-excited ($\theta > 0$, $P < 0$, $\cos \phi < 0$, $Q < 0$, $\sin \phi < 0$, inductive)

8.5.3   Power and Torque

On the stator side the machine takes up power, the stator power P_1. With neglect of the stator resistance, this is equal to the rotary field power P_{RF} that is transferred to the rotor via the air gap and which is converted to mechanical power in the synchronous machine.

$$P_1 = P_{RF} = 3U_{1,star} I_1 \cos \phi_1 \quad \text{for} \quad R_1 = 0 \tag{8.58}$$

With the reshaping of the equation by means of relationships, a different expression can be derived from the phasor diagram for the stator active power P_1 which is a typical notation for a synchronous machine. The same can be carried out for the reactive power Q_1. The following equations always apply for star voltages.

$$P_1 = -3U_{1,star} \frac{U_{R,star}}{X_1} \sin \theta \tag{8.59a}$$

$$Q_1 = 3U_{1,star} I_1 \sin \phi_1 = -3U_{1,star} \left(\frac{U_{R,star}}{X_1} \cos \theta - \frac{U_{1,star}}{X_1} \right) \tag{8.59b}$$

The powers are defined from the stator and synchronous internal voltage and the rotor displacement angle. From the active power the torque is determined as:

$$M = -M_{stall} \sin \theta \tag{8.60a}$$

$$M_{stall} = \frac{3U_{1,star} U_{R,star}}{2\pi n_1 X_1} \tag{8.60b}$$

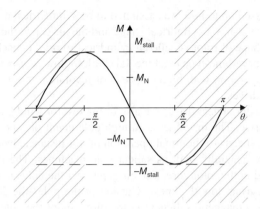

Figure 8.29 Torque–rotor displacement angle diagram of the synchronous machine; non-permitted operating region is hatched

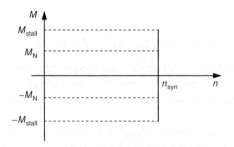

Figure 8.30 Torque–speed diagram of the synchronous machine at a fixed stator frequency

As can be seen from the formula, the torque is also dependent on the stator and synchronous internal voltage and the angle between them, the rotor displacement angle. If both voltages are fixed, then the torque relates sinusoidally to the rotor displacement angle (see Figure 8.29). The torque has a maximum value called the stalling torque M_{stall}. For reasons of static stability, only angle regions of the rotor displacement angle to 90° can be driven; in practice, due to the design of the machine, the region is even more limited.

The synchronous machine at fixed stator frequency, for instance at the grid, can only be properly operated at a speed n_{syn}, as it is only there that an appropriate team torque can be generated. At this speed, however, the machine can generate a desired torque in the permitted region. According to this, the result is a torque–speed diagram of the grid-supplied machine in Figure 8.30.

8.5.4 Embodiment of Externally Excited Synchronous Machines

The method of operation of the synchronous machine has been discussed in the previous section for the variant with external excitation. External excitation means excitation with direct current in the rotor that is fed by means of slip rings. Here, too, there are subvariants.

The machine discussed up to now was assumed to be symmetrical at the circumference of rotor and stator. The rotor has a cylindrical shape and the air gap is the same over the whole circumference. This shape of rotor is called a cylindrical rotor; the machine is called a cylindrical motor machine or turbine-type machine. It is built for low pole-pair numbers and is used in wind turbines for operation with higher speeds of revolution in this type of construction, thus when using gear drives.

However, variants without gear drives are also of interest for wind turbines. The drive train is denoted as direct drive. The machine has a high pole count of 50 or more and is a slow-running machine with speeds for nominal operation of, for instance, below 100 rpm.

These machines are generally built as salient pole machines. The poles are individually produced and installed in the rotor core. The air gap under the poles in the region of the flux path is designed with usual low thickness. Outside of the poles the air gap is greater. The machine therefore has a magnetic saliency; in the direction of the poles and in the pole gap there are different inductances due to the different air gaps. The course of the torque over the rotor displacement angle therefore changes somewhat compared with that of the sinusoidal shape for cylindrical rotor machines.

8.5.5 Permanently Excited Synchronous Machines

Permanently excited synchronous machines are increasingly being used in wind turbines. Their field of application was previously mostly in the field of servomotors with capacities under 100 kW. This machine variant is also being used successfully in wind turbines in the megawatt region.

Permanently excited synchronous machines differ from externally excited ones in that the magnetic excitation is generated by means of permanent magnets. Permanent magnets with high field strengths are required to operate the machines with limited magnetic material. Field strengths of $H_C = 1000$ kA/m and remanent inductions (flux densities) of up to $B_R = 1.5$ T are achieved. The materials used are rare earth metals such as samarium–cobalt and neodymium–iron. An important condition for use is that the magnets maintain their magnetic properties during heating and during possible counter-fields or faults in the current feed such as short circuits. The production of the machines and especially also the installation of the rotor into the stator must be especially carefully carried out due to the strong magnetic fields and the attractive forces.

The derivation of the formulae for the externally excited synchronous machine can be used for the determination of the electrical and mechanical operating behaviour of the permanent magnet machine. The equations given there (8.55 to 8.60) apply. It is should be noted that the magnetisation emanating from the rotor is constant when the saturation is neglected, as is assumed here. The synchronous internal voltage $U_{R,PM}$ in the voltage equation for the permanently excited machine is generated by means of the permanent magnet magnetisation.

$$\underline{U}_1 = R_1 \underline{I}_1 + jX_1 \underline{I}_1 + \underline{U}_{R,PM} \tag{8.61}$$

This applies correspondingly for the torque equation.

$$M = -M_{stall} \sin\theta \quad \text{with} \quad M_{stall} = \frac{3U_{1,star} U_{R,PM,star}}{2\pi n_1 X_1} \tag{8.62}$$

It must also be noted here that in contrast to the externally excited synchronous machine, the size of the rotor flux is fixed and for fixed speed thus the amplitude of the rotor internal voltage is also fixed, corresponding to the design of the machine. Correspondingly, the design also determines whether the machine in its operating points is rather over- or underexcited (higher or lower synchronous internal rotor voltage). The phasor diagram and the speed–torque curves of the externally excited synchronous machine also apply here.

8.5.6 Variable Speed Operation of Synchronous Machines

Variable speed operation occurs in that the machine is fed with stator voltage of variable amplitude and frequency via converters. Variable speed operation requires that the flux in the machine is kept constant and that the torque is adjusted via the stator current. If the stator flux linkage is to be kept to its rated value, then it follows:

$$\psi_1 = \psi_{1N} = \frac{U_1}{\omega_1} = \frac{U_{1N}}{\omega_{1N}} = \text{const.} \tag{8.63}$$

Thus a simplified control law applies for the variable speed, permanently excited synchronous machine. It states that the stator voltage must be adjusted proportionally to the stator frequency. This is shown in Figure 8.31. In the region of smaller infeed frequencies, the ohmic portion of the stator resistance acts more strongly and must be taken into account by means of voltage increases (for motorised operation, decreases for generator operation).

In the torque equation, the terms U_1/f_1 und $U_{R,PM}/n$ occur in the stalling torque as a dependency of the stator voltage and stator frequency and of the synchronous internal voltage. As the first term is to be controlled as a constant, it remains constant. The synchronous internal voltage $U_{R,PM}$ changes with the stator frequency as it is induced by the rotor with speed $n_1 = n_{1syn} = f_1/p$. Consequently the quotient also remains constant. Thus, the torque equation in this type of torque control is independent of the speed.

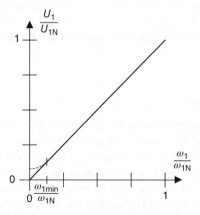

Figure 8.31 Control of the stator voltage of the synchronous machine for constant stator flux linkage for variable speed operation (dashed: required voltage increase for motorised operation)

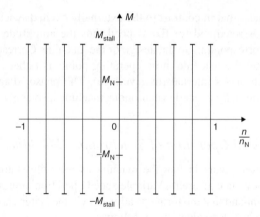

Figure 8.32   Torque–speed characteristic curve of the synchronous machine for speed control

$$M = \dfrac{3U_1\,\dfrac{U_{R,PM}}{\omega_1 L_1}}{2\pi n_1}\sin\theta \qquad\qquad (8.64)$$

This results in torque–speed curves of the variable speed synchronous machine as vertical lines in Figure 8.32. Continuous torque speed regulation is possible.

8.6   Converter Systems for Wind Turbines

With the current state of technology, wind turbines are generally equipped with variable speed generators. The capability of speed control allows the system of electrical power of the generator to be fed via an interconnected frequency converter into the electrical grid. The frequency converter is operated on the generator side with variable frequency of the generator, and on the grid side with the grid frequency of 50 or 60 Hz. The task of frequency converters is to convert frequency and amplitude of voltage appropriately between the two systems [1,2,8]. At the same time it is equipped with controls that regulate the power flow and carry out special grid and generator-side regulating requirements (grid codes) such as regulating the reactive power.

8.6.1   General Function

The amplitude and frequency conversion are realised in the frequency converter, in that current or voltage are switched on and off within the shortest of time intervals in the range of milliseconds [9–12]. The amplitude and frequency of the output quantities are controlled by means of the relationship of the switch-on time to the overall interval switch sector time. Switching occurs by means of suitable power semiconductors. In today's converters mostly IGBTs (isolated gate bipolar transistors) and also IGCTs (integrated gate commutated thyristors) are used. They can block voltages in the range of kilovolts and conduct currents in the range of kiloamps.

Because of the capability of these power semiconductors to independently switch the current on and off, this type of converter is known as self-commutated. Because of the formation of the output quantities by means of voltage pulses, the power converters are also denoted as pulse-width modulated inverters.

There are a multitude of topologies for these converters. In the field of wind turbines with capacities below approximately 5 MW at present the variant known as the two-level converter in basic topology is mainly used. This topology and its function are described below. For plants from approximately 5 MW, power circuits of the three-level converter variant or others are used. This will be briefly discussed.

The circuits of power electronics contain certain components for which here, for purposes of a clear overview and analysis of basic functions, ideal conditions are assumed as is usual. This means:

- Ideal power semiconductors, i.e. forward voltage and blocking current zero, no switching delay, switching time zero.
- Ideal inductive and capacitive elements, i.e. no losses, reactance coils with constant inductivity, capacitors with constant capacitance.

8.6.2   Frequency Converter in Two-Level Topology

8.6.2.1   Topology

Figure 8.33 shows a frequency converter for converting the electrical power of a three-phase system into another three-phase system; in the case of the wind turbine, for converting the power of the generator with variable frequency for feeding the electrical grid with a fixed frequency.

The frequency converter for feeding the electrical power of the variable speed generator into the electrical grid with constant frequency consists of two parts. The left, generator-side part-converter in the form of a three-phase to DC converter is connected to the generator terminals in a three-phase manner. It consists, as also the other part-converter, per phase of two on/off switchable power semiconductors designated here by the symbol of the IGBT (*insulated gate bipolar transistor*) and respectively anti-parallel diodes for freewheeling. It generates an

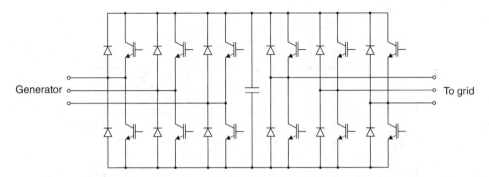

Figure 8.33   Frequency converter in two-level topology for converting the power of generator into the grid in wind turbines

output-side nearly direct voltage from the frequency-variable three-phase voltages. These are smoothed by means of capacitors. This part has the task of connecting and decoupling between the two part-converters and is called an intermediate DC link.

Connected to this direct voltage link is a DC to three-phase converter, the grid-side part-converter that generates a three-phase voltage with grid frequency and suitable amplitude and phase angle from the direct voltage in order to be able to feed power into the grid.

Both components are carried out in the same circuitry, the two-level circuitry. The dimensionings as regards voltage and current are slightly different. With the aim of unification, converters of the same dimensions can be installed on both sides.

Operation is by means of suitable control of the switching of the power semiconductors, called pulse-width modulation, as well as suitable controls of the currents and superimposed quantities.

8.6.2.2 Pulse-Width Modulation

Pulse-width modulation for the base circuitry of a DC to three-phase AC frequency converter in two-level topology is discussed in more detail here. Its circuit is shown in Figure 8.34 again but in more detail. This derivation applies for the machine-side as well as for the grid-side part-converter.

The circuits consist of six on/off switchable power semiconductors that are shown in Figure 8.34 with the symbol of the IGBT power semiconductor, for example V_1. Diodes, for example V_1', are positioned anti-parallel to each controllable power semiconductor and are necessary for the operation of the circuits, for guiding inductive current portions, called free-wheeling. From the direct current the circuit generates an amplitude and variable-frequency three-phase voltage. For this the power semiconductors are necessary, for instance, the power semiconductors V_1, V_1', V_4, V_4' for phase 1. At the input is the direct voltage U_d that is made available by the smoothing of a voltage by means of capacitors. For reasons of clarity in this depiction, intended as the basis for the following analysis, it is shown in two parts, each with half the voltage ($U_d/2$). On the three-phase side, reactors are required (not shown here) whose purpose is to reduce the pulsations of the three-phase current.

The frequency conversion is carried out by means of pulse-width modulation. The function of the conversion of direct voltage to three-phase voltage is shown in Figure 8.35 as the

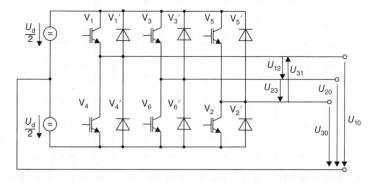

Figure 8.34 DC/three-phase frequency converters with two-level circuitry

example of a sine wave delta modulation for one phase. The sine wave component voltage reference value of adjustable amplitude and frequency, here shown related to a reference quantity and denoted as the modulation function $m(t)$, is compared with a higher-frequency delta signal $d(t)$ that determines the switching frequency.

The information of the sine wave value being greater than the delta value or the sine wave value being less than the delta value is used from the comparison of both functions. Where the sine wave value is greater than the delta value, the upper switches of the respective phase are switched on, for instance, V_1 for phase 1, while the lower switches, V_4 in case of phase 1, are switched off. With this the voltage $U_{10} = +U_d/2$ lies between the output point of phase 1 and the middle point of the intermediate circuit (0), independently of the current flow. If this is positive, then it flows through the controllable power semiconductor V_1, in the other case through the anti-parallel diode V'_1. If the sine value is less than the delta value, then the lower controllable power semiconductor V_4 is switched on, the upper one is switched off, and the output voltage is negative $U_{10} = -U_d/2$, also independently of the direction of the current flow. One phase can therefore generate two voltage values, hence being the reason for calling it two-level circuitry.

Both power semiconductors of a phase must not be switched on at the same time as this will lead to a short circuit of the DC link circuit. In order to ensure that a short circuit does not occur, a blocking time is arranged between the switching-on of the upper and the switching-on of the lower power semiconductors. Normally for converters in the power region of 1 MW and above, this time is set to microseconds.

The resulting output voltage of a phase is shown in the lower diagram in Figure 8.35. The sine wave modulated pulse widths can be clearly seen. The sliding middle value forms the required sine wave course. The other two phases are controlled in the same way, whereby the nominal sine values are each displaced by 120° and 240°.

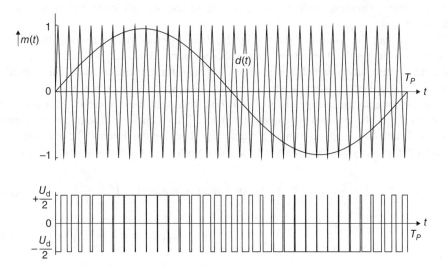

Figure 8.35 Pulse-width modulation by means of sine wave delta comparison for one phase of the DC to three-phase voltage converter; above: sine wave reference voltage value and delta function; below: resulting pulse series, voltage U_{10}, sine wave modulated

In order to achieve higher voltage values, the sine function can also be selected to be greater than the delta function. This is called overmodulation. This results in a more distorted voltage so it is usually avoided. The phase-to-phase three-phase voltages are the result of the difference of the voltage against the point 0, here as an example for the switching condition 1 (SC1, explained in the following) with the valves V_1, V_6 and V_2 switched on:

$$U_{12} = U_{10} - U_{20}; \quad \text{for switching condition 1}: \quad U_{12} = \frac{U_d}{2} - \left(-\frac{U_d}{2}\right) = U_d$$

$$U_{23} = U_{20} - U_{30}; \quad \text{for switching condition 1}: \quad U_{23} = -\frac{U_d}{2} - \left(-\frac{U_d}{2}\right) = 0 \qquad (8.65)$$

$$U_{31} = U_{30} - U_{10}; \quad \text{for switching condition 1}: \quad U_{31} = -\frac{U_d}{2} - \left(+\frac{U_d}{2}\right) = -U_d$$

Thus, the results for the phase-to-phase voltage are three voltage levels.

The relationship of the amplitude of sine wave reference value to the amplitude of the delta-shaped carrier function (half of the DC voltage) is designated as the degree of modulation m. The three-phase voltage u_{Pnm} between phases n and m for the pulse-width modulation is proportional to the degree of modulation up to its value 1.0, and in reverse proportional to the direct voltage:

$$m = \frac{U_{star}}{U_d/2}; \quad m = 0\ldots1(K1.15) \qquad (8.66)$$

for linear modulation without overmodulation. The value $m = 1.15$ applies when for the DC three-phase converter, instead of the pure rated sine wave function, use is made of one with an additional component of three-times frequency and 1/6 amplitude ($\sin\omega t + (1/6)\sin 3\omega t$). For systems without a neutral conductor, the direct phase components fed in with it have no negative effects.

The process of the control of the voltage has been clarified on the basis of the sine wave delta pulse-width modulation. However, another widespread method of voltage control is the space vector modulation, which is well suited as a basis for the realisation of control hardware with microcontrollers or other programmable circuits. Space vectors, for instance $\underline{U}(t)$ of the voltage, as complex time-dependent dimensions, represent the properties of a three-phase voltage or three-phase current or other dimension. They contain information on amplitude, phase angle and frequency also under dynamically changing conditions. The real components U_α and the imaginary components U_β of the space vector are determined according to:

$$U_\alpha = \frac{2}{3}U_{12} + \frac{1}{3}U_{23}; \quad \text{for switching condition 1}: \quad U_\alpha = \frac{2}{3}U_d \qquad (8.67a)$$

$$U_\beta = \frac{\sqrt{3}}{3}U_{23} + \frac{1}{3}U_{23}; \quad \text{for switching condition 1}: \quad U_\beta = 0 \qquad (8.67b)$$

The space vector is made up of real and imaginary components:

$$\underline{U}(t) = U(t)e^{j(\omega(t)t)} = U_\alpha(t) + jU_\beta(t) = U(t)(\cos\omega(t)t + j\sin\omega(t)t) \qquad (8.68)$$

Table 8.2 Circuit conditions for two-level direct current/three-phase converter

U_{10}	U_{20}	U_{30}	V_1V_4	V_3V_6	V_5V_2	No.	$\dfrac{U_\alpha}{U_d}$	$\dfrac{U_\alpha}{U_d}$	$|U|$	Arc (U)
$+U_d/2$	$-U_d/2$	$-U_d/2$	10	01	01	1	2/3	0	2/3	0°
$+U_d/2$	$+U_d/2$	$-U_d/2$	10	10	01	2	1/3	$\sqrt{3}/3$	2/3	60°
$+U_d/2$	$-U_d/2$	$+U_d/2$	10	01	10	6	1/3	$\sqrt{3}/3$	2/3	−60°
$+U_d/2$	$+U_d/2$	$+U_d/2$	10	10	10	7	0	0	0	0°
$-U_d/2$	$-U_d/2$	$-U_d/2$	01	01	01	0	0	0	0	0°
$-U_d/2$	$-U_d/2$	$+U_d/2$	01	01	10	5	−1/3	$\sqrt{3}/3$	2/3	−120°
$-U_d/2$	$+U_d/2$	$+U_d/2$	01	10	10	4	−2/3	0	2/3	−180°
$-U_d/2$	$+U_d/2$	$-U_d/2$	01	10	01	3	−1/3	$\sqrt{3}/3$	2/3	120°

The space vector of a sine wave symmetrical three-phase system is a circumferential moving vector with constant amplitude and frequency on the complex plane that describes a circle. This provides understanding of the properties of the space vector.

In pulse-width modulation, discrete space vectors are switched due to the various switching conditions of the converter. Table 8.2 shows the switching conditions for all possible circuit combinations of the power semiconductors of the circuit. Columns 4 to 6 show which valves are switched on (1: on, 0: off).

Columns 1 to 3 give the respective values of the voltage phase against the middle point of the DC input circuit. The circuit conditions are numbered in column 7 and the values of the space vectors are given in columns 8 to 11.

There are six active space vectors and two zero space vectors. Zero space vectors with the amplitude of zero are created when all upper or all lower valves are switched on. Two switched-on valves in one phase are not allowed as this means a short circuit. The circuit conditions can be followed in the circuit diagram.

The circuit conditions can be clarified in a diagram in the complex α,β plane as space vectors (see Figure 8.36). The conversion to space vectors was given further above. The six active space vectors (switching states 1…6) that lead to the outside corners of the hexagon and the two zero space vectors (switching states 0, 7) in the middle of the hexagon can be clearly distinguished.

The space vector, for instance, for sine wave symmetrical three-phase voltages with constant amplitude and frequency or for the pulse-type output voltage of the sine-wave converter, can be generated by means of rapid sequential switching-on of the nearby base space vectors. In the respective sector in which the rated space vector is situated, the right, the left and the neutral space vector must be switched on for a suitable time. For the first sector between base space vectors 1 and 2, this is shown as an example in Figure 8.37, and the equations resulting from this are summarised. This applies correspondingly for the other sectors.

Always one pulse period T_P with the time of a fraction of the period T of the rated sine function of the voltage is observed. The base space vectors 1, 2 and 7 or 0 are to be switched on for a certain period of time T_{on1}, T_{on2}, T_{on7} or T_{on0}, and together result in the rated voltage $\underline{U}$:

$$\underline{U} = \frac{T_{on1}}{T_P}\underline{U}_1 + \frac{T_{on2}}{T_P}\underline{U}_2 + \frac{T_{on0,7}}{T_P}\underline{U}_{0,7} \tag{8.69}$$

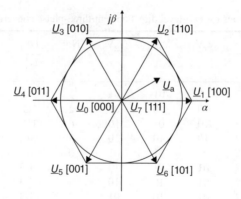

Figure 8.36 Basic voltage space vectors and example of a reference space vector for a two-level inverter on the complex plane

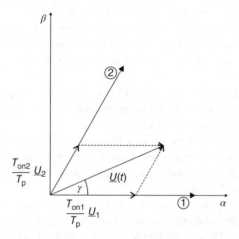

Figure 8.37 Formation of the rated space vector $\underline{U}(t)$ from the base space vectors (example)

The phase angle γ of the space vector to the real α axis is used for calculation of the switch-on times. The switch-on times of the space vectors 1, 2 and 7 or 0 then result in:

$$\frac{T_{on1}}{T_P} = \sqrt{3}\,\frac{U(n)}{U_d}\sin(60° - \gamma) \tag{8.70a}$$

$$\frac{T_{on2}}{T_P} = \sqrt{3}\,\frac{U(n)}{U_d}\sin\gamma \tag{8.70b}$$

$$\frac{T_{on0,7}}{T_P} = \frac{T_P - T_{on1} - T_{on2}}{T_P} \tag{8.70c}$$

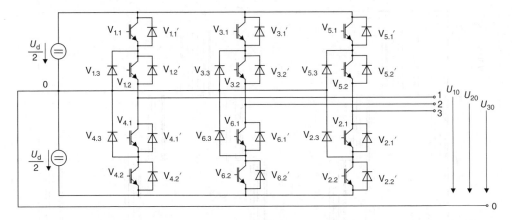

Figure 8.38   Three-level frequency converter, neutral-point-clamped topology

These equations are also used in the implementation of microcontrollers. In the control system, the reference values of amplitude of the phase-to-phase voltage $U_{pp}(n)$ and phase angle $\gamma(n)$ of the voltage are defined and prescribed. From the reference voltage value the switch-on times are determined by the frequency of the pulse frequency f_p and the reverse value pulse period T_p. Counters are used for the implementation in microcontrollers; these count the switch-on times and the switch signals are then passed on to the valves. The switch frequencies of the converters in wind turbines are in the kHz region (e.g. 2.5… 5 kHz) or below that for greater power.

Today, mostly IGBTs and diodes based on silicon are used as power semiconductors in converters for wind turbines of power in the middle megawatt region. The IGBT is a combination of MOSFET input and bipolar output stage.

8.6.3   Frequency Converter with Multi-Level Circuits

Converters with power semiconductors in series or parallel connection are used for wind turbines with a converter power in the higher megawatt regions. For this case the variant of the converter with multi-level circuits has been shown to be economical. These are circuits of power semiconductors in series. They are not switched at the same time but in a displaced sequence. With this method it is possible to switch finer voltage steps at the output and thus to reduce the required filter effort.

Mostly three-level converters in the neutral-point-clamped (NPC) topology [12] are used for wind turbines in the area of higher power. The circuit diagram is shown in Figure 8.38. Again the converter consists of three identical phases. In each phase the upper and lower branches are structured almost identically. Two power semiconductors, here depicted as IGBT with anti-parallel freewheeling diodes, are switched in series in each branch. The phase output is connected between the two branches.

Because there is a link to the middle point of the intermediate circuit via additional diodes within the series switching of the power semiconductors of a branch, each individual phase can generate the output voltage $U_d/2$, 0 and $-U_d/2$ against the middle point 0. For generating the output voltage $+U_d/2$ it is necessary, for instance, for phase 1 to switch on the power

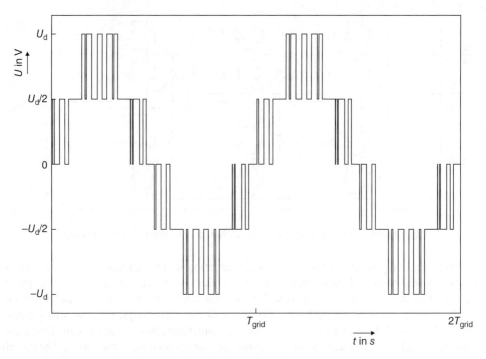

Figure 8.39   Three-level frequency converter; course of the phase-to-phase AC voltage

semiconductors $V_{1.1}$ and $V_{1.2}$, for the outlet voltage 0 the power semiconductors $V_{1.2}$ and $V_{4.1}$. Thus, three voltage levels can be generated for each phase, which has led to the designation 'three-level converter'. Figure 8.39 shows the phase-to-phase output voltage of the three-level converter. The phase-to-phase output voltage is calculated by the difference of two phase-to-midpoint voltages and forms a five-level shape.

8.7   Control of Variable-Speed Converter-Generator Systems

At present, four different variants of variable-speed generators for wind turbines are found in new installations. On the one hand, there are variants with short-circuit rotor induction machines and with permanently excited and externally excited synchronous machines, each with a fully rated converter whose structure is shown in the upper part of Figure 8.40. On the other hand, there is the variant with double-fed induction machines, thus an asynchronous machine with a slip-ring rotor, here with partly rated converters in the rotor circuit, as shown in the lower part in Figure 8.40.

With the aid of control systems [7,13,14], appropriately structured for each of the different generator variants, the generator is speed-controlled according to the demands of the operation, and feeds power into the DC link and the grid. The first function is normally carried out by the generator-side part converter, and the second function by the grid-side part converter.

The controls at the generator-side converter regulate the speed of the generator, and this occurs indirectly via the torque and therefore the power taken from the wind. The control of

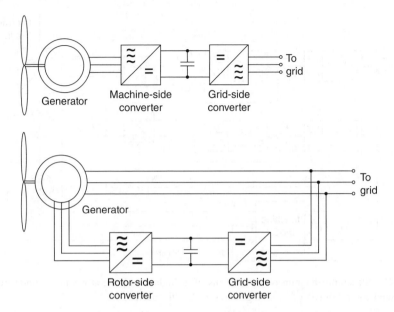

Figure 8.40 Structure of electrical drive trains state-of-the-art in wind turbines; top: generator with fully rated converter, bottom: induction generator with slip-ring rotor and converter at the rotor, with partly rated converter

the grid-side converter controls the power to be fed in to the electrical grid. This latter occurs indirectly in that the voltage in the DC link circuit is controlled to a constant value. Without control, it would rise or sink depending on greater or lesser feed from the generator.

These two control tasks are carried out separately; the coupling occurs via the system, with the flow of power in the DC link of the converters and its capacitor.

There are also concepts of the control system where the grid-side power converter regulates the speed and the machine-side one the DC link circuit voltage, thus the functions are distributed in reverse.

In the fully rated converter the whole of the flow of power passes through and is controlled by the converter [15]. In the variant with partly rated converter, only a part of the overall power passes through the rotor-side converter, but this also controls the overall power of rotor and stator [15].

The following presents a short example of the controls of the induction generator representing a generator control as well as the controls of the grid-side current converter. The controls of the other generator variants are briefly mentioned.

8.7.1 Control of the Converter-Fed Induction Generator with Short-Circuit Rotor

The standard field-oriented control for an induction generator is presented as an example. Figure 8.41 shows a structural picture of the control. The control is structured in two channels.

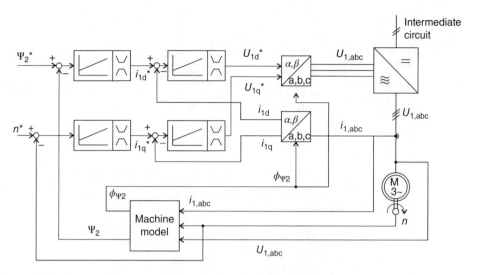

Figure 8.41 Structural diagram of the controls of an induction generator with a short-circuit rotor (field-oriented control of rotor flux linkage and speed, with speed sensor)

The concept comprises one control channel in which the flux in the machine, for instance the rotor flux linkage Ψ_2, is controlled and one channel in which the speed N, or alternatively the torque M, is controlled. In addition, the controls are cascaded in that a secondary subordinated current control in each channel is used. This has advantages in that the respective controllers can be set on the respective components, and also a limitation to the rated current value is possible in order to protect the converter.

The operation is described in brief. The actual values of the rotor flow linkage Ψ_2 and its reference value Ψ_2^* are compared and the difference is passed to the PI flux controller. This controller has the task of influencing the system via its output in such a manner that the reference and actual values are equal. The output of the flux regulator is the reference value for the flux-forming current i_{1d}^*. This current can be limited, as shown by the special symbol box behind the controller. It is compared with the measured and determined actual values of this current and the difference is passed to the PI controller for the flow-forming current and controlled. Its output value is the flux-forming reference value U_{1d} of the stator voltage. The channel for the speed control is structured accordingly. Input quantities are the signals of reference speed value n^* and actual value n, the intermediate step is the torque-forming current i_{1q} with reference and actual values, and the output quantity is the torque-forming part U_{1q} of the stator voltage. The two voltage parts are combined into the overall voltage and give the reference voltage amplitude and phase angle for the pulse-width modulation of the generator-side converter. In the pulse-width modulator that is seen here as a component of the converter, and not shown individually in the figure, the switching patterns for the converter are calculated in order to provide the required output voltage value.

The actual values required for the controls are the stator current, the stator voltage and the speed. Internal dimensions of the machine that are needed for the controls, such as the rotor flux linkage in the amount Ψ_2 and its angle $\phi(\Psi_2)$ as well as the torque M, if required, are calculated with the aid of the machine model. This contains the equations of the machine, or is executed by means of observers. In this way a problematic flux measurement and torque measurement are prevented.

In the technology of drives it is usual to make use of proportional integral controllers (PI controllers) that are able to regulate the control deviations, the differences between reference and actual values, to zero. The equation is:

$$y = K_p x + \frac{1}{T_1} \int x \, dt \tag{8.71a}$$

$$G(s) = \frac{y(s)}{x(s)} = K_p \frac{sT_1 + 1}{sT_1} \tag{8.71b}$$

In this equation, x is the input variable of the controller, thus the difference between rated and actual value, and y is the output variable, thus the actuating variable. $G(s)$ is the frequency transfer function of the controller, the relationship of output to input quantities over the frequency. K_p is the proportional amplification factor, and T_1 the integration time constant of the controller. These two parameters are always to be designed for the system and the other conditions for the controllers [7]. In drive technology, the controllers are designed according to two strategies: the technical optimum and the symmetrical optimum. In this way, good results are generally obtained with regard to dynamics, overswing and robustness. The adjustment rules can be found, for instance, in [7].

In order to determine the parameters for the controllers, the generator of the system to be regulated must be modelled for dynamic operation. Often, as is also shown in Figure 8.41, the dimensions are represented in space vectors and the system control is executed in space vectors, frequently in a special dq coordinate system circulating with the rotor flux. For this reason, conversion units are necessary to transform from the three-phase abc system into the two-phase dq system for the input variables, and in reverse for the output quantities. For this reason the modelling takes place in the space vector representation as has already been introduced for the converter. A space vector is a complex dimension that represents an equivalent to a three-phase system whose quantities are linearly dependent. The real α components and the imaginary β components of the space vector, here written for the voltage, are determined from the three-phase dimensions according to the following equations.

$$\underline{U}^{\alpha\beta}(t) = \frac{2}{3}(U_a(t) + U_b(t)e^{j2\pi/3} + U_c(t)e^{j4\pi/3}) \tag{8.72a}$$

$$U_\alpha(t) = \frac{2}{3}\left(U_a(t) - \frac{1}{2}U_b(t) - \frac{1}{2}U_c(t)\right) \tag{8.72b}$$

$$U_\beta(t) = \frac{2}{3}\left(0 + \frac{\sqrt{3}}{2}U_b(t) - \frac{\sqrt{3}}{2}U_c(t)\right) \tag{8.72c}$$

In this, the superscript $\alpha\beta$ indicates that the voltage is represented in this coordinate system. The return transformation from the space vector representation into fixed coordinates in the abc three-phase dimensions occurs according to the following equation:

$$\begin{pmatrix} U_a(t) \\ U_b(t) \\ U_c(t) \end{pmatrix} = \begin{pmatrix} 1 & 0 \\ -\dfrac{1}{2} & +\dfrac{\sqrt{3}}{2} \\ -\dfrac{1}{2} & -\dfrac{\sqrt{3}}{2} \end{pmatrix} \cdot \begin{pmatrix} U_\alpha \\ U_\beta \end{pmatrix}$$

(8.73)

The named control of electrical machines in a coordinate system that moves with an electrical quantity, such as the flux, can simplify the analysis. The equations are transformed here into a rotating dq coordinate system that, for instance, is oriented to rotor flux and that has the time-dependent angle $\theta = \theta(t)$ to the fixed $\alpha\beta$ system.

$$\underline{U}^{dq}(t) = \underline{U}^{\alpha\beta}(t)e^{j\theta(U_{dq};U_{\alpha\beta})} = U_d(t) + jU_q(t) = \underline{U}^{\alpha\beta}(t)(\cos\theta + j\sin\theta)$$

(8.74a)

$$U_d(t) = U_\alpha(t)\cos\theta + U_\beta(t)\sin\theta$$

(8.74b)

$$U_q(t) = -U_\alpha(t)\cos\theta + U_\beta(t)\sin\theta$$

(8.74c)

The reverse transformation is:

$$U_\alpha(t) = U_d(t)\cos\theta - U_q\sin\theta$$

(8.75a)

$$U_\beta(t) = U_d(t)\sin\theta + U_q\cos\theta$$

(8.75b)

The transformation is carried out with the turn operator $e^{j\theta(U_{dq};U_{\alpha\beta})}$. The features of the representation of the space phasors in a rotating coordinate system can be seen in Figure 8.42. The

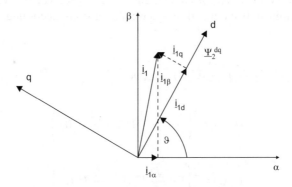

Figure 8.42 Rotor flux linkage and stator current in the stator-fixed $\alpha\beta$- as well as in the rotor flux-oriented dq coordinate systems, and partitioning of the stator current in these components

three-phase equations for the stator current I_1 are transformed firstly for this representation in the complex space vector equation for the stator-fixed $\alpha\beta$-coordinate system. Next, the rotor flux-oriented coordinate system is selected in which the control is to be carried out. In this case the d-component is placed on the rotor flux $\underline{\Psi}_2$. As shown in Figure 8.42, the rotor flux lies in the d-axis. The stator current is transformed in this coordinate system. Correspondingly it can be partitioned into the components i_{1d}, i_{1q} in both axes, thus the longitudinal component in the direction of the rotor flux and the lateral component orthogonal to it. For the controls this is a very advantageous representation, as the longitudinal portion of the stator current forms the flux and the lateral portion the torque, as shown by the torque equation (8.79a) below. These values can be used directly in the control, and form the already-depicted two channels.

It is a great advantage that for control in rotating coordinate systems, the dimensions for stationary operation are direct quantities and the control is therefore easy to carry out. Examples for these are the fluxes and currents in Figure 8.42 for the rotor flux-oriented coordinate system. However, control for three-phase machines can also be carried out in coordinate systems oriented to other quantities, for instance, referenced to the stator flux linkage. Also, a control in the stator-referenced system, thus in the $\alpha\beta$ system can be used.

The equations for the stator and the rotor voltage and the corresponding fluxes in modelling for dynamic behaviour of the induction generator can be derived from the known equation of the induction machine. These are transferred directly into the rotating coordinate system and the result is:

$$\underline{U}_1^{dq}(t) = R_1 \underline{i}_1^{dq}(t) + \frac{d}{dt}\underline{\Psi}_1^{dq}(t) + j\omega_{dq}\underline{\Psi}_1^{dq}(t) \tag{8.76a}$$

$$\underline{U}_2^{dq}(t) = R_2 \underline{i}_1^{dq}(t) + \frac{d}{dt}\underline{\Psi}_2^{dq}(t) + j(\omega_{dq} - p\omega_m)\underline{\Psi}_2^{dq}(t) \tag{8.76b}$$

$$\underline{\Psi}_1^{dq}(t) = L_1 \underline{i}_1^{dq}(t) + L_h \underline{i}_2^{dq}(t) \tag{8.76c}$$

$$\underline{\Psi}_2^{dq}(t) = L_2 \underline{i}_2^{dq}(t) + L_h \underline{i}_1^{dq}(t) \tag{8.76d}$$

In some parts these are known from the quasi-stationary point of view. The voltage equations contain a portion for the ohmic component drop and a portion for the component generated by variation of the flux. The respective term with the factor j result from the mathematical transformation into the rotating coordinate system, in detail due to the partial derivation of the respective flux term. If the equations in the components are dissolved here for the voltage, then they result in the characteristics of these terms:

$$U_{1d}(t) = R_1 i_{1d}(t) + \frac{d}{dt}\Psi_{1d}(t) - \omega_{dq}\Psi_{1q}(t) \tag{8.77a}$$

$$U_{1q}(t) = R_1 i_{1q}(t) + \frac{d}{dt}\Psi_{1q}(t) + \omega_{dq}\Psi_{1d}(t) \tag{8.77b}$$

These represent the rotating portion of the induced voltage (e.g. $\omega_{dq}\Psi_{1q}(t)$), a fictitious portion that exists due to the mathematical transformation. The portion with the derivation of the flux is called the transformative portion. The rotative portion is a coupling portion, that is, the portion of the lateral axis (q) acting on the voltage of the longitudinal axis (d) and in reverse.

The components i_{1d} and i_{1q} of the stator current that occur in the voltage equations are controlled in the current controller. The system of the machine can be interpreted for the design of this controller as a delay element of the first order, with the rotor stray time constant. In general the design according to the technical optimum is used [7].

In addition, the behaviour of the power converter is to be included. Power converters are dynamically characterised by a dead time that in a simplified manner is taken into account as a delay element of the first order [7].

$$G_{conv}(j\omega) = \frac{1}{1+sT_{conv}} \tag{8.78}$$

The time constant T_{conv} is determined, depending on the pulse pattern generation (single or double update, immediate implementation of the reference values), as a value approximately once or twice the reverse value of the pulse frequency ($T_{conv}=(1...2)/f_p$) [7].

For these derived dynamic equations, the dynamic equivalent circuit diagram of the induction machine as shown in Figure 8.43 can be assembled. It is equal largely to the stationary equivalent circuit diagram, but extended to take the rotary voltages into account.

The torque of the machine in the representation with rotor flux orientation results in a simplified formula where the simplification comes from the orientation of the coordinate system at the rotor flux (thus $\Psi_{2q}=0$):

$$m_{gen}(t) = \frac{3pL_h}{2L_2}\left(\Psi_{2d}i_{1q} - \Psi_{2q}i_{1d}\right) = \frac{3pL_h}{2L_2}\Psi_{2d}i_{1q} \tag{8.79a}$$

$$\underline{\Psi}_2 = \Psi_{2d} + j0 \tag{8.79b}$$

The variation with time of the speed or the angular frequency of the mechanical rotation of the generator is formed from the generator torque m_{gen} and the rotor of the wind turbine m_{rot}, which act on the moment of inertia Θ of the drive train.

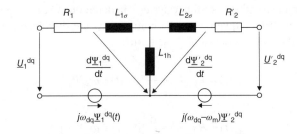

Figure 8.43   Dynamic equivalent circuit diagram of the induction machine

$$\frac{d}{dt}\omega_{mech} = \frac{1}{\Theta}\left(m_{rot}(t) - m_{gen}(t)\right) \tag{8.80}$$

The speed controller acts on this system. The controlled system has an integral behaviour, so that according to the rules in general, a design of the controller parameters according to the symmetrical optimum is selected [7].

As the power electronics generator systems in wind turbines are not operated in the field weakening region, this is not discussed here.

8.7.2   Control of the Doubly-Fed Induction Machine

The structural control diagrams for further types of generators are presented and explained here in order to give an idea of the method of control. Figure 8.44 shows the control structure for the doubly-fed induction machine. It corresponds in parts to the basic structure of the induction machine with a short-circuit rotor. It contains a speed and reactive power control in parallel channels as well as a subsidiary current control. In contrast to the induction machine with a short-circuit rotor, it is not the flux that is directly controlled, but the stator voltage via the flux, in amount and phase, and thus the stator reactive power. Also the torque control is implemented indirectly, as it is not the stator current but the rotor current that can be directly influenced.

The d channel and the q channel are shown here together in a single channel. Input dimensions of the controls are the stator voltage and flux as well as the speed, and in addition the rotor current that is required for the current control of the rotor-side converter. As the controls are in dq coordinates, also the corresponding abc/dq and dq/abc transformation blocks are required. In addition there is a reactive power calculation for the stator side that must be carried out dynamically [16] and a calculation of the angle γ_2 of the rotor flux position for the transformation blocks is required. The phase angle γ_1 of the stator voltage is determined by means of a phased lock loop (PLL).

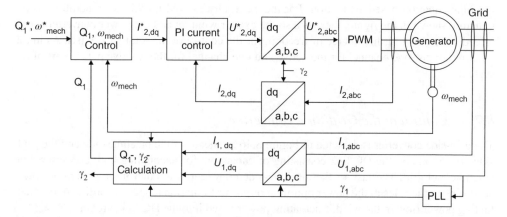

Figure 8.44 Structural diagram of the field-oriented control of the doubly-fed induction machine (speed and reactive power controls, rotor voltage-oriented regulation)

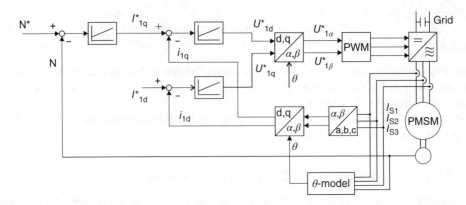

Figure 8.45 Structural diagram of the field-oriented controls of the synchronous machine (torque or speed and flux and reactive current control, rotor-oriented method)

8.7.3 Control of the Synchronous Machine

The control structure of the synchronous machine, in the externally excited as well as the permanently excited version, is structurally much the same as the induction machine with a short-circuit rotor. Figure 8.45 shows the control structure of a permanently excited synchronous machine. Speed control and flux control, here carried out as reactive current control, as well as the subsidiary current control and calculation of reference values for the converter, are features of the structure.

The controls are normally carried out with the rotor-circulating dq coordinate system which is oriented in the direction of the rotor flux. In the permanently excited synchronous machine, the rotor flux generated in the rotor is fixedly conditioned in the amplitude by the excitation with permanent magnets. However, due to flux in the stator coils also for this type of machine, flux portions can be generated in the direction with or against the flux of the permanent magnet (field weakening or strengthening).

The angle of the rotor for transformation of the variables is determined via observer or machine equation based estimation. For the externally excited machines, in addition to the structure shown in the figure, there is also the exciter winding and the power converter for its infeed, whereby a further degree of freedom for the control of the rotor flux exists. As in the other diagrams, one channel for the active and one channel for the reactive part is part of the system.

8.7.4 Control of the Grid-Side Converter

The grid-side converter is depicted in Figure 8.46 by means of the machine-side and the grid-side converter, and the DC link consisting of capacitors for decoupling both converters and machine and grid. The task of the machine-side converter is to control the generator power according to the rules in the power optimum or power limitation range. According to the fluctuating power flow in the wind, fluctuating power is fed into the DC link circuit.

The grid-side converter has the task of feeding the power from the generator, fed into the DC link, into the grid. This occurs indirectly in that the task of the controls is to keep the DC

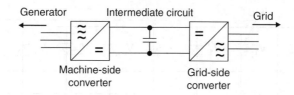

Figure 8.46 Machine-side and grid-side converter with intermediate circuit

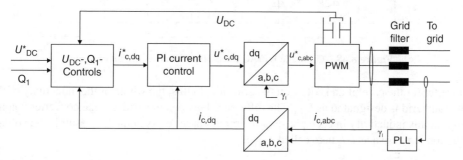

Figure 8.47 Structure diagram of the grid-side converter control (control of the DC link circuit voltage and the reactive power; oriented to the grid voltage)

link circuit voltage constant. This ensures that the power output from the generator is fed directly and properly into the grid.

The control structure of the grid-side converter, as shown in Figure 8.47, again corresponds to the basic structure of the other systems. The control of the DC link circuit voltage is super-imposed, which prescribes the d components of the reference converter current. Subimposed is the current control for the active current. By this, a proper feed-in of power from the generator from the DC link circuit into the grid is ensured. The reactive power control takes place in the second channel. If no reactive power is to be exchanged with the grid, then the reference value is set to zero.

Subordinately the control consists of a two-channel grid current control whose outputs form the reference value of the converter voltage. The control of the grid-side pulse power converter is generally carried out in rotating dq coordinates oriented to the grid voltage. The angle of the grid voltage is determined via a phase control circuit (PLL, phase locked loop) [7]. The actual values of the grid are transformed from the three-phase system into the corresponding dq system, and the reference voltages for the pulse-width modulation from dq into the abc system. The PWM block contains the pulse-width modulation and the power section of the converter.

The power to the electrical supply grid is fed in via a grid filter that is required for maintaining the limiting values for the harmonics from the standards [14]. Here an L-type filter is shown; grid filters can also be designed as LCL filters. The result is then a reduced filter effort but at the expense of increased demands on the controls due to the capability of the filter to oscillate. The detailed converter circuit with L-type filter is shown in Figure 8.48.

The dynamic behaviour of the controlled system of the grid-side converter will be shown as an example. It is the dynamic modelling of the controlled system of the grid current sine wave

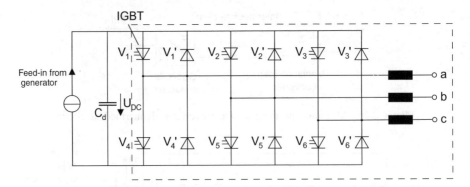

Figure 8.48 Grid-side DC/AC three-phase inverter: circuit diagram

converter for the case of an L filter on the grid side. In this, the whole of the inductivity of the filter and grid is designated as L_{grid}. The difference between the grid and the converter voltage drops at this inductivity in each phase of the three-phase AC system. The result is the voltage equation in phase a as (U_c here called U_{conv}):

$$U_{grid,a} = L_{grid}\frac{di_{conv,a}}{dt} + U_{conv,a} \tag{8.81}$$

and analogues for those in phase b and c. After the transformation into the space vectors and the transformation of this equation into a coordinate system rotating with the grid voltage as well as the dissolution for the differential quotient, the result is:

$$\frac{di_{conv,d}}{dt} = \frac{1}{L_{grid}}\left(U_{grid,d} + \omega_{grid}L_{grid}i_{conv,q} - U_{conv,d}\right) \tag{8.82a}$$

$$\frac{di_{conv,q}}{dt} = \frac{1}{L_{grid}}\left(U_{grid,q} - \omega_{grid}L_{grid}i_{conv,d} - U_{conv,q}\right) \tag{8.82b}$$

Thus, the dynamic behaviour of the grid-side converter is described by means of the current dynamics, which is necessary for the design of the parameters of the controls. Here, too, cross coupling occurs, with the influence of the q-components of the converter current on its d-components and vice versa. As regards the behaviour of the dynamic system, the effects of the converter voltage U_{conv} on the converter current i_{conv} result in the behaviour of an integrator. A PI control is usually applied. The control parameters are to be defined according to the symmetrical optimum [7,14].

The superimposed control of the DC link circuit voltage U_{DC}, which is equal to the voltage U_{Cd} at the intermediate circuit capacitor, works on a system that basically consists of this DC link circuit capacitor C_d. This capacitor is fed in by the current i_{Cd}, consisting of the active generator current $i_{gen,d}$ minus the active current $i_{con,d}$ led off by the grid-side converter.

$$U_{Cd} = \frac{1}{C}\int i_{Cd}\,dt + U_{Cd0} \tag{8.83a}$$

$$\frac{dU_{Cd}}{dt} = \frac{1}{C} i_{Cd} = \frac{1}{C}\left(i_{gen,d} - i_{conv,d}\right) \tag{8.83b}$$

Here, integral system behaviour for the DC link is found. In general, a PI controller is applied for the DC link voltage control. The design of the control parameters according to the symmetrical optimum is recommended.

8.7.5 Design of the Controls

The controls are required to adhere to the general requirements such as dynamics, stability and robustness, and many requirements that are typical for wind turbines such as:

- smoothed generator torque;
- low voltage pulsations in the intermediate circuit;
- smoothed grid power;
- suitable control reaction to harmonics in the grid voltage;

and this in both stationary and dynamic operation. The design of the controls is carried out generally in accordance with the usual rules for drive technology of the technical optimum and the symmetrical optimum [7], as briefly described for the individual controls.

In addition, wind turbine generator systems, and here especially their power electronics, must ensure a high standard of requirements for grid support in stationary operation as well as dynamics for voltage failures and interruptions (low-voltage ride-through, LVRT). This is achieved with special control design or by means of additional control measures.

8.8 Compliance with the Grid Connection Requirements

On the one hand, proper operation of the whole electrical and mechanical system is to be ensured for the general operation of the wind turbine. This is achieved by a suitable converter-generator system and by means of a suitable control operation.

On the other hand, the respective stationary grid connection requirements as regards low-frequency electromagnetic interference must be met, thus the permissible harmonics in current and/or voltage must be limited. This is achieved by means of a corresponding selection of pulse width and space vector modulation in connection with suitable L or LCL filters on the grid side, and possibly special converter topologies (for example, with interleaved switching). Suitable single-frequency filter circuits may also be used in order to reduce individual harmonics in the current.

Further important factors are the grid connection requirements [17–20]. An important requirement is the grid support in stationary operation. The grid connection requirements require active power-influencing and reactive power-feeding, dependent on the condition of the grid. This must be implemented in the controls and the power components must provide power reserves – for the case of an additive active and/or reactive feed-in.

A further grid connection requirement is the grid support in dynamic operation. For instance, the grid connection rules require an uninterruptible operation in the case of grid faults with

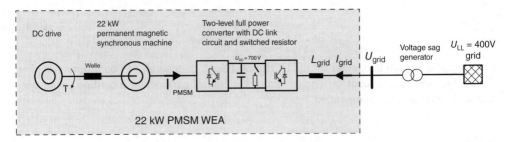

Figure 8.49 Structure of a laboratory system of a wind turbine drive system with a frequency converter and permanent magnet excited synchronous machine; with voltage sag generator for the analysis of low-voltage ride-through

low or zero grid voltage. If the grid voltage collapses, then the grid-side converter must remain in operation and feed in reactive current. This must be ensured by the control concept, the controllers and the power section. In addition, the generator continues to generate power that is fed by the generator-side converter into the DC link circuit. This leads to an increase in the DC link voltage. Protective measures must be taken to protect the converter from damage due to overvoltage [2,21]. This is achieved in general via a switched resistance in the DC link circuit, if control measures are insufficient.

The power electronics generator systems are to be designed for these requirements of the grid connection conditions on the hardware and control side. It must be ensured that the requirements for grid feed-in are constantly refined and sharpened with constantly expanding knowledge and experience from operation with decentralised renewable energy. Thus, requirements in this field on newly installed wind turbines grow continuously.

In order to clarify these properties, an example of the sequence of a grid under-voltage and wind turbine reaction is presented in Figure 8.49. Use is made of measurements in a 22 kW system in the laboratory. It corresponds to the concepts presented in the previous chapters.

The measurements were carried out on a system with a permanently excited synchronous machine. This system feeds power into a grid via a two-level DC/AC converter and a filter inductance. Installed in the DC link circuit of the converter is a resistor with a direct DC/DC converter that becomes active above a certain voltage, absorbing power and limiting the DC voltage. Connected with the grid is a voltage sag generator that performs the grid voltage sags. The rotor power is fed by means of a DC machine that can emulate a wind power profile. Measurement results for a three-phase voltage sag with a voltage down to 12% of the nominal voltage and a duration of 300 ms are shown in Figure 8.50. In this, the measured quantities (grid voltage, grid power, DC link power, stator flux, speed) are depicted, taken up by means of the hardware of a control system and visualised with Matlab.

In stationary operation with full grid voltage in the time period of 0 s to approximately 0.08 s, the grid-side converter feeds a current of $I_{grid}=6\,$A into the grid. At the time of 0.08 s the grid voltage collapses to 12% of the nominal grid voltage. Transient overcurrents of up to 30 A exist on the grid side. Due to the low grid voltage and the converter currents limited to their nominal value, only very limited power can be fed into the grid. In order to feed as much of the power from the DC link circuit into the grid as possible, the grid power flow is controlled

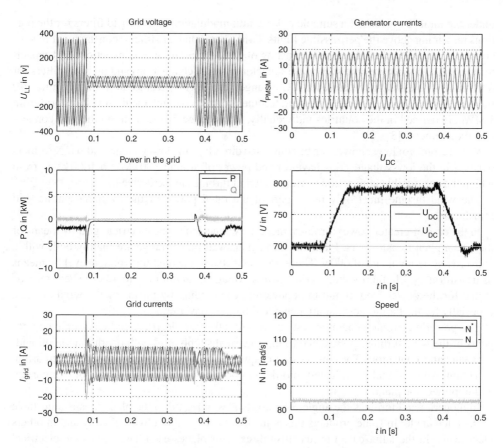

Figure 8.50 Sequence of the low-voltage ride-through in the system with a permanently excited synchronous machine and full power converter in the laboratory (22 kW); DC/DC converter with resistor and voltage limit controller in the DC link; grid voltage sag down to 12% of the grid nominal voltage

up to the system limits (here $I_{grid,max} = 10\,A$). The generator power flow and thus the generated power remain constant. Because of the power equalisation, the DC link circuit voltage U_{DC} rises. The DC/DC converter and the resistance in the DC link circuit are activated and limit the DC link voltage there to $U_{DC} = 790\,V$. The grid voltage collapse is mastered well. At the time 0.38 s the full grid voltage is restored. The grid flow collapses for a short period. In the time period 0.39 s to approximately 0.46 s the grid currents remain at the limiting value of 10 A and unload the DC link circuit capacitor to its rated voltage of 700 V. Then a regular stationary operation occurs again.

8.9 Further Electronic Components

On the grid side, the permissible voltage and current harmonics are defined in standards or guidelines of the grid operators. The harmonics generated by the converter, dependent mostly on the power converter circuits, pulse-width modulation and operating point, must be kept under the limits by means of a suitable pulse-width modulation and via grid filters for the portions remaining above the permissible limits. Grid filters can be L-filters (consisting of reactor coils), LCL filters in T connection (consisting of grid reactor in series connection, grid capacitor in parallel connection and second grid reactor in series connection) and possibly filters for remaining single frequency currents (LC arrangements). The permissible limiting values for the current harmonics, generally for wind turbines connected to the medium voltage grid from 1 MW, are defined in the countries individually, for instance, in Germany in the grid connection requirements [3].

Wind turbines of lower power can be connected directly to the low-voltage grid. Transformers existing in the low-voltage grids have a rated power of around 300 kVA to 1000 kVA, from which the permissible connection power of a wind turbine can be derived.

The wind turbine must only use a part of the rated power of the available public grid transformer.

In the case of greater power turbines, these are connected to the electrical grid by means of their own plant transformers. Plants of lower megawatt power range are mainly built with a converter system voltage of 690 V. These are connected to the medium voltage level by means of transformers. Medium voltage grids have a voltage of about 6kV to 40kV. They must be suited for the power feed in and the converter type feeding by means of the wind turbine, especially for the harmonics that are generated by the converter.

The power electronic generator system, as with all other electrical and electronic components of the wind turbine, is switched to the electrical supply grid with operation via on-and-off switchable contactors. The contactors are equipped with electronic and thermal overcurrent detection, whose signal is evaluated in order to open the contactor in the case of overcurrent and to protect the plant.

In modern wind turbines the angle the rotor blades in regard to the blade axis is adjustable to the pitch angle. In some products this adjustment is carried out hydraulically and in others electrically. In the latter case the (usually) three rotor blades each have their own electrical pitch drive. Systems such as geared motors can be used. An emergency power supply is necessary for the pitch drives in order to turn the rotor blades out of the wind into a safe position in the case of wind turbine system or grid voltage failure.

Also, additional measuring arrangements are necessary for the operation of the wind turbine. On the one hand this is a wind velocity measurement (anemometer) in order to select the type of operation from the actual value of the wind velocity and to prescribe the reference value of the velocity (power optimising range) or power (power limiting range). On the other hand, an arrangement for wind direction measurement is necessary in order for the plant to follow the nacelle to the wind direction on the basis of its measured value

8.10   Features of the Power Electronics Generator System in Overview

The power electronic generator systems of the wind turbines in operation and production are based on various basic concepts that are dealt with here. In addition there are always manufacturer-specific features. Table 8.3 summarizes the most important basic features of the various concepts, which should be self-explanatory.

Table 8.3 Features of power electronics generator systems in wind turbines

Features	System IM-short circuit rotor	System IM-slip ring rotor	System SYM-external excitation	System SYM-permanent excitation
Machine type, special features	Stator standard; rotor short-circuit squirrel cage	Stator standard; rotor with coil; slip rings for rotor current	Stator standard; rotor with coil; slip ring for exciter current	Stator standard; rotor with permanent magnets
Displacement factor generator stator current in rated operation	Approximately 0.85	Approximately 0.95	Approximately 0.95	Approximately 0.95
Efficiency of machine	Medium	High	High	High/very high
Converter model, rated apparent power S_N	Standard, full power converter, approx. 118% P_n	Standard, part power converter, approx. 30% P_n	Standard, full power converter, approx. 105% P_n	Standard, full power converter, approx, 105% P_n
Efficiency of converter	Not so high due to high reactive current	High and only part power via converter (low losses)	High	High

Up until now the use of gears has not been discussed in detail in this chapter. In general, as regards gears at present, the three variants for wind turbines in the upper megawatt region are found with corresponding consequences for the speed of the generator:

- Plants with drives with gears of approximately 1:100 transmission and generators with medium speeds of about 1500 rpm.
- Plants with drives with gears of approximately 1:10 transmission and generators with medium speeds of about 150 rpm.
- Gearless plants with gears of 1:1 transmission and generator with extremely low speeds of about 15 rpm.

The non-use of gear drives in the latter variant reduces the load on the mechanics substantially and leads to savings of cost, volume and weight for the mechanical part. However, the opposite is the case for the electrical part: electrical machines with lower rated speeds are more voluminous and more expensive. Obviously, a certain trend in the past few years to gearless plants has been observed.

8.11 Exercises

Three-phase systems: A wind turbine feeds power into a three-phase grid. The grid voltage is 690 V (effective value of the phase-to-phase voltage). A rated current of 1500 A is fed in at a displacement factor of cos $\phi = 0.8$ capacitive.

1. Determine the active, reactive and apparent powers.
2. What reactive power, e.g. from a parallel-fed reactive power current converter, is required in order to compensate (extinguish) the fed-in reactive power?

3. To what value can the grid current be reduced when the same active power of part (1) would be fed with cos $\phi = 1$?

Transformer: A wind turbine of low-voltage design feeds into the medium voltage grid via a three-phase transformer.

Data for the transformer:
Rated apparent power $S_N = 3\,\text{MVA}$
Rated primary voltage $U_{1N} = 10\,\text{kV}$
Rated secondary voltage $U_{2N} \approx 690\,\text{V}$
Circuit type Dyn5
Ohmic relative short-circuit voltage $u_R = 0.03$
Reactive relative short-circuit voltage $u_X = 0.05$
Ratio of primary to secondary winding numbers $w_1/w_2 = 14.925$.

The transformer is operated on the primary side at the 10 kV grid. A simplified short circuit equivalent circuit diagram is to be assumed for deriving the solutions.

1. Draw the equivalent circuit diagram of the transformer. On the secondary side the transformer is operated in idling. The voltage drop at the primary side reactance should be neglected due to the low magnetising current. Determine the secondary voltage for this idling point.
2. On the high-voltage side of the transformer, a pure active power of 2 MW at 10 kV, no reactive power, is now fed into the grid. Determine the required current (RMS-value).
3. The active current from part (2) is fed via the transformer into the 10 kV grid with fixed voltage of 10 kV. What is the voltage drop at the transformer impedance in amount and phase, referenced to the high-voltage side and the low-voltage side? What is the voltage at the low-voltage side?
4. In addition, a reactive current of 30% of the active current is fed in, once capacitively and once inductively. What is the voltage in these cases on the low-voltage side of the transformer?

The following points apply: the complete equivalent circuit diagram, only without iron resistance, is to be taken for deriving the following solutions. From the primary side idling measurement it is known that there is a current of 2% of the rated current (at nominal primary side voltage U_{1N}).

5. Draw the equivalent circuit diagram. The data of the five impedances are to be determined from the above information. The conditions that the primary and the secondary side reactance, real and imaginary part, are assumed to have the same value, apply.
6. On the primary side 2 MW active power is to be fed into the grid again. The voltages and the currents in the transformer are to be determined using the impedance values determined above. Draw the phasor diagram of the voltages and currents, and determine the losses.

Induction generator and converter: Figure 8.51 shows the measured Heyland diagram of an induction machine with short-circuit rotor. The machine is fed by a three-phase grid with 400 V/50 Hz (phase-to-phase voltage). The stator of the machine is star-connected. The machine is a 4-pole type (pole pair number $p = 2$).

In the following it is assumed that the primary and secondary leakage numbers of the machine are identical ($\sigma_1 = \sigma_2$). In addition, the ohmic losses of the stator and the iron losses can be ignored. *Remarks*: (1), (2) and the first part of (3) can be solved independently of each other.

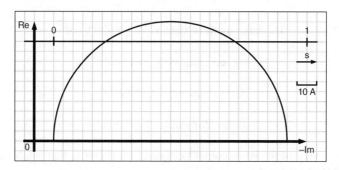

Figure 8.51 Heyland diagram of an induction machine with short-circuit rotor

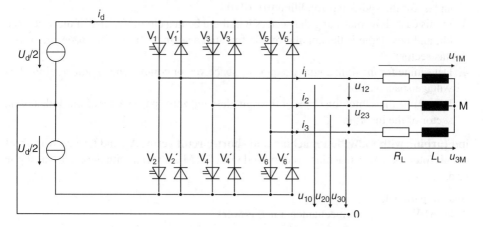

Figure 8.52 DC/AC three-phase two-level inverter

1. Calculate the following elements of the single-phase equivalent circuit diagram of the star-connected machine with the aid of the Heyland diagram: primary leakage inductance $L_{1\sigma}$, primary main inductance L_{1h} and stator referred secondary leakage inductance $L_2'\sigma$. Draw the basic stator current phasor from calculation into Figure 8.51.
2. Draw the stator current $I_{1,\text{opt}}$, for the optimum operating point, the phase angle ϕ_{opt} and the slip s_{opt} into Figure 8.51 and determine the values of the three quantities. Then calculate the power factor $\cos\phi_{\text{opt}}$, the electrical active power $P_{1,\text{opt}}$, the efficiency η_{opt}, the mechanical speed n_{opt}, the mechanical power $P_{\text{mech,opt}}$ and the torque of the machine M_{opt} at this operating point.
3. Determine the stall slip s_{stall} with the help of Figure 8.51 and calculate the stall speed n_{stall} as well as the rotor resistance R_2' of the single-phase equivalent circuit diagram of the machine.
4. Calculate the stall torque M_{stall} of the given machine.

Converter with space vector modulation: With the aid of a pulse controlled inverter, at an electrical load a sinusoidal symmetrical three phase AC system shall be emulated as well as possible. The equivalent circuit diagram of the converter used for analysing the modulation is shown in Figure 8.52.

$$\text{DC link circuit voltage}: U_d = \frac{3}{2} \times 400\,\text{V} \times \sqrt{2}$$

$$\text{Load voltage}: \hat{u}_{LM} = 150\,\text{V}$$

$$\text{Grid frequency}: f_N = 50\,\text{Hz}$$

$$\text{Pulse frequency}: f_P = 10\,\text{kHz}$$

1. Make a drawing of the basic output voltage space vector u_s that can be generated with the inverter and name it.
2. What is the maximum instantaneous value of the inverter output voltage? Calculate for this the amount of the corresponding space vector out of the space vector components u_α and u_β for the switching condition [01 10 01].
3. In this case: how many angular sectors will there be in a circumference of the space vector, and how large is the annular region for such a segment (time, the space vector is in this sector)?
4. Determine for the 45th angular sector the space vector components u_α and u_β of the load voltage space vector.
5. Determine the switch-on times of the right, left and zero space vector of the 45th angular sector of the inverter.

Wind turbine with induction machine and short-circuit rotor: A wind turbine with short-circuit rotor induction machine is connected via a PWM converter with the electrical supply grid.

System parameters:

$P_N = 2\,\text{MW}$	(mechanical rated power)
$c_p = 0.5$	(power coefficient)
$\rho = 1.2$	(density of the air)
$r = 50\,\text{m}$	(rotor blade length)
$N_N = 1800\,\text{rpm}$	(rated speed)
$p = 2$	(pole pair number)
$\ddot{u} = 100$	(transfer ratio of the gears)
$U_{grid} = 690\,\text{V}$	(grid phase-to-phase voltage)
$\text{Cos}\,P_{grid} = 1$	(displacement factor)

1. Is this a fixed or variable speed system? Reason?
2. What must be the given wind velocity in order to generate the rated power P_N for the plant? ($P = 0.5 c_p \rho A_v v^3_{wind}$; $A_v = $ area covered by the rotor)
3. With what speed of revolution N_{rotor} do the rotor blades rotate in the wind when the generator is operated with the rated speed?
4. What is the rated torque M_N at the rotor of the machine when the plant generates its rated power?
5. Calculate the grid current, the active and the reactive power that is fed into the grid for the case that the wind turbine generates the rated power.
6. Draw the speed–torque diagram of the machine at the grid and mark the rated operating point. How does the diagram change when the stator voltage of the machine is changed proportional to the speed?

7. Draw the block diagram of the power electronic–generator system with generator, converter, gear and grid. Why are gears necessary?
8. Draw the simplified UI characteristic of the converter and the machine. Mark and name the limits of the operation.

Drive train with doubly-fed induction machine: A small wind turbine is equipped with a doubly-fed four-pole pair induction machine and a converter in two-level circuit.

Data for the doubly-fed asynchronous machine:
Rated stator power $P_{1N}=25\,\text{kW}$
Rated stator voltage $U_{1N}=400\,\text{V}$
Stator resistance $R_1=0.15\,\Omega$
Stator referred rotor resistance $R_2=0.12\,\Omega$
Stator inductance $L_1=L'_2=50\,\text{mH}$, equal to stator referred rotor inductance
Stator referred main inductance $L_{1h}=49\,\text{mH}$
Rated stator frequency $f_{1N}=50\,\text{Hz}$
Pole pair number $2p=4$

The induction machine is operated at the stator side at the grid and on the secondary side at the inverter and is controlled with constant stator flux linkage.

1. Create the phasor diagram for idling operation with synchronous speed and with 1.3 × synchronous speed. In both cases create the phasor diagram when, in addition, in the first case only the active rotor voltage and in the second case only the reactive rotor voltage is increased by 5%.
2. The stator reactive power is to be controlled to zero. At the highest active power the machine produces 22 kW at a speed of 1950 rpm, connected to the converter. Determine the torque, the slip, the speed, the stator flux, the displacement factor, the active stator current, the reactive stator current and the stator voltage, the rotor current and the rotor voltage.

The data for the inverter apply for the following operating points: $U_{DC}=1.35\times\sqrt{2}\times400\,\text{V}$.

3. For what rated current, for which rated voltage, for which rated apparent power and for what displacement factor is the converter to be designed?
4. At which maximum degree of modulation is the converter to be operated? Can overmodulation be prevented?

Control of the converter: The grid-side converter of a wind turbine with three-phase machine and full power converter feeds into a grid. The current controls are to be designed. The equations for the dynamic behaviour of the grid can be found in the corresponding chapter. The grid data are:

$U_{grid,N}=690\,\text{V}$, rated grid voltage
$u_{SCgrid}=12\%$, grid-side relative short-circuit ratio

A proportionally integral controller is to be used:

$$y=K_P x+\frac{1}{T_1}\int x\,dt; \quad G(s)=\frac{y}{x}=K_P\frac{sT_1+1}{sT_1}$$

Design equation for the technical optimum rule:

$$T_I = T_{max}; \quad K_P = \frac{T_I}{2\sigma_I V_S}$$

T_{max}: largest time constant; σ_I: sum of the smaller time constants; V_S: amplification factor of the controlled system.

The converter operates with a pulse frequency of 2.5 kHz. Two pulse periods are assumed for the delay by the converter.

1. The differential equation for the d-portion of the converter current is to be taken from the corresponding chapter and the disturbance components (U_{Grid}, I_{1q}) must be neglected for the controller design. The time constants T_1, system and the amplification factor V_S of the system are to be determined from the remaining equation.
2. The proportional amplification factor K_P and the adjustment (integrator) time constant T_I as control parameters of the proportional integral controller are to be determined with the aid of the equations for defining the parameters for the technical optimum.

Control of Generator, Controller and Grid Feeding:

1. Draw the phasor diagram of a synchronous machine connected to the grid that works in phase shift operation (stator-side only, capacitive reactive current).
2. Sketch the block circuit diagram of the general control circuit and designate all the part-systems. Name three types of controllers.
3. Draw the complete circuit for field-oriented speed control of the induction machine. Describe the important advantage of the field-oriented method.
4. What coordinate system is often used as reference system for field-oriented control? Name the reasons and explain the value of synchronising the control to specific electrical quantities.
5. Draw the equivalent circuit diagram (topology) of the grid-side and generator-side converter in the often-used circuit.
6. Name and describe the steps of control synthesis. Describe the aims of the control design 'technical optimum'.
7. A wind turbine with a permanent magnet excited synchronous generator fed via a full-power converter operates in power-limitation operation. Sketch the control structure. Name all the participating circuits of the power section of the wind turbine and allocate them to the control circuit tasks.

References

[1] Liserre, M., Cárdenas, R., Molinas, M. and Rodriguez, J. 2011. Overview of multi-MW wind turbines and wind parks. *IEEE Transactions on Industrial Electronics*, **58**: 1081–95.
[2] Lohde, R., Wessels, C. and Fuchs, F.W. 2009. Leistungselektronik Generatorsysteme inWindenergieanlagen und ihr Betriebsverhalten, Fachtagung 3: Direktantriebe und Generatoren, ETG Congress October 2009, Düsseldorf.
[3] BDEW: Bundesverband der Energie- und Wasserwirtschaft, information available at www.bdew.de
[4] Norotny, M.D., Lipo, T.A., Jahns, T.M. 2010. Introduction to Electrical Machines and Drives. Wisconsin Power Electronic Research Center, Madison.

[5] Fitzgerald, A.E., Kingsley, C., Umans, S.D. 2000. Electric Machinery Tata McGraw Hill Education.

[6] Camm, E.H., Behnke, M.R., Bolado, O., *et al.* 2009. Characteristics of wind turbine generators for wind power plants. IEEE Power & Energy Society General Meeting.

[7] Leonhard, W. 2001. Control of Electrical Drives. Springer, Berlin.

[8] Zhe, C., Guerrero, J.M. and Blaabjerg, F. 2009. A review of the state of the art of power electronics for wind turbines. *IEEE Transactions on Power Electronics*, **24**: 8, 1859–75.

[9] Erickson, R.W., Maksimovic, D. 2001. Fundamentals of Power Electronics. Springer.

[10] Hagmann, G. 2009. *Leistungselektronik – Grundlagen und Anwendungen in der elektrischen Antriebstechnik.* Aula, Wiesbaden.

[11] Mohan, N., Undeland, T.M. and Robbins,W.P. 2003. *Power Electronics: Converters, Applications, and Design.* John Wiley and Sons, Ltd., Chichester.

[12] Schröder, D. 2007. *Elektrische Antriebe: Grundlagen.* Springer, Berlin.

[13] Kazmierkowski, M.P., Krishnan, R. and Blaabjerg, F. 2002. *Control in Power Electronics: Selected Problems.* Academic Press, Amsterdam,

[14] Teodorescu, R., Liserre, M. and Rodriguez, P. 2010. *Grid Converters for Photovoltaic and Wind Power Systems.* John Wiley and Sons, Ltd., Chichester.

[15] Khan, M.A. and Pillay, P. 2005. Design of a PM wind generator, optimized for energy capture over a wide operating range. *IEEE International Conference on Electrical Machines and Drives, Proceedings*, 1501–6.

[16] Akagi, H., Watanabe, E.H. and Aredes, M. 2007. *Instantaneous Power Theory and Applications to Power Conditioning.* John Wiley and Sons, Ltd., Chichester.

[17] BDEW. 2008. *Technische Richtlinien für Erzeugungsanlagen am Mittelspannungsnetz; Bundesverband der Energie- und Wasserwirtschaft.*

[18] International Electrotechnical Commission. 2008. *IEC 61400–21: Wind turbines, Part 21: Measurement and assessment of power quality characteristics of grid connected wind turbines.* IEC.

[19] SDLWindV. 2009. *Verordnung zu Systemdienstleistungen durch Windenergieanlagen* (System-dienstleistungsverordnung – SDLWindV. Part 1, No. 39. SDLWindV, Bonn.

[20] Tsili, M. and Papathanassiou, S. 2009. A review of grid code technical requirements for wind farms. *Renewable Power Generation*, **3**(3): 308–32.

[21] Fujin, D. and Zhe, C. 2009. Low-voltage ride-through of variable speed wind turbines with permanent magnet synchronous generator. Industrial Electronics, 2009, 35th Annual Conference of IEEE, pp. 621–6.

9

Control of Wind Energy Systems

Reiner Johannes Schütt

Wind turbines (WTs), or several WTs gathered in a wind farm (WF), are complex systems whose operation requires extensive control of the overall system as well as of the subsystems. In this it is expected that wind turbine systems (WTSs) fulfil at least the requirements of conventional power stations with regards to reliability, efficiency and operational management. In contrast to conventional power stations, the supply of energy cannot be influenced but varies very strongly and quickly due to the wind speed. For independent and secure operation and the extensive adaption of the operation of WTSs for different application conditions, it is necessary to have a complex control technology that distributes the tasks to different subsystems.

Modern WTSs supply the electrical grid at the medium or high-voltage level by means of transformers. In order to optimise the feed-in power and to limit this to the rated power, WTs are used with and without gear drives, full power or part converters with variable speeds, and variable blade angles. The automation of WTSs includes the measurement, control, automation and monitoring of important characteristics within the WT as well as the WT and WF operational management. It includes remote monitoring and visualisation and the information and communication linkages to other systems, and here especially the integration into the primary grid management systems. The automations for this consist of the technical equipment (hardware), the associated programs (software) and the required communication systems. Despite the different WT systems, the automations have an increasing commonality, especially because in future the WTs, independent of the manufacturer, must be integrated into primary control systems in the same manner.

This chapter deals with modern WTSs and the basic control and automation circuits within the WTs, the WT and WF operational management, the linking of WTSs to primary systems

Understanding Wind Power Technology: Theory, Deployment and Optimisation, First Edition.
Edited by Alois Schaffarczyk. Translated by Gunther Roth.
© 2014 John Wiley & Sons, Ltd. Published 2014 by John Wiley & Sons, Ltd.

and the so-called SCADA systems. It provides an overview of the important aspects of the automation of WTSs; comparable discussions are found in [1–4].

9.1 Fundamental Relationships

The aim of the WTS automation is a safer and more efficient operation of the WT, which is extensively independent of people. For this purpose different tasks must be fulfilled, which in principle comprise the measuring, controlling, automation, monitoring and visualising. These individual functions are all combined in the primary term 'control'; in German the term "automation" is used. To make the content readable, there are some symbols that are not used as usual in British or English-language publications. The main differences are the wind speed v, the voltage u or U, and the torque M.

9.1.1 Allocation of the WTS Automation

Several hundred physical inputs and outputs must be processed in order to operate a WT. A WF can consist of several hundred individual WTs and in the grid control system, besides the WFs, also further generators, loads and grid resources must be monitored and coordinated. The large number of automation functions, the requirement of speed of reaction and the quantity of data to be processed can only be mastered when the tasks are distributed and arranged in a hierarchy. In plant automation this hierarchy is depicted in the form of a pyramid.

Figure 9.1 shows an often-used automation pyramid with three main levels, whereby the term automation or control level has become customary for the middle level. The tasks and requirements of reaction time and data quantities shown in the figure can also be transferred to the automation of WTS. Thus, the switching times of the converter in the field level move in the microsecond range, the reaction times of the programmable controls in the automation level in the millisecond range, and the required reaction times of the WT to the requirements

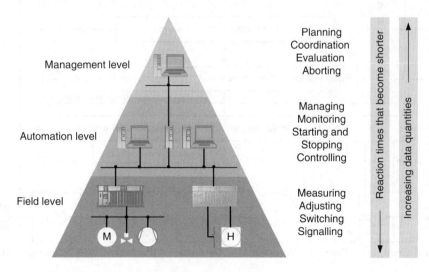

Figure 9.1 Three-level model of plant control with important functions

of the power reduction from the grid management system in the second range. If in the field level only a few analogue and digital inputs and outputs are processed by a device, then in the management level the data series acquired over several years are all saved and evaluated in a single unit connected to the supply area.

The three main levels are often divided for separating the functions. The division into four to six levels is typical; unified naming of these levels has not been agreed upon.

Figure 9.2 shows the important subsystems of the WTS automation arranged in six levels. The designations in the following correspond to the main functions of the respective levels.

Sensor-actuator level: In the lowest level, the drive train of the WT is depicted with the input dimensions wind speed v_W and wind direction γ_W. The marked output dimensions are the three-phase grid voltage u_n and grid current i_n, the grid frequency f_n and the phase angle ϕ_n between the current and voltage of the three-phase system. The rotor speed n_R is influenced by the variable yaw angle γ_G and the variable blade angle β_B.

Different types of generator are used in WTs; in general these are double-fed induction generators (DFIGs), direct current excited synchronous generators (DCSGs), foreign excited synchronised machines (PSM), permanent magnet synchronous machines (PSM) and asynchronous machines with squirrel cage induction generator (SCIGs). They are operated partly with and partly without gears and feed the generator power either fully or partly into the grid by means of a converter. The figure shows the drive train of a WT with full-power converter feed PMSG with and without gears.

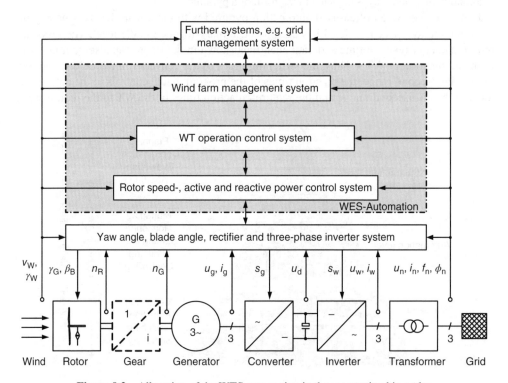

Figure 9.2 Allocation of the WTS automation in the automation hierarchy

Mechanical power of the rotor P_R is transferred by the drive train to the generator with the generator speed n_G. The generator power P_G is fed into the electrical grid with the aid of a full-power converter via a transformer. The full-power converter consists of the generator-side rectifier with the generator-side three-phase current i_g and voltages u_g, the direct current intermediate circuit with the direct voltage u_d, as well as the grid-side inverter with the grid-side three-phase current i_w and voltages u_w.

The rectifier and inverter are carried out as six-pole bridge circuits with insulated gate bipolar transistors (IGBTs), that are driven via six control signals, s_g on the rectifier side and s_w on the inverter side respectively. With the aid of a transformer, the voltage on the inverter output side is adapted to the grid voltage.

Adjustment level: The adjustment actuators of the individual drive systems are situated above the sensor-actuator level. The yaw angle drive system, often called the azimuth system, adjusts the direction of the nacelle according to the direction of the wind. The blade angle drive systems, often also called the pitch systems, set the blade angle according to the desired reference values. The generator-side electrical power P_g is influenced by the pulse width modulation control signals s_g of the pulse rectifier. The inverter-side active power P_w and reactive power Q_w are influenced via the pulse width modulation control signals s_w of the pulse inverter. The intermediate circuit voltage u_d remains constant when the active power on the rectifier- and inverter-side is identical.

Closed control loop level: The automation to the reference values of the active power P_n and grid reactive power Q_n as well as the limitation of the rotor speed occur above the automation level allocated to the setting level. Depending on the mode of operation of the WT, various controllers calculate the reference values for the yaw and the blade angles as well as for the rectifier and the inverter.

Operation control loop level: The controls receive their reference values from the primary operation management that is allocated to the control level. The operation management controls the WT in the required operating condition depending on the speed of the wind, the reference values of a grid management system or a wind farm management system, and external requirements. Besides the requirements of the reference values for automation, the operation management also controls further actuators in the WT such as the brakes, protections, heating and ventilation or the obstruction lighting system, and processes additional sensor signals such as temperature, air moisture or oil pressure. The operation management exchanges the information directly or via a management system with remote monitoring and control of the manufacturer or operator, as well as with the management system of the grid operator.

Process control level: If several WTs are operated in a wind farm, then the WF controls take priority over the individual WT controls. The system allocated to the process control level has a management function and distributes the required power reduction and reactive power feeds in accordance with the reference values given by the grid operator to the individual WTs and ensures the required participation in the short-circuit current in the grid. Also the wind farm management system ensures an orderly and continuous start-and-stop process of the wind farm so that there are no power jumps.

Management or planning level: WT operating management or WF management systems obtain their reference values from the management level, also called the planning level, and return the relevant actual values back to it. The grid management system as well as the accounting system or the commercially-oriented management system of the plant operator, can be assigned to the uppermost level.

9.1.2   System Properties of Energy Conversion in WTs

The aim of the management, operation and control systems of the WT is to optimise the energy conversion from the kinetic energy of the wind into electrical energy of the grid, and to operate the plant securely within the prescribed limiting values.

Below the rated wind speed of a WT v_{WN}, the power fed into the grid is maximised; above this wind speed it is limited to the rated power P_N, the electrical rated power P_{nN} at the grid connection point (GCP). In order to adhere to the grid connection conditions, the WTS must limit the active power to the requirements and with rising grid frequency. The reactive power must be changed on demand in the case of grid voltage fluctuations. For time-limited voltage interruptions, the WTs must support the grid by feeding in a defined reactive current.

In order to develop the basic control and automation functions of a WT, first the required system properties important for the description of the energy conversion in WTs must be collected. The controls have basic similarities despite the differences in the drive trains (with and without gears, high-translation, low-translation), the generator system (SCIG, DFIG, DCSG, PMSG) and the pitch and azimuth drives (hydraulic, electrical).

Figure 9.3 shows the block diagram of energy conversion in which the subsystems are linked corresponding to the described parameters of their power current. The direct voltage intermediate circuit, the grid-side inverter and the transformer are combined into a single block. The possibilities of influencing the power current are described on the basis of this figure.

9.1.3   Energy Transformation at the Rotor

From the point of view of a WT (a), Figure 9.4 shows the section of a blade (b) and the plan view (c) of the orientation of the dimensions v_w, n_R characterising the energy transformation, the rotor diameter D, γ_G and β_B. A good summary of the dependency of the rotor power on these dimensions is found in [1]. At this juncture the decisive connections for control and automation are summarised.

In the section view the circumferential speed of the blade u_B and the active speed v_u acting on the blade cross-section are depicted.

The circumferential blade tip speed is designated by u.

$$u = 2\pi n_R D / 2 \tag{9.1}$$

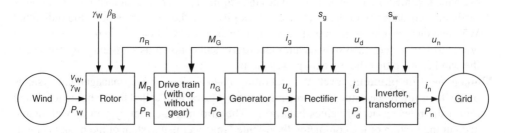

Figure 9.3   Schematic depiction of the power flow in a wind turbine

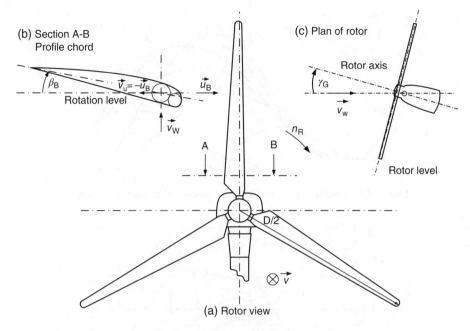

(b) Section A-B
Profile chord

β_B
$\vec{v}_u = -\vec{u}_B$
$\vec{u}_B$
Rotation level
$\vec{v}_W$

A   B

(c) Plan of rotor

Rotor axis

γ_G

$\vec{v}_W$

n_R

Rotor level

D/2

$\otimes \vec{v}$

(a) Rotor view

Figure 9.4   Characteristic dimensions of the rotor for calculating the rotor power

The kinetic energy of the undisturbed current of air is partly converted into mechanical energy of the rotor by the rotor. The relationship of the mechanical power P_R to the power of the wind P_W is called the rotor power coefficient, or in short, power coefficient c_{pR}:

$$c_{pR} = \frac{P_R}{P_W} = \frac{2\pi n_R M_R}{\frac{1}{2}\rho(\pi D^2/4)v_W^3} \qquad (9.2)$$

In this, ρ is the density of the air and D the diameter of the rotor. The power coefficient is determined by the number of blades N, the geometry of the blade, the angle of the blade β_B, the yaw angle γ_G, the wind speed v_W and the rotor revolutions n_R.

$$c_{pR} = f(N, \textbf{blade geometry}, \alpha_G, \beta_B, v_W, n_R) \qquad (9.3)$$

The rotor power is at a maximum when the yaw angle, which gives the deviation of the wind direction from the direction of the rotor axis, is zero. With increasing yaw angle the relationship to the normal component of the rotor surface in the direction of the wind becomes smaller so that the power in a first approximation sinks proportionally to the cosine of the yaw angle. WTs lead the nacelle to follow the direction of the wind, thus the effect of the yaw angle on the rotor power is ignored in the following discussion.

The relationship of the blade tip speed to the wind speed v_W becomes the TSR (tip speed ratio) λ and is an important parameter to the power controls.

$$\lambda = u/v_W = \pi D n_R/v_W \qquad (9.4)$$

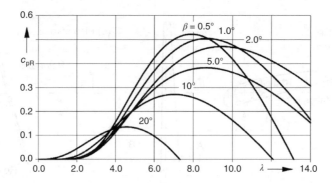

Figure 9.5 Course of the rotor power coefficient at constant blade angle

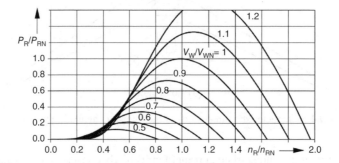

Figure 9.6 Course of the rotor power coefficient at constant wind speed

For a given number of blades and blade geometry and with constant yaw angle, the introduction of the tip speed ratio makes the two-dimensional depiction of the power coefficient of the rotor power or the rotor torque possible. It has been shown that at constant blade angle the course of the power coefficient in relation to the tip speed ratio is at a maximum. In the case of given blade geometries, the characteristic fields of $c_{pR} = f(\beta, \lambda)$ can be calculated or dimensioned for implemented WTs.

The family of characteristic fields can be emulated by means of analytical functions that lead via parameter adaptions to a satisfactory agreement with measured values. Building on the proposal in [5], the following form is selected for Figures 9.5 and 9.6:

$$c_{pR} = c_1 (c_2/\lambda_i - c_3\beta - c_4\beta^x - c_5)\, e^{-c_6/\lambda_i} \quad \text{with} \quad \frac{1}{\lambda_i} = \frac{1}{\lambda + 0.08\beta} + \frac{0.035}{\beta^3 + 1} \tag{9.5}$$

The constants are: $c_1 = 0.6$, $c_2 = 116$, $c_3 = 0.4$, $c_4 = 0.001$, $c_5 = 5$, $c_6 = 20$, $x = 2.0$.

The maximum rotor power coefficient applies to very small blade angles, and for the courses shown, for an optimum tip speed ratio of 8. Below the rated wind speed the rotational speed is adapted to the changing wind speed in order to convert the maximum power of the wind into rotor power.

It is interesting for the power automation and limitation how the rotor power for the optimum blade angle is dependent upon the rotational speed of the rotor. The rotor power referenced to the rated rotor power P_{RN} is defined by the use of the power coefficient in the rated operation c_{pRN} as follows:

$$\frac{P_R}{P_{RN}} = \frac{c_{pR}}{c_{pRN}}\left(\frac{v_W}{v_{WN}}\right)^3 = \frac{c_{pR}}{c_{pRN}}\left(\frac{n_R}{n_{RN}}\frac{\lambda_N}{\lambda}\right)^3 \tag{9.6}$$

With the power equation and the reference values characterised by the index N, there results the sequences for $\beta = 0.5°$ and $\lambda_N = 8$ shown in Figure 9.6 for different wind speeds.

The courses of the curves show that the optimum rotor revolution rises proportionally with increasing wind speed. If the power above the rated wind speed at constant rotor revolutions is to be limited to the rated power, then the blade angle must be increased.

9.1.4   Energy Conversion at the Drive Train

It is decisive for the control of WTs how the rotor power P_R is converted into the generator power P_G. The drive train is a complex system that can vibrate and can be seen in simplified form in the two-dimensional Figure 9.7. The characteristic properties are the moment of inertia of the rotor J_R, the stiffness of the shaft k_{shaft}, the damping constant d_{shaft}, the gear translation i, which is equal to 1 for gearless plants, and the moment of inertia of the generator J_G. In this depiction the gears are assumed to be frictionless and weight-free.

The rotor torque can be determined with Equation (9.2) from the rotor power.

$$M_R = c_{pR}\frac{\frac{1}{2}\rho(\pi D^2/4)v_W^3}{2\pi n_R} \tag{9.7}$$

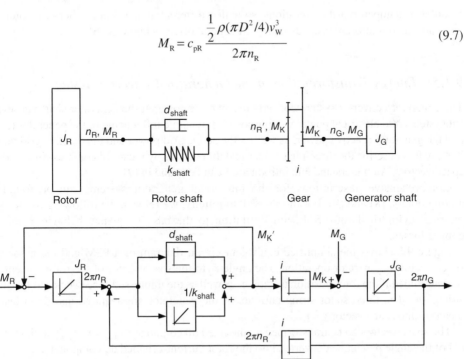

Figure 9.7   Two-dimensional model of the drive train and associated block diagram

With the introduction of the torque M_k applied at the generator coupling and the coupling torque M'_k on the rotor-side, as well as with the use of the translation of the ideal gear drive $I = n'_R/n_g = M_k/M'_k$, there result the following equations for rotor and generator torques:

$$M_R = J_R 2\pi \frac{dn_R}{dt} + M'_K \tag{9.8}$$

with $M'_K = d_{shaft} 2\pi(n_R - n_G i) + k_{shaft} 2\pi \int (r_R - n_G i) dt$

$$M_G = M_K - J_G 2\pi \frac{dn_G}{dt} \quad \text{with} \quad M_K = iM'_K \tag{9.9}$$

According to this for the two-dimensional model, the block diagram in Figure 9.7 results in a system capable of vibrations. By means of long start-up times of the WT in the seconds range and continuous changes of the generator torque, the deviation between cut-in and cut-out rotational speed can be held low. Then, in a simplified manner, one can assume a rigid shaft in which only one rotational speed need be taken into account. For a rigid shaft with $i = n_R/n_G$ the well-known acceleration equation is obtained:

$$\frac{dn_G}{dt} = \frac{1}{i}\frac{dn_R}{dt} = \frac{(M_R/i - M_G)}{2\pi(J_G + J_R/i^2)} \tag{9.10}$$

In order to dampen out the vibrations, so-called differential speed controllers for rotor and generator revolutions are used, as known in general drive technology [6].

9.1.5 Energy Transformation at the Generator-Converter System

The generator-converter system converts mechanical power at the generator shaft via several subsystems into electrical power. For purposes of dynamic influencing of the generator torque and the generator reactive power, as well as the active grid power and the reactive grid power, it is usual to describe the three-phase electrical dimensions by means of complex time-varying space vectors. An understandable introduction can be found in [7].

Comprehensive descriptions for the individual generator systems can be found in, among others, [8] and [9]. For the sake of simplicity, we refer in the following to the symmetrical cylindrical rotor-SM with limitation to the base frequency behaviour and the copper losses.

Figure 9.8 shows the simplified equivalent circuit diagram of a PSM and the associated space vector in generator operation. The angle δ_g to the space vector of the rotor current can be determined if the rotor position is known, as well as the impressed space vector of the stator voltage u_g. The three stator string currents of the generator are acquired and described by means of the space vector $i_s = i_g$.

The space vector can be transformed on the fixed $\alpha\beta$ components $i_{g\alpha\beta} = i_{g\alpha} + j\, i_{g\beta}$ and, with the aid of the angle δ_g, can be converted into the rotor current-oriented dq components $i_{gdq} = i_{gd} + j\, i_{gq}$. The component i_{gd} shows the direction of the permanent magnetic current Ψ_{PM} and the component i_{gq} is at right angles to it.

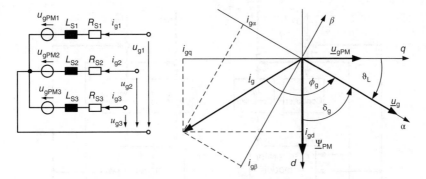

Figure 9.8   Simplified equivalent circuit diagram of the PSM and space vector in generator operation

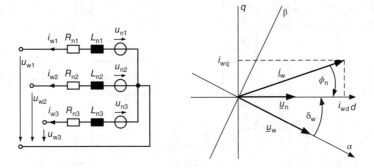

Figure 9.9   Simplified equivalent circuit diagram and space vector in generator operation

The absolute value of the space vector of the induced voltage for a generator with the polar pair p from the induction law results in:

$$u_{gPM} = 2\pi n_G p \Psi_{PM} \tag{9.11}$$

The amount of the space vector is fixed such that in the stationary case it corresponds to the complex phasor of the amplitude of the geometric sum of the string dimensions, and is therefore two-thirds of the geometric sum of the individual current. This results in active and reactive power as well as generator power and generator torque as follows:

$$P_g = \frac{3}{2}(u_{gd}i_{gd} + u_{gq}i_{gq}) \quad \text{and} \quad Q_g = \frac{3}{2}(u_{gq}i_{gd} + u_{gd}i_{gq}) \tag{9.12}$$

$$P_G = 2\pi n_G M_G = \frac{3}{2}u_{gPM}i_{gq} \quad \text{and} \quad M_G = \frac{3}{2}p\Psi_{PM}i_{gq} \tag{9.13}$$

Figure 9.9 clearly shows that for the PSM the maximum power for minimum current $i_{gd} = 0$ occurs. The space vector of the voltage u_g for a constant torque can be influenced by a changing current i_{gd}.

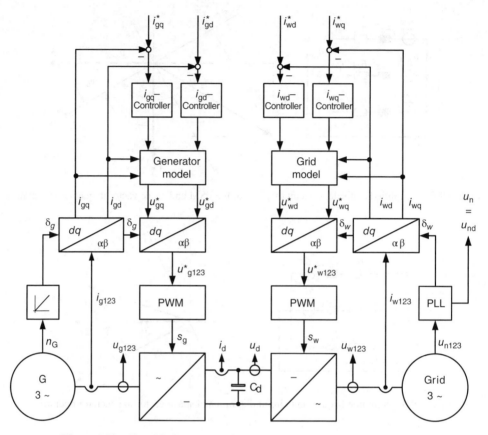

Figure 9.10   Simplified structure of the generator-converter system for PSM

The grid-side inverter can be dealt with in a similar manner. For this purpose the grid-side voltages can be transformed to the inverter side. Figure 9.9 shows the simplified equivalent circuit diagram of the grid referenced to the inverter side and the associated space vectors in generator operation. Frequency and phase position of the grid voltage are acquired with the aid of a phased lock loop (PLL) and thus the space vector of the grid current is divided into the grid-voltage-oriented d components and the vertical q components.

If the grid component $i_{nd} = i_{wd}$ points in the direction of the space vector $\underline{u}_n$, then active and reactive power can be influenced in a simple manner with the aid of the current components.

$$P_n = \frac{3}{2}(u_n i_{wd}) \quad \text{and} \quad Q_n = \frac{3}{2}(u_n i_{wq}) \tag{9.14}$$

Figure 9.10 shows the simplified structure of the generator-converter system for a PSM with full power converter and grid dimension transformed on the inverter-side. The reference values transferred by the controls are marked by *. The highly dynamic control of generator torque and voltage is possible when the current components are regulated to the reference values i_{gq}^* and i_{gd}^* with the aid of a controller.

The output of the current controller determines the voltage components to be set using precontrol values that are calculated with the aid of a simple generator model. The voltage components are transformed with the aid of a pulse-width modulation (PWM) into the required control signals of the six-pulsed rectifier.

Also with the grid-side inverter it is possible for the reference values of P and Q to be highly dynamically impressed with the aid of the current reference values i_{wd}^* and i_{wq}^*, of the current controller and the PWM of the six-pulsed inverter. Here the pre-control values are determined by means of the grid model.

Direct and three-phase converters are connected by means of the direct voltage intermediate circuit. The voltage at C_d is determined for loss-free assumed direct current and three-phase converter by integration of the difference of the momentary power $p_g - p_w$ with $p_w = p_n$.

$$C_d = \frac{du_d}{dt} = \frac{P_g}{u_d} = \frac{P_w}{u_d} \quad \text{with} \quad u_d(t) = \sqrt{u_d(t=0)^2 \times \frac{2}{C_d} \times \int_0^t (p_g(t^*) - p_w(t^*))\, dt^*} \quad (9.15)$$

If, in the stationary case, the active generator and active grid power are the same, then the intermediate circuit voltage does not change.

9.1.6   Idealised Operating Characteristic Curves of WTs

Because of the system properties in the energy conversion of the rotor, four different operating ranges can be determined for a WT. These are described on the basis of the idealised sequences of the characterised dimensions in Figure 9.11.

The operating ranges have the following properties:

Underload (A): Below the switch-on speed (cut-in speed v_{cin}) the blade adjustment angle is reduced with increasing wind speed so that the WT starts up. The WT does not yet feed the grid and the rotational speed of the rotor increases.

Partial load (B): In the range from v_{cin} to v_{wN} the WT is run with optimum fast speed so that the rotor rotational speed with constant blade adjustment angle rises linearly. The power fed into the grid rises cubically.

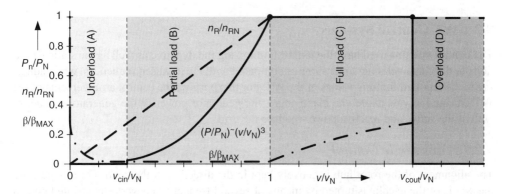

Figure 9.11   Idealised operating characteristic curves of WTs

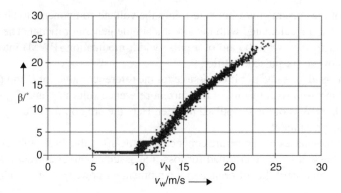

Figure 9.12 Measured blade adjustment angle of a WT

Full load (C): Above the rated speed up to the switch-off speed (cut-out speed v_{cout}) the grid power is limited to the rated power in that the blade adjustment angle is increased and thus the rotor rotational speed is kept constant with rising wind speed.

Overload (D): Above the switch-off speed the blade angle is increased so that the WT is broken to the zero revolution. The WT feeds no power into the grid.

Across the entire operating range the nacelle follows the direction of the wind so that the yaw angle is always zero.

In fact, these characteristic dimensions of the WT deviate from the idealised sequences. Short-term wind speed changes, as they occur especially in gusts, lead to a change of power that has an effect on the characteristic curves depending on the properties of the drive system and the control arrangements.

Figure 9.12 shows an example of the measured blade adjustment angle of a WT with a cut-in wind speed of 4 m/s, a rated wind speed of 12.5 m/s and a cut-out speed of 25 m/s. The deviations from the idealised characteristic curve in the transition from partial-load to the full-load range are clearly seen. The figure shows the short-term average value divided into minutes.

The actual self-adjusting operating characteristic curves, the deviations from the idealised characteristic curves and the width of the fluctuations depend greatly on the control system in use and the automation parameters.

9.2 WT Control Systems

The control systems used have the aim of minimising the dynamic as well as the static deviations of the characterising dimensions as compared with the idealised sequences, while adhering to the required limiting values of the WT. For this purpose the control arrangements of the WT change the yaw angle, the blade angle, the generator power or the generator torque, as well as the active and reactive power fed into the grid.

9.2.1 Yaw Angle Control

The alignment of the rotor hub is actively kept in the direction of the wind. For this purpose the whole of the nacelle is turned by means of several frequency converter-controlled geared motors (azimuth drives) and a geared wheel fastened to the tower. Brakes at the tower ring

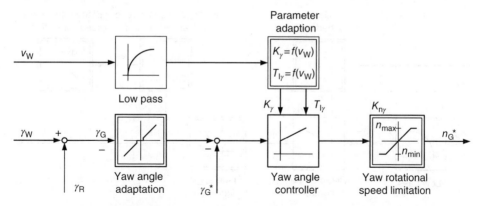

Figure 9.13 Block diagram of the yaw angle automation

ensure that the nacelle is fixed when the motors are not driving it. The cut-in times, revolutions and direction of turning of the azimuth drives are controlled and regulated so that, on the one hand, the yaw angle remains small with fluctuating wind directions, and on the other hand, the number of adjustment movements is not too great.

Figure 9.13 shows the block diagram of a yaw angle automation that optimises this behaviour with the aid of the parameter adjustments for different wind speed ranges. For tracking purposes the wind direction γ_W is acquired and compared with the alignment of the rotor hub γ_R.

If the yaw angle $\gamma_G = \gamma_W - \gamma_R$ exceeds a minimal value (some degrees) for a longer period of time, then the brakes are released and the nacelle alignment is regulated. The time for which the deviation must be in existence is reduced with increasing wind speed and increasing yaw angle. Strengthening and integration time constants of the PI yaw angle controls are adapted with dependence on the wind speed.

The nacelle is not aligned for very low wind speeds ($v_w \ll v_{cin}$). With increasing wind speed, the rotational speed of the azimuth gears increase with an increasing yaw angle. The rotational speed is limited so that the nacelle only moves slowly. Typical maximum speeds are well below one revolution per minute.

One possibility of parameter adaption by determining various operating ranges dependent on the wind speed and the yaw angle is described in [3]. Because of the imprecise formulation of the control behaviour, the use of fuzzy controls is also proposed.

9.2.2   Blade Angle Control

Power and rotational speed limitation in the full-load range occur by means of blade angle automation with the aid of electrical or hydraulic actuators. The three actuators situated in the hub increase the blade angle above the rated speed in order to keep the rotor power constant with increasing wind power.

In order not to have to react to the fast and imprecisely determined wind speed, use is made at the start of the blade angle adjustment of the deviation of the rotor rotational speed from the rotor rated rotational speed. In the case of positive deviation, the blade angle is increased by the PI controller.

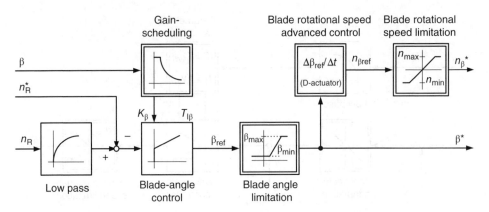

Figure 9.14   Blade angle control with parameter adaption

The behaviour of the course is non-linear. For small deviations of the wind speed from the rated wind speed, a large blade angle change is necessary in order to keep the power and thus the revolutions constant. For high wind speeds a small change of blade angle is sufficient to limit the power and revolutions. Well-known controls therefore use a PI blade angle controller with adaption of the strengthening factor (gain-scheduling). In this the strengthening factor decreases with increasing blade angle, so that the control deviation of the revolutions for high wind speed does not become too great and the closed control circuit avoids becoming unstable.

Figure 9.14 shows a possible structure of the blade angle controls in which both the blade angle reference value as well as the blade angle revolution are passed on to the pitch drives. Normally, electrical servo-drives then use an internal cascade automation of a superimposed angle automation, a rotational speed control and a sub-imposed torque or current automation. Among others, detailed description of the design can be found in [10] and [11].

The limitation of the rotational speed is designed in such a manner that even in the case of an emergency stop the maximum blade angle of almost 90° is reached in less than 10 s, which means angular speed of several degrees per second. In this the torque or power controls are substantially slower so that in the case of dynamic speed changes there are marked changes in the rotational speed.

In order to obtain a continuous change from variable rotation to constant rotation operation, the blade angle control begins before reaching the rated speed and rotational speed.

An increase in the strengthening and reduction of the integration time can reduce the deviation of the rotational speed. However, especially with the change from the variable to constant speed of revolution ranges of the WT, such an automation setting leads to large adjustment activities in the actuator, which can have an effect especially as an increased level of noise in this wind speed region. The design is therefore often adapted in the test operation. Because of the imprecise formulation of the control behaviour, the use of fuzzy controls is also proposed here.

9.2.3   Active Power Control

Active power control and limitation also occurs without direct evaluation of the wind speed. As the optimum power in the partial-load range is reached at optimum fast rotational speed, the rotor speed is also used for the active power control. In partial-load mode it adjusts itself freely in accordance with drive and generator power.

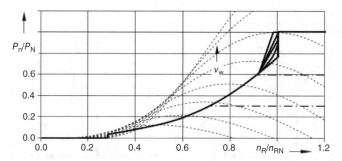

Figure 9.15   Schematic reference value curves for power

The control to the maximum achievable power is reached in that the rated power P_n^* in dependency of the rotor rotational speed is prescribed and adjusted via the converter with a characteristic curve. If the power required by the characteristic curve is smaller than the power provided by the wind, then the rotor turns faster. If the power required by the characteristic curve is greater than the power provided by the wind speed, then the rotor turns slower. In the stationary case there is a maximum power factor for each wind speed in the partial load range, and an optimum rotor rotational speed.

In Figure 9.15 the schematic rated curve of the power characteristic curve is entered into the characteristic curve field. Recognisable is the cubed trajectory for low revolutions and the constant power for large powers. The dot-dashed lines are the characteristic curves for the required power limits imposed by the energy supplier, here the typical 60° and 30° limits.

If the power provided by the wind exceeds the limiting characteristic curves, then the increase in rotational speed has the effect of adjustment of the blade angle setting and therefore limitation of the power. For optimising the change-over between variable and constant speeds of revolution, modifications in the characteristic curve path are used. A comprehensive discussion can be found in [12].

In the stationary case, and with neglect of the losses in the generator-converter system, the intermediate circuit voltage is constant, the generator power and active grid power are the same. The control of the generator power or the generator torque can therefore occur either directly or by means of active grid power, if a controller for the intermediate circuit voltage ensures that this remains almost constant. Thus, there are two possibilities for active grid control:

a) controlling the grid-side inverter to the maximum achievable power and controlling the active power of the rotor-side rectifier so that the intermediate circuit voltage remains almost constant;
b) controlling the rotor-side inverter to the maximum achievable power and controlling the active grid power so that the intermediate circuit voltage remains almost constant.

Figure 9.16 shows the solution for variant (a). Because of the rotor rotational speed, the reference value of the active power is pre-defined. According to the deviation between the rated and actual values of the grid power P_n, the active power-forming rated power value i_{wd}^* of the inverter is changed with the aid of a PI controller.

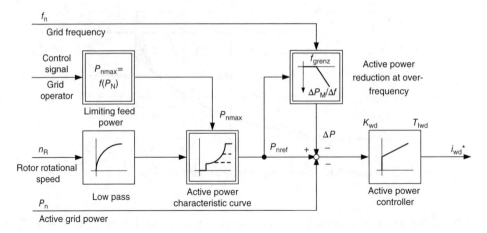

Figure 9.16 Active power control by means of grid-side converter

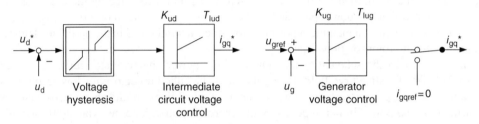

Figure 9.17 Control of the generator-side current

In addition, in the picture the active power reductions are supplemented due to the prescription of the grid operator [13–15]. On the one hand, this is the stepwise reduction of the active power referenced to the rated power at the request of a control signal, for the protection of overloading of individual grid sections of the transfer grid. On the other hand, it is the continuous reduction of the power available at the time of the request when the grid frequency is too high.

The control of the generator-side current is shown in Figure 9.17. The torque-forming rated current value i_{gq}^* of the generator-side converter is prescribed in such a manner that the intermediate circuit voltage remains constant. If the intermediate circuit voltage exceeds an upper limit, then the torque-forming current is reduced. If the intermediate circuit voltage falls below a lower limiting value, then it is increased. The rotor-oriented current i_{gd}^* is either set to zero or used for control of the optimum voltage $u_s = u_g$. If the voltage is too low, then the rotor-oriented current is increased.

In Section 9.1.4 it is shown that the rotor rotational speed can fluctuate about a middle value because of the flexible drive shaft. In [9] it is proposed that the fluctuations of the rotor rotational speed can be damped, in that the reference value of the intermediate circuit voltage is prescribed dependent upon the rotor rotational speed deviations, in order to impress a torque-forming current that actively counters the fluctuations. The voltage hysteresis must then fall away.

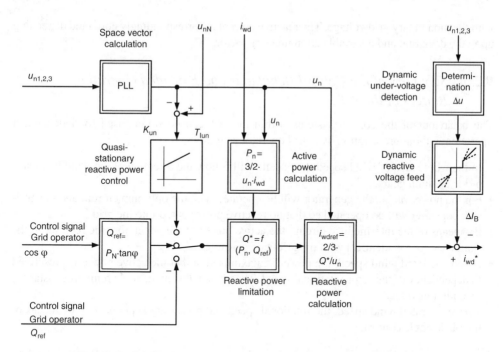

Figure 9.18   Quasi-stationary and dynamic control of the grid reactive power

9.2.4   Reactive Power Control

The grid feed of the WT is normally operated so that the reactive power at the connection node is zero. Thus, the active power is fed into the grid with the minimum required grid current. For efficient and secure grid operation, the WTs must contribute in the quasi-stationary operation as well as dynamically to voltage automation, so that they feed an inductive or capacitative reactive power into the grid. For purposes of voltage increase in the connection nodes, the capacitative reactive force is usually impressed; the corresponding grid connection guidelines speak of over-excited operation. For purposes of voltage reduction, an inductive reactive current is impressed in the so-called under-excited operation.

Figure 9.18 shows a block diagram that permits the reactive power control by means of impressing the reactive-forming current component $i^*_{wd} = i^*_{nd}$. In order to take possible requirements of the grid operator into account, the block diagram shows a quasi-stationary and a dynamic reactive power feed according to the automations in [13] and [14]. The requirement for the quasi-stationary reactive power feed can occur in various ways: by means of a control signal, the grid operator can directly prescribe the reactive power to be fed in or prescribe the power factor cos ϕ. Alternatively, the reactive power feed can be carried out continuously depending on the deviation between the actual voltage value and the reference value. In order to limit the required over-dimensioning of the grid-sized converters, the grid operator prescribes the required limiting curves for the reactive power.

In the case of dynamically significant voltage deviations, especially for voltage decreases due to grid errors, WTs must be able to independently impress a reactive flow for voltage support. For this purpose, the voltage before the error is used as a reference value and is compared with the momentary grid voltage. The reactive current is correspondingly corrected depending upon the deviation and a variable strengthening factor.

9.2.5   Summary of the Control Behaviour and Extended Operating Ranges of the WT

The behaviour of the control systems within the WT can be summarised for independent operation without prescription by a grid operator as follows:

- In the whole area of wind speed the nacelle will follow the direction of the wind by means of the azimuth gears.
- For the power range, the generator will be operated with the optimum current and the grid-side converter will be operated so that no reactive power is fed into the grid.
- By means of the intermediate circuit, the active power of the generator-side rectifier in the stationary case is identical with the grid-side converter.
- Below the rated wind speed, the power is maximised in that the active power is prescribed in dependency of the rotor rotational speed. The blade adjustment angle remains constant at its optimum value.
- Above the rated wind speed, the rotational speed and power are kept constant by means of the blade angle control.

Rotational speed, blade adjustment angle and power are interdependent. Besides the four types of operation shown in Figure 9.11, modern WTs possess further significant transfer ranges. Figure 9.19 shows the typical sequences in relation to the wind speed.

For a continuous transfer of the system sizes of partial loads into full loads, the control characteristic curve affects the power so that the rotor rotational speed does not cause it to rise in proportion to the wind speed already below the rated wind speed. Therefore, the WT is no longer operated with the optimum fast rotational speed from v_2. At v_3 the blade angle controls set in even when the rated power has not yet quite been reached. With this, the rotor power factor sinks additionally. The optimum power factor is thus reached by a WT below v_N, so when the rated wind speed is reached the power factor has already sunk substantially.

9.3   Operating Management Systems for WTs

The technical management system is used to ensure the automatic, efficient and secure operation of the WT. It processes the input signals of the WT as well as the signals of the operating equipment and further primary systems such as the management system. From these input signals, the operating management determines the output signals for the WT, for the display arrangement and the primary systems. The important tasks of the operation management include:

- control of the operating sequence and the requirements for the reference values of the WT control systems;
- monitoring of critical dimensions and activation of corresponding security measures; and
- acquisition, saving and exchange of relevant information with the primary systems.

The technical management system can be divided into three parts according to the main tasks.

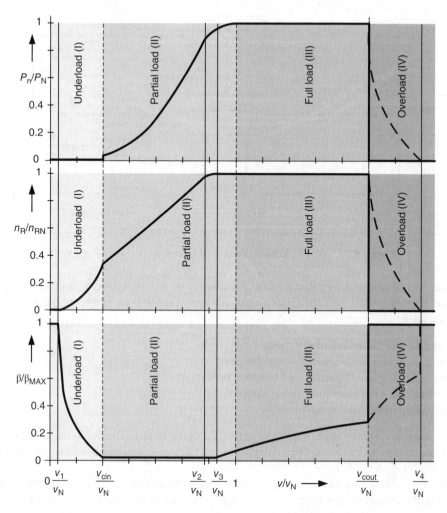

Figure 9.19   Extended operating ranges of a WT with operating characteristic curves

9.3.1   Control of the Operating Sequence of WTs

The operating condition is determined with the aid of the operating sequence, and the transfer between operating conditions is coordinated. It depends upon the type of operation selected (automatic, manual) and the place of operation (remote, local).

In automatic operation the WT can only be switched on and off, all other functions being carried out by the controls. In the manual operation that is necessary for start-up, testing, servicing or maintenance, the change of operating conditions as well as the activation of output signals for the WT can be manually influenced. WTs are usually remote-controlled and monitored by means of management systems. Only in manual operation can the WT be operated with the aid of operating devices on site (local control). In this the local control has priority over remote control.

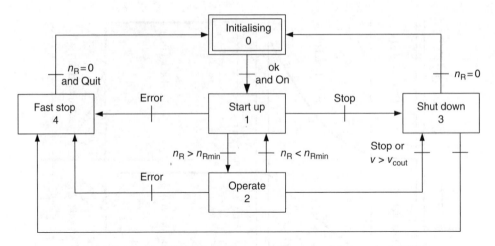

Figure 9.20 Simplified base structure of the sequential control of a WT

The automatic operating sequence can be displayed by means of a sequential function chart in which the following important operating conditions of the WT are differentiated in a simplified manner:

- initialisation and testing of the plant;
- starting up the plant without feeding the grid;
- operating the plant in partial- and full-load operation;
- shutting down the plant in a normal case; and
- quick-stop in the case of malfunction.

The sequence is shown in the base structure of the controls in Figure 9.20. The change from one condition to the other takes place depending on the transition conditions that are shown at the lateral line on the linkage.

The five operating conditions and the transitions are shown in Figure 9.21, taking into account the extended operating ranges from Figure 9.19 in greater detail. Individually the following tasks are to be fulfilled in the steps:

Initialising: After switching on the plant, all the partial controls initialise themselves and the communications between the subsystems and the primary controls are taken up. After the initialisation phase, all the status signals and measuring values are recorded and saved.

Testing the plant: After initialisation, all the important actuators and sensors are checked. Brakes are vented and closed again, pitch and yaw drives, as well as their displacement transducers and rotor and generator rotation transducers, are tested by means of a short test operation. The environmental conditions such as temperature and air moisture, wind speed and direction, as well as the numerous installed vibration and force sensors and current and voltage sensors, are checked for adherence to their limiting values before the plant switches to a state of operational readiness with a stationary rotor.

Stationary: As long as the wind speed is lower than the minimum limiting value v_1, the WT waits with stationary rotor and non-activated pitch and azimuth drives. Only when the wind speed exceeds this value does the WT change over into the start-up condition of the plant.

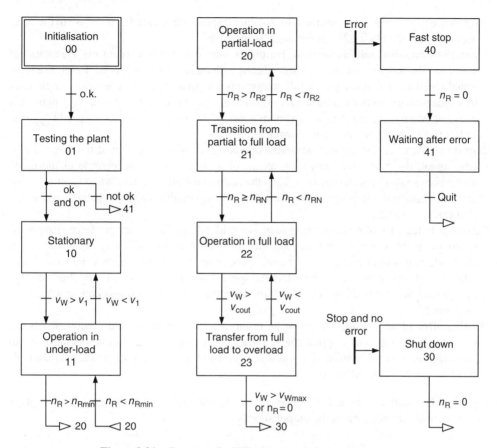

Figure 9.21   Sequence for WT with extended operating ranges

Operation in under-load: When starting up, the plant is aligned to face the wind with vented brakes. The blade adjustment angles are slowly reduced until the rotor starts to rotate. After exceeding a minimum required rotor rotational speed, the generator-side rectifier builds up a torque and the WT starts to feed power into the grid via the converter.

Operation in partial load: In the partial-load range the active power is maximised by means of adapting rotational speed at the optimum blade adjustment angle. If the wind speed increases so that the rotor exceeds the characteristic value n_{R2}, then the WT changes into the transition range between partial and full load.

Transition from partial to full load: In this region the rotational speed of the rotor and the blade adjustment angle are changed. The rotational speed of the rotor rises slightly due to the required power characteristics curve; the power fed into the grid has not yet reached the rated power.

Operation in full load: Above the rated wind speed the WT feeds the rated power into the grid. The power and rotational speed are kept constant by the control of the blade adjustment angle.

Transition from full load to overload: On exceeding the limiting wind speed v_{cout}, the power of the plant is continuously reduced to zero (storm control). For this purpose both the reference power and the reference rotor rotational speed are reduced in response to the wind speed, so that the blade adjustment angle is increased.

Transition from full load to overload: For plants with so-called storm controls, the rotational speed and the torque are continuously reduced with increasing wind speed. Reference rotational speed and reference power fall continuously so that the blade adjustment angle rises to a substantially higher gradient than in full load operation. In the case that the plant still does not stop when exceeding a maximum permissible operating rotational speed v_{Wmax}, the WT automatically goes into shut-down mode.

Shut down: After exceeding the maximum operating rotational speed or at the demand of a stop signal, the plant shuts down by means of a ramp-type, slow increase of the blade adjustment angle to revolution zero. Then the brakes take effect again. When shutting down at the demand of the stop signal, the normal operating condition always changes over to the shut-down condition.

Fast stop: In the case of a malfunction such as a grid voltage failure or very large gusty wind speed, the plant is shut down in a controlled manner by means of a very rapid increase in the blade adjustment angle to rotational speed zero. For safety reasons the rotor brake releases and acts again even if the rotational speed has not gone down to zero after a certain time period. When a malfunction occurs, the operating condition always changes to the fast-stop mode.

Waiting after malfunction: An automatic restart after a security-critical malfunction is not possible. A WT that was stopped due to a fast stop can only be placed into operation again when the cause of the malfunction has been found and is no longer applicable and the malfunction can be acknowledged.

For the tasks mentioned here, the adjustment arrangements for the adjustments and displays can be derived for each step of the output signals.

9.3.2   Safety Systems

Besides the control of normal operations, safe operation of the WT must also be ensured. The technical operating management must monitor the operation of the plant and avoid malfunctions, recognise errors and changes in the plant at an early stage, and automatically initiate safety measures and maintenance work.

As for any other plant, the WT must also fulfil basic requirements of safety. A good overview of general requirements for prevention and mastering failures, as well as organisational demands on operation of plants with programmable control systems, can be found in [16]. The basic features of the safety systems for WTs can be found in [17].

For purposes of safe operation, a safety system is superimposed on the technical management system. So-called failsafe control and circuit components are used, and additional safety systems are installed that acquire the signals from the controls and carry out immediate safety measures.

In the case of exceeding limiting values, malfunctions or disturbances in the sequence of operation, the controls act early in order to ensure safe operation of the plant. The WT is brought into a safe condition during extraordinary wind speeds, environmental conditions such as extreme outside temperatures, conditions of grid supply or unforeseen failure of individual components, for instance due to lightning.

The following events first lead to a reduction of power before triggering a fast stop upon exceeding the limiting values:

- exceeding the high values of rotational speeds for forces, torques and vibration amplitudes of tower, blades or drive train;
- exceeding threshold values for the generator current, for the intermediate circuit voltage and the grid current; and
- exceeding the thermal limiting values in the nacelle, in the generator or the converter, or the limiting values of other climatic dimensions such as moisture.

The following events lead to an immediate stop due to blade displacement and the subsequent triggering of the rotor brake:

- failure of the measuring system for rotational speed, wind speed, current and voltage;
- failure of individual adjusting devices such as yaw and azimuth drives;
- failure of the communication systems within the WT between the controls and adjustment devices;
- exceeding the grid voltage limiting values or falling below the limiting curves for the permissible grid undervoltage; and
- exceeding the maximum or falling below the minimum grid frequency limiting values.

In addition, the operating condition of important operating media of the WT is acquired by means of so-called condition monitoring systems (CMS). These systems automatically introduce the required measures, maintenance work or replacement of operating media before damage occurs.

9.4 Wind Farm Control and Automation Systems

Wind farms are characterised by a large number of WTs, spread over a large area, connected to the public grid via a grid connection point (GCP). The connection line of such wind farms exceeds 100 MW, and they are often connected directly to the transmission grid.

The WTs in such WFs often have different capacities. Thus, in large wind farms, selected WTs are often completely still because of routine maintenance, and with others the power fed in is not identical because of the different wind speeds within the farm. The wind speed of the WTs that are first exposed to the wind is much greater than for WTs situated on the lee side of the wind farm. In new, large wind farms, the installed power of the wind farm is larger than the rated connection power so that for WTs that are stopped for maintenance and with sufficient wind speed, the rated connection power can be fed into the grid.

Individual control of a single WT for adhering to the grid connection guidelines no longer makes sense. Therefore wind farm control and automation systems are now installed that are primary to the WT control and automation systems and which prescribe individual reference values. These WF control and automation systems have the following tasks:

- controlled increase of the power when switching on, with limited dP/dt (switch-on controls prescribing the ramp-up time);
- controlled reduction of the power when switching off with limited dP/dt (switch-off control prescribing the ramp-down time);

- prescription of the reserve power;
- adherence to the active power limit and quasi-stationary frequency support at GCP; as well as
- adherence to the reactive power, quasi-stationary voltage stabilising and dynamic voltage stabilising, in the case of voltage failure due to reactive current feed at the GCP, prescribed by the grid operator.

Section 9.2 describes the control system for an individual WT. When administering a large WF, the tasks for adhering to the grid connection conditions are carried out by a farm control system. Proposals for this were made already in [18] and [19] and these have now been introduced in new farm controls on a step-by-step basis. Figure 9.22 shows a wind farm control and automation system that coordinates the tasks for active and reactive feed according to the requirements of the grid operator.

The blocks for distributing the active and reactive power to the individual WTs can be recognised. The distribution is optimised by taking into account the WT operating conditions, the wind speed and the actual values of active and reactive power.

The dynamic and voltage stabilisation, in the case of voltage failure due to reactive power feeding, is only possible when the reference values are transferred quickly and there is a fast reaction by the WTs. It may be necessary that additional units are only extended for reactive power feeding in a WF.

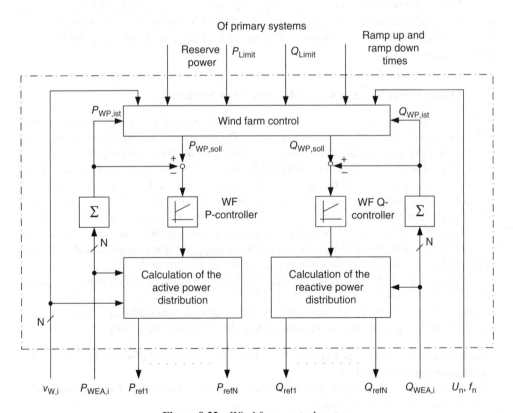

Figure 9.22 Wind farm control system

9.5   Remote Control and Monitoring

Both the data of the technical operating management as well as the data of the operation condition monitoring are acquired, displayed, evaluated and saved in modern remote-controlled operating, maintenance and monitoring systems. These systems for remote control are abbreviated as SCADA for 'supervisory control and data acquisition'. In contrast to operating management, control and automation, the SCADA system does not access the process of energy conversion of the WT directly in real time.

SCADA systems do not control the individual WTs but often all WTs delivered by one manufacturer or all the plants from different manufacturers in use by one operator. In addition, the SCADA system also provides selected data for other applications. Thus, selected data for information purposes is made available to shareholders, or the operating companies can include selected data in their business data processing systems (accounting systems).

The technical tasks of the SCADA systems for WTs and WFs can be classified as follows:

Control tasks

Communication and identification of the WTs: The SCADA system connects to the WT and WF and identifies the position, type and further required plant information.

Access and types of operating control: The SCADA system controls the access to the plant and the data in relation to user rights and access controls. It coordinates on-site and remote control, the switch-over from manual to automatic operation and the necessary locking.

Function checks and parameterising: With the aid of the SCADA system in manual operation, the various actuators can be checked for their functions. Important plant properties can be re-parameterised with the SCADA system.

Alarming: The SCADA system triggers automatic warnings, alarms and possible servicing times and spare parts acquisition by various means such as e-mail, fax or SMS.

Tasks for data acquisition, analysis and archiving

Data acquisition: In fixed cycles, the SCADA system reads and stores the relevant measuring values of the WTs such as wind speed, wind direction, rotational speed, power, current, voltage, frequency, power factor, temperature, moisture and pressure. Operating conditions, system interventions and maintenance activities are also acquired.

Data analysis: The SCADA system evaluates the data and from this it determines characteristic dimensions such as average values, minimum and maximum values, standard deviations, wind speed distributions, power distribution, production and availability statistics, and collates the information for reporting or event protocols.

Archiving: The SCADA system stores and saves the data.

Display: The SCADA system clearly displays site, time and operating conditions and alarms. It shows wind speed and direction, rotor rotational speed and further WT characteristic curves such as power and energy.

Visualising: The SCADA system makes available overview pictures of wind farm, WT and WT components in a hierarchical arrangement. It can present important time sequences as time series.

Operation: The SCADA system offers standard operating possibilities such as start, stop, reset and a selection of types of operation. Using password-protected login/logout functions, it permits user management and access control.

In order to fulfil the tasks, the remote control and monitoring system consists at least of a man–machine interface at the control stand (workplace terminal), a computer for processing, sending and receipt of data, data storage, communication infrastructure, and possibilities for local interaction in the WT and WF arranged system with the control room. Different solutions for linking the communication and automation infrastructure with different communication protocols and data formats are being implemented.

9.6   Communication Systems for WTS

Different communication systems are used for the exchange of necessary information with primary systems and are determined basically by the transfer rate, the number of inputs and outputs, the required reaction times, the distances between the systems, and the security of the transfer. The communication structures and protocols for WTSs are unified with the increasing use of the international standard for '*Communications for monitoring and control of wind power plants*' [20] and for '*Communication networks and systems for power utility automation*' [21].

Figure 9.23 shows a typical WT and WF communication structure and the linkage into a grid system, in the WTS remote monitoring and control as well as further monitoring and diagnostic systems.

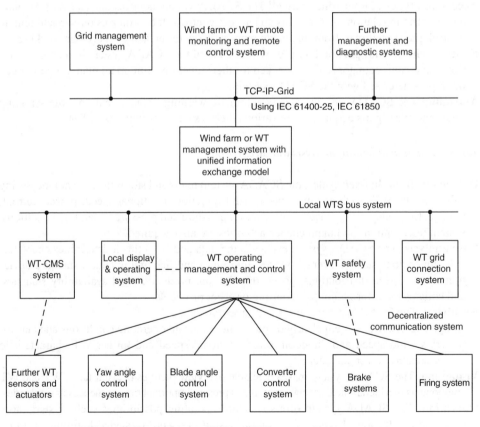

Figure 9.23   Diagram of WT and WF communication structure

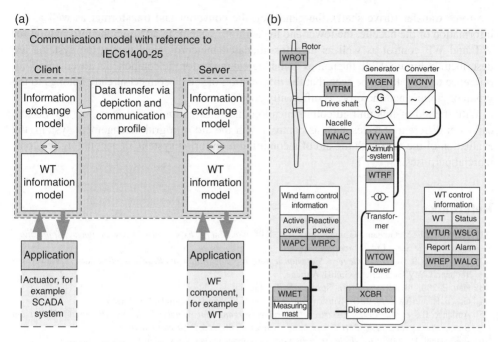

(a)

Communication model with reference to IEC61400-25

Client

Information exchange model

Data transfer via depiction and communication profile

Server

Information exchange model

WT information model

WT information model

Application

Application

Actuator, for example SCADA system

WF component, for example WT

(b)

Rotor
WROT
Generator Converter
WGEN WCNV
WTRM
Drive shaft G 3~ ~ ~
Nacelle
WNAC WYAW
Azimuth-system
WTRF
Wind farm control information
Active power | Reactive power
WAPC | WRPC
Transfor-mer
WTOW
Tower
WT control information
WT | Status
WTUR | WSLG
Report | Alarm
WREP | WALG
WMET
Measuring mast
XCBR
Disconnector

Figure 9.24 Unified WT communication model (a) and information model (b)

Within the WT, the sensors, actuators, the WT controls and controller exchange the required information via different decentralised communication systems. The various WTs in a WF are linked via a local WTS bus system with the WF management system, which must fulfil the real-time requirements for time-critical processes. Glass-fibre cables are used for achieving higher transmission speeds for long transmission lines. Communication to the outside then occurs via the TCP-IP network using standardised protocols and data formats.

The sequential and information exchange as well as the depiction of the WT information is unified in order to integrate specific information for the wind turbines of different manufacturers into the primary system, and to be able to control them in a manufacturer-independent manner. Figure 9.24 shows (a) the WT communication model and (b) the information model of a WT with reference to [20].

Communication is based on the well-known client–server relationship. The WT components exchange their information with the WT information model on a server that prepares this for different clients. Various applications such as the SCADA system can use these data when the client possesses the corresponding rights. The actual data exchange between client and server occurs via unified information exchange models for loading, setting, reporting, controlling or protocoling and, with the data transfer for reading and writing, with the aid of unified communication profiles.

The WT information model provides the content that is required for information exchange between client and server. For this purpose, the components of the real WT write into a virtual WT with unified arrangements, designations, data types, resolution and functionalities of the information. Figure 9.24b shows the important components of the virtual WT. The organisation of the model is hierarchical and comprises the information for the drive train with rotor, the force transfer (drive shaft), the generator, the converter and transformer as well as the information of the nacelle, the tower and the azimuth system. In addition, the information for WT and WF control as well as data for allocated meteorological measuring systems are described. In Figure 9.24b the names of the logical nodes for the components are entered with reference to [20]. It can be seen that the information of the disconnect already belongs to the group of circuits and therefore is differently coded.

With the aid of this unified communication and information model, it is now possible to carry out different services such as controls with the aid of a grid management system, the evaluation of the data with the aid of a condition monitoring system, or accounting for commercial purposes.

References

[1] Manwell, J.F., McGowan, J.G. and Rogers, A.L. 2008. *Wind Energy Explained – Theory, Design and Application*. John Wiley & Sons, Ltd., Chichester.
[2] Heier, S. 2009. *Windkraftanlagen – Systemauslegung, Netzintegration und Regelung*. Wiesbaden, 5. Vieweg und Teubner, GWV Fachverlage GmbH, Auflage.
[3] Hau, E. 2003. *Windkraftanlagen*. Springer-Verlag, Berlin.
[4] Gasch, R. and Twele, J. 2010. *Windkraftanlagen*. GWV Fachverlage GmbH, Wiesbaden.
[5] Amlang, B., *et al.* 1992. *Elektrische Energieversorgung mit Windkraftanlagen. Technische Universität Braunschweig: s.n., 1992. BMFT Forschungsvorhaben 032-8265-B Abschlussbericht*.
[6] Riefenstahl, U. 2008. *Elektrische Antriebssysteme*. Teubner Verlag, Wiesbaden.
[7] Lubosny, Z. 2003. *Wind Turbine Operation in Electric Power Systems*. Springer-Verlag, Berlin.
[8] Nuß, U. 2010. *Hochdynamische Regelung elektrischer Antriebe*. vde-Verlag, Offenbach.
[9] Michalke, G. 2008. *Variable Speed Wind Turbines – Modelling, Control and Impact on Power Systems*. Dissertation, Fachbereich Elektrotechnik und Informationstechnik, Technische Universität Darmstadt.
[10] van der Hooft, E.L. 2001. DOWEC blade pitch control algorithms for blade optimisation purposes. ECN Wind Energy Report ECN-CX-00-083, Petten, NL.
[11] Hansen, M.H., *et al.* 2005. *Control design for a pitch regulated, variable speed wind turbine*. Risø National Laboratory, Risø-R-1500(EN), Roskilde.
[12] Bianchi, F.D., Battista, H. de and Mantz, R.J. 2007. *Wind Turbine Control Systems – Principles, Modelling and Gain Scheduling Design*. Springer Verlag, London.
[13] Transmission Code. 2007. Netz- und Systemregeln der deutschen Übertragungsnetzbetreiber. Verband der Netzbetreiber VDN e.V. beim VDEW, Berlin.
[14] BDEW-Richtlinie. 2008. *Technische Richtlinie Erzeugungsanlagen am Mittelspannungsnetz: Bundesverband der Energie- und Wasserwirtschaft e.V.*
[15] Verordnung zu Systemdienstleistungen durch Windenergieanlagen. 2009. Veröffentlicht im Bundesgesetzblatt Jahrgang Teil I Nr. 39, Bonn.
[16] Funktionale Sicherheit sicherheitsbezogener elektrischer/elektronischer/programmierbarer elektronischer Systeme. 2010. VDE Verlag GmbH, Berlin.
[17] IEC DIN EN 61400-1 and -3. 2010. *Wind turbines: design requirements*; and *Wind turbines: design requirements for offshore wind turbines*. VDE Verlag GmbH, Berlin.
[18] Holst, A., Prillwitz, F. and Weber, H. 2003. *Netzregelverhalten von Windkraftanlagen. 6. GM/ETG Fachtagung "Sichere und zuverlässige Systemführung von Kraftwerk und Netz im Zeichen der Deregulierung"*. VDI/VDE, Munich.
[19] Soerensen, P., *et al.* 2005. *Wind farm models and control strategies*. Risoe National Laboratory Risoe-R-1464(EN), Roskilde.
[20] DIN EN IEC 61400-25. 2007. *Communications for monitoring and control of wind power plants*. Beuth-Verlag, Berlin.
[21] DIN EN 61850. 2006. Communication networks and systems for power utility automation. Beuth-Verlag, Berlin.

10

Grid Integration

Sven Wanser and Frank Ehlers

Modern wind turbines (WTs) are mostly connected to the medium-voltage or high-voltage grids of a grid operator. The grid integration must therefore possess important properties in order to make safe operation of the grid possible and to maintain the voltage quality of the public grid. This chapter deals with the important properties of public electrical grids, the possibilities of influencing the grid by means of operating media such as WTs, and the conditions under which WTs can even be connected to the public electrical grid. The end of the chapter deals with grid integration of WTs that have an effect on future developments such as Supergrids and smartgrids.

10.1 Energy Supply Grids in Overview

This section will deal with the basics of power transmission. The advantages and disadvantages of the different grid structures and the use of different voltage levels are described.

10.1.1 General

Because of current heat losses, which depend upon the square of the current, and because of the voltage changes dependent on the current (voltage decrease or voltage increase), there is also a rise of the required grid voltage generally with the distance and the power to be transmitted. The grid impedance rises with increasing distance or length of the grid; with constant power to be transmitted, the current is reduced by the selection of a higher voltage level. The flow of load set in the electrical power supply grid, made up of active and reactive power,

Understanding Wind Power Technology: Theory, Deployment and Optimisation, First Edition.
Edited by Alois Schaffarczyk. Translated by Gunther Roth.
© 2014 John Wiley & Sons, Ltd. Published 2014 by John Wiley & Sons, Ltd.

depends upon the grid impedances (ohmic inductive and capacitative portion) and the load behaviour of the connected users or generating plants. The capacitive reactances of the supply grid in low-voltage grids and in short medium-power grids (length less than 50 km) have no relevant effect on the power flow, but this is different for high and highest voltage grids. A rule-of-thumb formula for electrical power supply is that for the transmission of power via three-phase systems, 1 kV voltage is required per 1 km distance.

10.1.2   Voltage Level of Electrical Supply Grids

Electrical power supply grids are usually divided into different rated voltages. The limits of the individual voltage levels are defined according to IEC [1] as follows:

Low voltage:	$U \leq 1\,\mathrm{kV}$
Medium voltage:	$U > 1\,\mathrm{kV} \leq 52\,\mathrm{kV}$
High voltage:	$U > 52\,\mathrm{kV}$

In Germany a high-voltage grid is usually designated with a rated voltage of 60–110 kV. Highest voltage grids have a rated voltage of at least 220 kV and serve as transmission grids. In general they are confined to the voltage levels 220–380 kV and their job is to transmit electrical power into the subordinate distribution grids. In special cases, 110 kV grids can also have the function of a primary grid (exception).

Distribution grids, in contrast, serve within a restricted region for the distribution of electrical power for feeding local grid stations and customers. In distribution grids, the flow of power is basically determined by the power requirement of the customers. In Germany, low-, medium- and high-voltage grids (110 kV) are used by more than 900 grid operators as distribution grids. This includes small town and community works up to large regional suppliers. Before the boom of the decentralised power plants (DPPs) based on renewable energies and especially wind turbines, the flow of electrical power was clearly prescribed from the highest to the lowest voltage level (top-down) and the daily time-dependent power flows could be well prognosticated. With the rapid increase in wind turbines in particular, the clear power flows have been fundamentally changed. Thus, today there are feeds at all voltage levels and recovery systems into the respective voltage levels by means of the interposed transformers. Figure 10.1 provides an overview of the individual voltage levels and the WTs connected there.

10.1.3   Grid Structures

Electrical supply grids must be suited in their planning and implementation to the respective feeds that they are to take over. Besides the technical requirements, the commercial needs such as the one-off construction costs and the continuous operating costs must be especially taken into account. Decisive for the operating costs are the annual costs that result from the lost power during transfer over the grid. The technical requirements depend on what power density (MW/km²) is to be covered, how large the supply radius is around the supplying transformer, and what voltage changes (voltage reduction or voltage increase) apply to the grid. In the classical power supply without integration of WT, a central theme of the supply grids was the voltage drop from the supply point to the load. Today, because of the multitude of connected

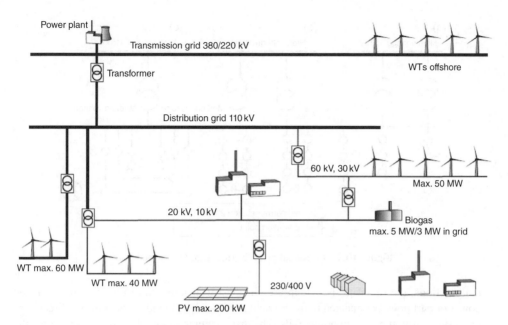

Figure 10.1   Structure of the power grid

wind turbine plants (as well as photovoltaic and biogas plants), one likes to talk of so-called disposal grids. These grids are subjected to voltage increases due to the reversal of the load flow direction.

The permissible voltage change is described as a quality characteristic of grid voltage in the DIN EN 50160 standard [2] with its tolerances, and as a deciding planning criterion. A further planning criterion is the reliability of the supply. This is understood to be the capability of an electrical system to fulfil its supply tasks under prescribed conditions within a particular period of time. The reliability of supply is also a measure of how well the limited power supply system is in a position to fulfil the given supply tasks, i.e. the covering of the load or the transmission of generated power spatially, and also to ensure this under disadvantageous operating conditions. The reliability of supply to an individual power customer is defined by the frequency and duration of interruptions. The supply reliability is different depending on the connection point of the power customer (voltage level and position) as well as requirements of the availability of the required power, so that there are different grid structures with one or more feed points. Feed points are those places in the grid at which fed-in or taken-out power is transformed, converted, or is unchanged and is branched to one or more lines. Generally, electrical grids are differentiated into two base structures: unmeshed or meshed grids.

Radial grids: In a radial grid (Figure 10.2a) the users or suppliers are connected by means of spur lines (radially) to a transformer substation or to a node point. Radial grids are often used for smaller power supplies and/or feeders with no special requirements for supply reliability in the low- and medium-voltage level. For acceptance of wind turbine power, in the first instance for reasons of costs, radial grids have been used in medium and partly even in high voltage, for so-called spur connections of transformer plants. However, with grid

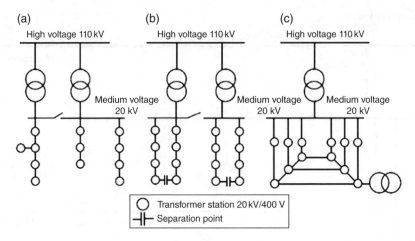

Figure 10.2   (a) Radial grid; (b) ring grid; (c) meshed grid

malfunctions, the radial grids, which are cheapest to install, lead to the failure of all the power or part power connected to the branch. For so-called 'wind grids' without direct supply tasks, 'only' the feed-in power fails. The disadvantage of radial grids, especially for the grid connection user at the end of the line, is the relatively high voltage drop for supply or the large voltage rise with feeding in.

Ring grids: Ring grids (Figure 10.2b) are a special case of the two-sided feed lines, as the lines at the ends are returned to the feed-in point. In this type of grid the feeding takes place from two sides. Ring grids and their derivatives are the most used grids. They offer the operators a good overview because of the simple form. For reasons of cost the individual take-off or feed-in points are not provided with circuit breakers or grid protection devices. The ring contains so-called separation points that are normally open so that two branches exist. The point of separation should, as far as possible, be positioned in the electrical middle (lowest line losses). Operational requirements such as accessibility of the grid stations, however, often require compromises. In cases of interruption the position is localised and disconnected. The point of separation is then closed again and in a short space of time all the users are again provided with power.

Meshed grids: A meshed grid is a grid whose lines run between various nodes and are connected to them (Figure 10.2c). The more complete a meshed grid, the lower are the line losses and the voltage changes. Meshed grids offer a high degree of supply reliability, but the short-circuit flows are also increased. Because of the large number of switching arrangements and protective devices, the monitoring of the grid is greater and more costly. High-voltage grids are often constructed as meshed grids. Thus, for instance, in Schleswig-Holstein, the whole of the 110 kV grid of the EON Netz GmbH is operated as a meshed grid.

10.2   Grid Control

The following section describes the grid regulation of the power supply grid. It will discuss the preparation of the momentary required power and maintaining the frequency of the grid. Also, the system services that are to be fulfilled for safe grid operation are described.

10.2.1   Controlling the Power Range

Electrical power must be generated at the same instant that it is needed by the user. The control of power ensures the supply to the current user of exactly the required quantity of electrical power even in the case of unforeseen events in the grid. Added to this, short-term power adaptations in regulation-capable power stations can be implemented; fast-start-up power stations (for instance, gas turbine power stations) can be started or pumped storage power stations can be used. Alternatively, certain current customers with load control can be separated from the net on a short-term basis.

Controlling the power range is a part of the equalisation services that are required within the scope of the power preparation for covering losses and for equalising of differences between supply and use. Often the term 'control energy' is used for the equalisation power.

In addition, the transmission grid operator, for particular operating conditions and for maintaining the system safety, can automatically or per switch command remove loads from the grid or allocate rated values to power stations. In this way the supply grid can be stabilised and, in the extreme case, prevent load shedding and resulting smaller current failures or regional current failures from occurring [3].

10.2.2   Compensating Power and Balancing Grids

A balancing circuit usually consists of the current supplier and his customers. The current supplier is responsible for estimating the amount of power that his balancing grid requires. He informs his power supplier (power stations) of this estimate so that this supplier can organise the ordered quantity. This procedure is called the scheduled supply. The current supplier can source his power from many different sources of energy, such as hydropower stations, coal-fired power stations, nuclear power plants, wind turbines, biomasses or photo-voltaic, and makes the balancing circuit available. The size of a balancing circuit can vary; a city, for instance, can have several balancing circuits. The balancing circuit pulls together a power trader and power supplier, and all input and withdrawal positions within a regulated zone, as well as scheduled supplies from other balancing grids. Balancing circuits are not only arranged for power traders and sales departments of power supply undertakings, but also, for instance, for large industrial enterprises that look after their own power acquisition (e.g. Deutsche Bahn AG).

On the basis of possibly exact prognoses, the person responsible for the balancing circuit must ensure that it is equalised within each 15 minutes. The balancing circuit is thus the sum of the take-offs, on the one hand, and the sum of the feed-ins on the other hand. Deviations from the schedule are charged to the person responsible for the balancing circuit in the case of undersupply or paid for in the case of oversupply. The accounting takes place on the basis of the costs accruing to the transmission consumer by the use of control energy. Deviations of the actual power offering from that prognosticated occur, for instance, in the case of power station shut-down, non-adherence to the reference profiles of large users, errors in the prognosis in the wind turbine feed or for power failures (loss of users). The larger the balancing circuit, the lower is the relative requirement for control energy. By means of the allocation of the balancing deviation of a balancing circuit to another balancing circuit, it is possible to minimise the deviation as an effect of the resulting mix. Because of the legal obligation for taking up power from renewable energies, the transmission grid operators carry out EEG-wide [Renewable

Energy Law] balancing circuits. Of special importance here is the balancing circuit for wind energy. The feed-in wind power is a dimension that is very difficult to define. Despite improved wind prognosis, the power fed in by wind is the greatest cause for the use of control energy for equalising the balancing circuits [4].

With increasing use of wind energy there is an inevitable increase in the required control energy; it particularly increases the requirement for negative control energy (absorption of production peaks). The renewable energy law forbids the technically easy solution, in the case of wind peaks, for overproduction at the source by reducing the power feed of wind turbines. Rather, it is a legal requirement that the whole of the available wind current is fed into the grid and is paid for. An exception here is the possibility of downward control of the feed-in power when otherwise the grid would be damaged due to overloading. The actual control energy provided has, however, remained constant in recent years or has even been reduced slightly [5]. The actual increased requirement of control energy can hardly be exactly quantified due to the superimposition of normal control energy needs, despite the accuracy of the prognosis system for wind turbine feeds being improved in recent years. Because of the relatively low overall power provided up to now, there is no prognosis program for the feed from photo-voltaic. Because of the added feed with the peak at noon times, the photo-voltaic feed can be seen as dampening the requirement from medium-load and very expensive peak-load power stations, and is only secondary to the control power that is especially needed in the middle of the day.

10.2.3 Base Load, Medium Load and Peak Load

The power requirement of the user and thus also the transport of the current over the grid is dependent upon the time of the year and the time of the day. The use of current reaches peak values in the morning, midday and evening. By contrast, it is particularly low after midnight. Different types of power stations cover these fluctuations. If, for instance, one considers the curve of current demand for one day (Figure 10.3) then it quickly becomes clear that below the minimum value, a certain amount of power is required around the clock. This demand is called the base load. In Germany it is about 40 GW (2005) [6] in comparison to the annual peak of 75–80 GW [7].

A base-load power station is a power station that operates for technical and commercial reasons possibly uninterruptedly and as near as possible to its full capacity. Fixed costs (mostly capital costs) and low power-generating costs mostly play a deciding role in this. For this

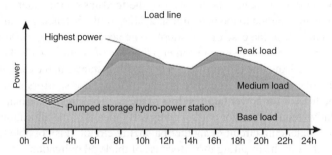

Figure 10.3 Typical load power requirements for one day

reason the power supply undertaking will attempt to estimate the base load requirement in advance for the long term. A corresponding reaction must occur when falling below the estimated value, either by switching on extra users (pumped storage hydro-power station, night storage heating), by transferring power to other power grids, or by throttling back the base power station. Demands on quick controls are not made. Base power stations include brown coal-fired power stations, run-of-the-mill power plants or nuclear power plants. Wind turbines provide a portion of the base power depending on the weather conditions. In the case of great amounts of wind power fed into the grid, the base-load power stations will need to throttle back their output as the preferred take-up of wind energy in Germany is ensured by the Renewable Energy Law (EEG).

In order to cover peak loads, power stations that can be brought up to full power in a few minutes are used. These are, for instance, pumped storage hydro-power stations and gas turbine power stations or compressed air storage power stations. They must be able to follow any change in power needs and thus possess a high degree of dynamics. Gas turbine power stations achieve change speeds of up to 20% of the rated power per minute and are distinguished by fast start-up times of a few minutes. The capacity can be regulated between 20% and 100% [8]. They are used to compensate for the fluctuations in power requirements that cannot be smoothed out by other power stations, or for those where it is not commercially sensible. As peak power stations are normally in use for just a few hours in the day, the power they generate is much more expensive than for other types of power stations.

Between the short-term peak load and the continuous base load is the region of the medium load of the daily load diagram. The medium-load power stations working in this range are switched off or at least are run down to a substantially lower power output at times of particularly low loading, for instance nights. The hourly loading of the grid can be foreseen and is primarily covered by coal power stations. Medium-load power stations are easier to regulate than base-load power stations [9].

10.2.4   Frequency Stability

In all European power stations the generators rotate at exactly 50 times per second and generate the alternating or three-phase current with a frequency of 50 Hertz (Hz). If the frequency in the grid sinks or rises, then the functions and partly the life of numerous electrical devices, such as clocks (in which the grid frequency is used as a reference of the time cycle), computers, motors or compensation installations are influenced. Also the generator themselves can be damaged insofar as the frequency sinks below 47.5 Hz. For this reason in the European linked grid the grid frequency may only vary very slightly from the rated value. The grid controls automatically intervene at a deviation of 0.01 Hz.

For instance, if all the machines in the industrial plants start in the morning at the same time, or if in the evening all the TV sets are switched on at the same time for the news, then the load on the generators rises and for a short moment they run a bit more slowly. This causes the frequency to sink. The automatic intervention of the power frequency controls in the grid ensure that the power stations receive new rated power values and thus the turbines receive more steam and the generators rotate again at 50 Hz. Contracting power stations in the UCTE (Union for the Coordination of the Transmission of Electricity) are participants in this and keep an agreed amount of their generating power as a reserve [10].

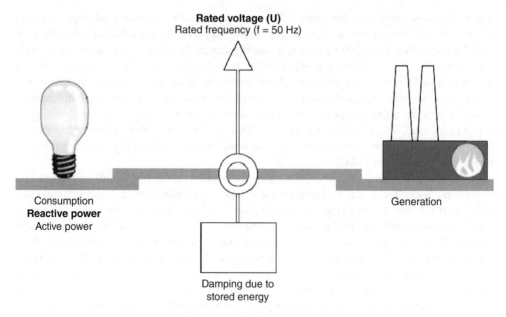

Figure 10.4 Diagram of electrical energy between generation and consumption

The generation and utilisation of power must be equalised constantly so that the frequency does not deviate greatly from the rated value of 50 Hz. The interplay between generation and use is shown in Figure 10.4.

10.2.5   Primary Control, Secondary Control and Tertiary Control

If equalisation between generation and consumption is not provided, then the equalisation reserve or control power cuts in. As there are several balance circuits in which fluctuations occur at the same time, they can balance each other out. However, often balance circuits cannot equalise themselves out. This occurs when all balance circuits are undersupplied or a power station fails. The frequency of 50 Hz can then no longer be maintained and the transmission line operator must provide control power as equalisation. In the UCTE the control power is provided in three control steps [3]:

- primary control
- secondary control
- tertiary control

The three types of control powers differentiate themselves by their speeds of activation and change. Primary and secondary control power is automatically called up by transmission operators from controllable power stations and is quasi-continuously used in different amounts and directions. Tertiary control is requested by means of telephonic instructions by transmission operators to the suppliers.

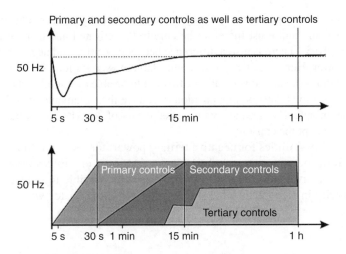

Figure 10.5 Primary, secondary and tertiary controls

The purpose of the primary control is to equalise any unevenness in the physical power supply and demand in the whole European linked grid with the aim of recreating a stable grid frequency. Each grid operator within the linked grid must provide, within 30 seconds, 2% of his momentary generation capacity as a primary reserve for this purpose. Not every power station participates in this primary control (for instance, wind farms, photo-voltaic installations). It is immaterial in which region of the European linked grid a fluctuation occurs, as the momentary grid frequency changes in the whole grid region due to load fluctuations. This is equalised for the proportional primary controllers of the participating power stations with the rated frequency of 50 Hz. If a deviation occurs, then the primary control power in each participating power station (almost all power stations with more than 100 MW rated capacity) are activated according to the control characteristic curve and the frequency (for a jump in the load) or a further increase in frequency (for load decrease) is prevented. The primary control power automatically comes into force in 5 seconds and must increase or decrease the power output within 30 seconds and be able to maintain this for 15 minutes. Also the purpose of the secondary control is to recreate the equalisation between physical current supply and demand after a difference has occurred. In contrast to the primary control, here only the situation in the respective regulation zone, including the current exchange with other regulating zones, is considered. For this purpose, the planned and the actual power flows to other regulation zones are compared and equalised. It must be ensured that the secondary and primary controls always work in the same direction, which is achieved by monitoring the grid frequency.

Primary and secondary controls can start at the same time; the secondary control should be achieved after at the latest 15 minutes. This is important so that the primary controls are released again for new control events. The secondary control power must thus be prepared to its full extent in 15 minutes (Figure 10.5). Pumped storage hydro-power stations or gas turbine power stations are used for this purpose.

If the control power has not been cleared up within 15 minutes, then the secondary control is relieved by the tertiary control. It is used for longer-lasting application of secondary power, especially after power station failures, in order to relieve the secondary control and make it available for new control processes. Tertiary controls are usually used as scheduled deliveries for full quarter-hours and must therefore be able to be activated and deactivated within 15 minutes (Figure 10.5). There is also a difference between positive and negative control power with tertiary power. Here use is made of controllable power stations such as pump storage hydro-power stations or coal power stations. In order to be able to smooth out the load fluctuations, the power stations must be in a position to change their gradients by at least 2% of the rated power per minute. For instance, with a rated power of 800 MW, this would be ±16 MW/min by which to adapt the capacity.

There are two possibilities for negative tertiary power: for rises in frequency, additional loads in the form of pump storage hydropower stations, or night storage heating, for example, can be activated in the grid. Besides this, it is possible to reduce the generated electrical capacity in steps within the shortest space of time, or to switch power stations off [3].

10.2.6 Voltage Stability

Whereas the grid frequency is literally determined by the active power flow, the voltage is mainly influenced by the exchange of reactive power. Depending on the primary active and reactive resistances of the grid, the reactive power behaviour of the user or the generating plant decides upon the amount of the change of voltage in the grid or the grid connection nodes. If the behaviour of a classic user in the past led to component failure, then the position with wind turbines is different today as the direction of the load flow has changed. With the reactive power region in which today the normal wind turbines are operated, the voltage at the grid connection nodes is normally increased.

In the reactive power take-up of the wind turbine (inductive behaviour) the rise in voltage is much lower than the reactive power feed (capacitative behaviour). In the high and highest voltage grids, the voltage stability is kept within permissible limits by the so-called reactive power bands of power stations. Here one talks of voltage reactive power control. Due to the massive expansion of wind energy, wind turbines must also take part in the reactive power exchange and thus in the voltage stability.

10.2.7 System Services by means of Wind Turbines

System services in electrical supply are designated as the aid services necessary for the functionality of the system, and are provided by the generator of electricity for his customers in addition to the transmission and distribution of electrical energy, and which thus determine the quality of the power supply. Included in the system services are [11]:

- (active power) control reserve (relevant for the stability of the frequency);
- primary power, secondary power and tertiary power;
- voltage stability;
- compensation of the active losses;
- black start/island operation capability;
- system coordination;
- operational measurements.

By 2001, EON Netz GmbH, at that time an operator of transmission grids, had carried out a short-circuit calculation with a simulated three-pole short-circuit in the 380 kV switching installation in Dollern. Physically a resistance area forms at the point of the malfunction. At this point of malfunction the voltage collapses to 0%. Due to the impedances of the grid, the voltage increases with an increase of the distance from the point of the malfunction. Thus, the voltage in the 380 kV grid at the Danish border was approximately 70%. The voltage collapse transmits itself more or less damped through the transformer impedances also in the lower levels of high-, medium- and low-voltage grids.

The BDEW guidelines of the time, *Generating plants at medium voltage grids* and *Generating plants at low voltage grids*, were based on the fundamental idea of minimising the reactions of generating plants on the distribution grid and thus on maintaining the supply quality. Also, with their requirements, they ensured that a rapid uncoupling of the generating plants from the grid would occur in the case of disturbances in the distribution grids. After the above short-circuit simulation it was determined that wind turbines with an installed capacity of around 2700 MW would be connected to the grid within the voltage funnel $U \leq 70\%$ U_n. If such a simulated malfunction would occur and if there was a corresponding wind capacity available, then a very high feed capacity due to uncoupling protective devices would separate the wind turbines from the grid immediately. At this scale the stability limit of the UCTE link grid is endangered.

Because of the increase in generating plants that are connected to the grid and in use on the basis of the Renewable Energies Law in a primary manner, there are other demands than those to be met by the behaviour of these plants in normal operation in the case of grid malfunction and in order to ensure stable and proper system operation. Newly installed WTs must participate actively in the stability of the voltage and frequency. For instance, the consequences of a malfunction in the grid due to a failed power feed must be limited in order to prevent uncontrolled spread of the malfunction; a rapid uncoupling of the generating plant in the case of malfunctions in the high and highest voltage grids must not occur unselectively. In the future, generating plants of the medium-voltage grids will also participate in supporting the grid. They must not, as previously, immediately disconnect from the grid in case of a malfunction, and even in normal operation must contribute their portion to the voltage stability of the medium voltage grid.

From this knowledge, wind turbines must fulfil the following general requirements that are summarised under the term 'wind system services':

- Static voltage stability (by means of a controllable reactive power region).
- Dynamic grid support by means of voltage stability in the case of voltage collapse in high and highest voltage grids. This means the capability:
  - not to disconnect from the grid in the case of grid malfunction.
  - in the case of a grid malfunction to support the grid voltage by feeding in reactive power.
  - after clearing the malfunction, not to take inductive reactive power as before the malfunction.
- Operate with reduced feed-in capacity.
- Active power reduction for excess frequency in the grid.

The relevant guidelines for connection of generating plants of high and highest voltage grid as well as medium- and low-voltage grids have been reworked or are being reworked on the basis of compulsory requirements. In this they collate the important points of view that are to be taken into account in the connection to the respective grids, so that the safety and reliability of the grid operation can be maintained in accordance with the demands of energy management law, considering the growing proportion of decentralised generating plants and the limiting values of the voltage quality formulated in DIN EN 50160 [12].

10.3   Basic Terminology of Grid Integration of Wind Turbines

This section describes the basic electrical terminology that is necessary for grid integration. Emphasis is placed here on the voltage quality that is to be noted especially in grid integration.

10.3.1   Basic Electrical Terminology

10.3.1.1   Reference Arrow System

The user reference arrow system is used by power station and grid operators to denote the direction of current and voltage as well as the phase angle.

In the following the user reference arrow system will be used for the reference customer plants connected to the grid as well as for the generating plants. Flows and voltages in the direction of the arrow are counted as being positive. In order that there is no misunderstanding of the terms 'under-excited operation' and 'over-excited operation', as well as 'inductive' and 'capacitative' behaviour for users, it is important to agree to a unification of the arrow system. General use is made of the user reference arrow system in the connection guidelines as well as the technical guidelines of wind energy for the measurement and evaluation of electrical properties, as well as for the reference customer plants connected to the grid. A power circle is used (Figure 10.6) for depicting the individual operating conditions in quadrants. The different 'operating conditions' are depicted in the four quadrants I to IV.

The current indicator always lies on the real axis (3 o'clock) whilst the position of the voltage indicator corresponds to the virtual power and the phase angle. Angles are counted anticlockwise as positive. The phase angle is defined as the angle of the current indicator to the voltage indicator.

In quadrants I and IV an active power (P+) is withdrawn by the user plants, whereby in quadrant I reactive power is also withdrawn (so-called inductive behaviour) and in the quadrant IV reactive power is delivered (so-called capacitative behaviour).

The quadrants II and III, in contrast, describe the behaviour of a generating plant that feeds active power (P–). In quadrant II the generating plant withdraws reactive power from the grid whereas in quadrant III it makes reactive power available to the grid.

10.3.1.2   Single Pole Equivalent Circuit Diagram

In general the circuit diagram in the single pole depiction is sufficient for the evaluation of connection possibilities and for showing the overall layout. The overview circuit diagram of the whole electrical plant contains all the relevant data on the used operating media, the customer's own medium-voltage connections, cable lengths and switching plant, as well as an overview of the protection. Further, it shows where measurements are taken and upon which switches the protection is effective.

The all-pole circuit diagram, in contrast, goes into greater depth, detailing the plant and the components. They are used for cabling tasks and for possible searches for errors.

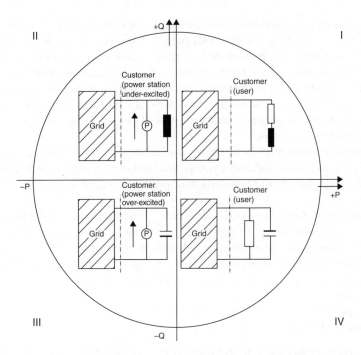

Figure 10.6 Depiction in the user reference arrow system

10.3.1.3 Operating Media

Operating media are all components, modules or devices of an electrical plant. Thus, they are all 'objects' that are used in whole or in part for generation, transport, distribution, storage, measuring, data transfer, and so on, in the sense of the utilisation of electrical energy.

WTs and other DPPs can cause a higher loading of the lines, transformers and other operating media of the grid. For this reason a check of the load capacity of the grid operating media with a view to the connected generating plants according to the normal measuring regulations is necessary. In contrast to the plant operating media for the user plants, continuous loading must be assumed here (degree of loading = 1) instead of the often-used EVU [electricity supply utility] load (degree of loading = 0.7).

10.3.1.4 Availability

The availability of a technical system is the probability or the measure that the system will fulfil certain requirements by, or within, an agreed time period. It is a quality criterion/characteristic of a system [13]. The availability can be defined on the basis of the time that a system is available:

$$\text{Availability} = \frac{\text{Overall time} - \text{overall failure time}}{\text{Overall time}} \qquad (10.1)$$

In electricity supply, availability is understood to be the relationship of the time period for which the system is available for its actual purpose (operating time), to the agreed-to time (overall time). The operating time can be limited by regular maintenance or repairs.

In wind energy technology the term availability has a different meaning. Wind turbines as technical units are subject to the abovementioned view and normally possess an availability of more than 95%. This means that in more than 95% of the overall time possible, they could produce power. Whenever possible, maintenance work is carried out at times of low winds in order to minimise the loss of yield.

A well-known manufacturer in Germany, for instance, provides the plant operators with warranty of 97% technical availability for its wind turbines. A plant is technically available when the technical condition of the wind turbine permits operation under normal environmental conditions.

The availability of the wind as a driving force is a different matter for the wind generators. Wind is not always sufficiently available and also cannot be simply switched on. The availability of wind depends on the respective meteorological conditions. WTs at good sites can produce 2000–3000 full load hours a year. The advantage of the high degree of availability can therefore not be fully utilised due to the low availability of the wind.

10.3.1.5   Supply Quality

The supply quality of an electrical grid is the sum of all quality-determining conditions from the point of view of the power customer and is defined by the following three components:

- supply reliability;
- voltage quality;
- service quality

The supply reliability as a part of the supply quality is the ability of an electrical system to fulfil its supply tasks under prescribed conditions during a particular period of time. The supply reliability of the individual power customer is determined by the frequency and duration of the supply interruptions. Supply reliability is different depending on the structure of the grid and the position of the power customer in the grid. In order to evaluate the whole grid of a grid operator, average values of corresponding power customers are taken. For instance, the average failure frequency and duration as well as the product from both dimensions are considered.

10.3.1.6   The (n – 1) Criterion

The (n – 1) criterion is a safety standard that is often used in planning and operating the management of grids. This criterion ensures that the grid (n) can ensure the supply even with the failure of the strongest line (−1) or the largest transformer. This occurs in high and highest voltage grids using automatic re-energisation. In meshed medium-voltage grids there are automatic switches, but often also manual switches are used. In low-voltage grids the (n – 1) criterion is often ensured by the meshed grid topology or the use of subsidiary grid plants (emergency power installations) in the case of failures or for planned work. The reference-oriented (n – 1) supply reliability is normal in grid operation.

Because of costs, wind turbines themselves are normally not connected to the grid of the general supply in a (n − 1) manner as in the case of malfunction, 'only' the feed falls away and the supply to grid connection entities is not endangered.

If the connection power of a WT or a wind farm exceeds the (n − 1) case – thus in the case of a grid malfunction the permissible power – then the generating plant in the (n − 1) case is limited in its capacity or must be switched off.

10.3.1.7 Short-Circuit Current/ Power

With the operation of a wind turbine or a wind farm, the short-circuit power of the grid, especially in the neighbourhood of the grid connection point, is increased by the short-circuit current of the generating plant. Information about the expected short-circuit current of the wind turbine and the grid connection point is therefore indispensible for grid planning and calculation.

The following values can be approximately assumed for the determination of the short-circuit current contribution of a WT (generating plant):

- for directly coupled synchronous generators, 8×
- for asynchronous generators and double-fed asynchronous generators, 6×
- for generators with inverters or direct current intermediate circuits, 1×

For an exact calculation of the rated current, the impedances between generator and grid connection points (customer transformer, lines, and so on) must be taken into account.

If, because of the wind turbine(s), the short-circuit current in the medium-voltage grid is increased above the rated value, then suitable measures such as limiting the short-circuit current from the wind turbine (e.g. by means of the use of short-circuit current limiters) must be agreed upon.

In consideration of the short-circuit current capability of all electrical operating media, attention must be paid, not just to the starting short-circuit alternating current I_k'' that is responsible for the thermal load, but also especially to the impact short-circuit current I_p, which substantially affects the operating media (attractive and repulsive forces).

The short-circuit power S_k'', or start short-circuit alternating current power according to DIN EN 60909-0 (VDE 0102) '*Short-circuit currents in three-phase a.c. systems*', can be found by:

$$S_k'' = \sqrt{3} U_n I_k''$$ (10.2)

where U_n is the rated voltage. In the design of medium-voltage switching circuits, the impact short-circuit current is in general 2.5 times greater than the short-circuit current capability for the starting short-circuit alternating current. Typical values for the short-circuit current capabilities are therefore:

$$I_k'' = 16\,\text{kA} \quad \text{(for 1 sec.)} \quad I_p = 40\,\text{kA}$$
$$I_k'' = 20\,\text{kA} \quad \text{(for 1 sec.)} \quad I_p = 50\,\text{kA}$$
$$I_k'' = 30\,\text{kA} \quad \text{(for 1 sec.)} \quad I_p = 75\,\text{kA}$$

10.3.1.8   Reactions

A grid reaction is understood as the mutual influencing of operating media on the grid and also the influence on the grid by these operating media. The disturbances from electrical operating media act upon the form, height and frequency of the supply voltage, and for three-phase systems also possibly upon the voltage symmetry [14].

10.3.1.9   Compatibility Level

The value of a disturbing value defined by a system that will be exceeded with such low probability that there is no danger for the electromagnetic compatibility of the equipment.

10.3.2   Grid Quality

The basis for determining a suitable connection point for a generating plant (e.g. wind turbine or wind farm) depends upon, on the one hand, the demand for a reliable grid operation according to the requirements of energy management law and, on the other hand, the formulated limiting values of voltage quality in EN 50160 *(Voltage characteristics of electricity supplied by public distribution networks)*. This European standard defines, describes and specifies the important characteristics of the supply voltage at the transfer point of the grid user in low-, medium- and high-voltage supply grids under normal operating conditions. With this standard the supply voltage is described as a product with fixed characteristics. Every grid connection consumer, whether user or generator (feeder), influences the voltage quality, as the grid is not rigid. A rigid grid would presuppose an infinitely high short-circuit power.

The quality characteristics are different depending on the grid level. Mutual interaction must be taken into account here. For this reason there follows the division of the connection guidelines according to voltage levels, as the specific requirements are too variable to summarise them in a single guideline. The requirements of EN 50160 are respectively considered, in that the following characteristics of the supply voltage are defined:

- frequency;
- voltage level;
- voltage curve form;
- symmetry of the conductor voltages.

Influences on these characteristics of the supply voltage have the following effects:

- harmonics;
- grid flicker;
- voltage changes;
- grid-signal transmission voltages;
- frequency stability;
- uneven load distribution on the individual conductors.

10.3.2.1 Harmonics

Constant, periodic deviations of the grid voltage from the sine wave form (voltage distortion) are caused by additionally superimposed fluctuations whose frequency is an integer multiple of the grid frequency. These so-called harmonics in the grid voltage are caused by operating media (devices and plants) with a sine wave current uptake. A harmonic voltage is the result of a corresponding harmonic portion of the current and the grid impedance for the corresponding harmonics.

The important harmonics generators are, for instance (see [14]): operating media (devices and plants) of power electronics (e.g. power converter drives, rectifier installations, dimmers, or mass appliances with rectifier supplies such as TV sets, compact lighting fluorescent lamps with built-in electronic pre-switch devices, IT devices) or operating media with non-linear current-voltage characteristics such as induction and arc ovens, gas-unloading lamps, motors, small transformers and throttles with iron cores.

High harmonics in the grid voltage can lead to impairments of the grid operation as well as in the operating media (devices and plant) for grid users, such as:

* reducing the life of capacitors and motors due to thermal overload;
* acoustic disturbances of operating media with electromagnetic circuits (throttles, transformers and motors);
* coupling on the harmonics in reporting and information equipment;
* functional disturbances in electronic devices;
* incorrect functioning of ripple control receivers and protective equipment;
* making earth fault neutralisation more difficult in grids.

The current regulations (e.g. the European standard EN 50160 *Voltage characteristics of electricity supplied by public distribution networks*) define the adherence to fixed limiting values for harmonics in grids. Limits are determined for individual harmonic flows as well as for the totality of all harmonic flows. In low-voltage levels the voltage distortions add up all superimposed voltage levels. The permissible harmonic voltage in the respective grid levels are used up to a great extent already by the connected user devices. The additional harmonic voltage values generated by wind turbines must therefore be limited to permissible values. For this reason, possible measures for the reduction of harmonics are necessary. An evaluation as regards harmonics is mostly required only when feed occurs via converters or inverters.

10.3.2.2 Grid Flicker

Changes of loads triggered by usage equipment or generating plants cause changes in the voltages at the grid impedances. The voltage changes lead to changes in the density of light in lamps, especially incandescent lamps. These light density changes, noticeable to the eye, are called flickering. As the eye reacts very sensitively to flickering, changes in voltage must be kept within limits.

A noticeable light density change is only found to be disturbing at a certain repeat rate. The disturbing effect grows quickly with the amplitude of the fluctuations. At certain repeat rates, even small amplitudes can be disturbing. Basically the voltage changes caused by a device are limited to 3%. The human eye is most sensitive to 18 changes per second. Voltage changes of only 0.3% can lead to complaints in this region. The flicker strength is used as a measure for flickering.

Voltage changes can occur due to, for instance, the switching-on of larger loads (e.g. motors, capacitors), controlled loads (fluctuation packet controls) and due to variable feeders (e.g. WTs). In the case of wind turbines the fluctuations of the wind (e.g. gusts) can lead to a change in the feed-in power. The power change, for its part, affects the impedances and can lead to impermissible flicker values.

10.3.2.3 Permissible Voltage Changes

Tolerances in the operating voltages in low-voltage grids are compulsorily defined in the DIN IEC 60038 standard (CENELEC standard voltages) and EN 50160 (*Voltage characteristics of electricity supplied by public distribution networks*). Throughout Europe the rated voltage is 400 V between the external conductors, and correspondingly 230 V between the external and neutral conductors or earth. The tolerance limit of the operating voltage is $\pm 10\%~U_n$ and must be adhered to by the grid operator for fluctuations between weak and strong loads as well as for fluctuations in the feed from wind energy.

The grid operator selects the operating voltage of the medium-voltage grid and the level setting of the grid transformers such that the operating voltage for the most distant customer installation is still above the lower tolerance limit and a customer installation nearer to the transformer substation receives a voltage that is not too much above the rated operating voltage. The control of the grid transformers from the high-voltage to the medium-voltage level occurs by means of a large number of steps so that the overall operating voltage region of the high voltage can be compensated. Thus, the medium voltage at the collector bus of a transformer substation fluctuates within a level (usually between 1% and 1.5%).

Steady-state voltage stability is understood as the voltage stability for the normal operating case in which the slow voltage change in the distribution grid is stable within tolerable limits. Wind turbines must participate in the steady-state voltage stability by means of their reaction power behaviour.

10.3.2.4 Frequency Stability

The frequency is controlled to such a fine extent in the large power stations connected to the UCTE grid that the operating frequency has fluctuated in the past only by a few mHz. With the increase in wind turbines, the stability of the frequency has become much more difficult. For this reason, WTs must also participate in maintaining the frequency when in operation, in that from a frequency of more than 50.2 Hz they reduce the momentary active power with a gradient of 40% of the momentary available power of the generator per Hz (Figure 10.7). Only when the current frequency has sunk to a value of $f \leq 50.05$ Hz may the active power be increased again, and only when the frequency falls below 47.5 Hz may the WT independently separate itself from the grid.

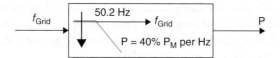

Figure 10.7   Example of active power reduction in the case of excess frequency $p=20\,P_{\mathrm{M}}(50.2\,\mathrm{Hz} - f_{\mathrm{grid}})/50\,\mathrm{Hz}$ at $50.2\,\mathrm{Hz}<f_{\mathrm{grid}}<51.5\,\mathrm{Hz}$; P_{M} is the momentary available power, p is the power reduction, f_{grid} is the grid reduction; in the region of $47.5\,\mathrm{Hz}<f_{\mathrm{grid}}<50.2\,\mathrm{Hz}$ no limitation; at $f_{\mathrm{grid}}=47.5\,\mathrm{Hz}$ and grid$=51.5\,\mathrm{Hz}$ separation from the grid

10.4   Grid Connections for WTs

With the enormous increase in wind turbines, there has been a paradigm shift at the turn of the century in the grid connection conditions for wind turbines. If in the past an unselective separations from the grid applied for grid malfunctions, it is now necessary for the stability of the protection concept that the connection conditions for wind turbines (but also other forms of power generation) changes fundamentally. The requirements of the currently valid EEG as well as the ordinance for system services by wind turbines (System Services Ordinance – SDLWindV) are taken into account, just as for the latest knowledge from stability calculations of high and higher voltage grid operators.

The grid connection guidelines or user regulations are the basis for the technical connection conditions of the grid operators and summarise the important points that must be considered in the connection of generating plants to the grid. They especially define the processing duties of the grid operators, the erecters, planners, as well as power customers. In Germany the technical connection conditions (TCCs) of a grid operator apply together with part 19 of the energy management law (EnWG), *Technische Vorschriften* [*Technical requirements*], and must be published. In general, the connection guidelines or the rules of application are valid for new generating plants to be connected, as well as for those where major changes are being carried out (e.g. repowering).

Wind turbines must be able to participate in the static and dynamic grid support in grids of the highest voltage, high-voltage and medium-voltage. With the static frequency stability in normal operating cases, the slow changes in the voltage can be kept to tolerable limits by the provision of reactive power conditions. The dynamic grid support prevents the undesirable switching-off of large power feeds, and the resulting grid collapses in the case of component failures in the high and highest voltage grids. As wind turbines with connections in the medium-voltage grid have an immediate or longer influence on the high and highest voltage grids, these plants must also be in a position to fulfil the system services described in Section 10.2.7.

In the future, plants feeding the low-voltage grids must also participate in the static voltage stability. As they are obliged during normal grid operation to make their contribution to maintaining the voltage of the low-voltage grid, this has an immediate effect on the design of the plants. Dynamic grid support by generating plants that feed into the low-voltage grid is not required.

The verification of the required electrical properties of WTs and wind farms must be provided by independent accredited institutions. For this purpose the project-specific electrical properties and the guideline-conforming behaviour of the sum of all wind turbines connected at the grid connection point, the internal wind farm connection, additional components such as compensating installation and the connection lines to the grid connection point (also the complete connection installation), must be verified.

For a connection evaluation as described in the next section, it is necessary to carry out a test of the voltage changes, the grid reaction, the short-circuit strength and the design of the grid operating media. In addition, long-term flickering, harmonics, reactions of tone frequency and ripple control systems and the dynamic grid support must be checked.

10.4.1 Rating the Grid Operating Media

This section describes the design of the grid operating media on the basis of an example. The following grid structure (Figure 10.8) for the connection of a wind farm with a medium-voltage grid is also used for viewing the voltage increase in Section 10.4.2 and the reactive power control in Section 10.4.3.

The Maxwind wind farm is to be connected to a transformer substation that is fed by two grid transformers and which has two separated 20 kV collector rails (CRs), connected by means of an 8 km long aluminium cable (NA2XS2Y $3 \times 1 \times 500^2$).

In addition, a reference plant is to be connected to the planned connection point of the wind farm. For this reason this point is also designated with a V point of common coupling.

These planned wind turbines are eight double-fed asynchronous generators of the Vielwind Company TYPE WEA 2000 with 2 MW rated virtual power per unit $S_{re} = 2.2$ MVA.

The displacement factor required by the grid operator at the connection point V is to be between 0.95 inductive and 0.95 capacitative.

In order to check the rated power P of the operating media, first the greatest expected power I of the wind farm is calculated. For this purpose the 'most disadvantageous' displacement

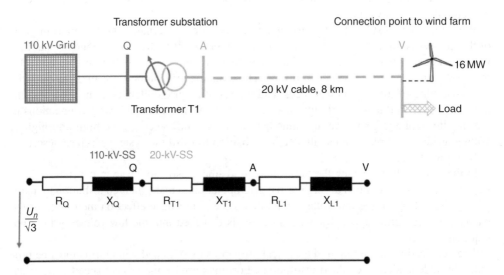

Figure 10.8 Example for calculations: simplified equivalent circuit diagram

point of 0.95 is assumed. For the active power to be evaluated, The determination of inductive or capacitive power is irrelevant for the active power to be evaluated.

$$P = U_n I \sqrt{3} \cos \phi$$

$$I = \frac{16\,\text{MW}}{20\,\text{kV} \times \sqrt{3} \times 0.95} = 486\,\text{A} \qquad (10.3)$$

For the sake of accuracy, the real operating voltage at the connection point V should be calculated. The calculation according to this formula, however, is sufficiently accurate. The exact values of the current will be slightly less as the voltage at point V is somewhat greater than 20 kV. Thus, the calculation is on the 'safe' side.

The transformer possesses a rated virtual power of $S_{rT} = 20\,\text{MVA}$. With a displacement factor of $\cos \phi = 0.95$, the virtual power of the wind farm is:

$$S_{rT} = \frac{P}{\cos \phi} = \frac{16\,\text{MW}}{0.95} = 16.8\,\text{MVA} \qquad (10.4)$$

The transformer is therefore sufficiently dimensioned. The 20 kV output fields in the transformer substation are designed for 603 A and the transformer feed field for 1250 A (future extension possibilities). The planned wind farm can therefore be connected to an output field.

According to the manufacturer's information, the 20 kV VPE cable is designed for a rated current I_r of 610 A. However, this rated current applies for defined environmental conditions such as soil temperature and heat conducting capability of the soil at a degree of loading of $m = 0.7$ of the so-called EVU [electric supply utility] load. The degree of loading is the result of the quotient of the surface under the load curve and the overall surface of the rectangle. In Figure 10.9 the degree of loading for an EVU and a continuous load are shown.

However, due to the operation of the wind farm, the degree of loading of 1.0 (continuous load) must be applied. Because of the non-available cooling phase, reduction factors for the current loading are to be applied. The permissible current loading of the cable is calculated according to DIN VDE 0276 [15]:

$$I_z = I_r f_1 f_2 \prod f = 610 \times 0.93 \times 0.87 \times 1 = 493\,\text{A} \qquad (10.5)$$

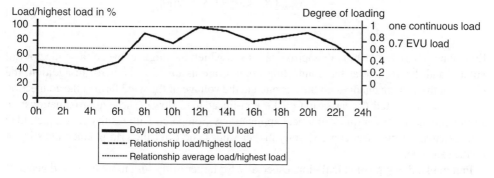

Figure 10.9   EVU and continuous load behaviour

where I_z is the maximum permissible loading on the cable, f_1, f_2 are reduction factors depending on the degree of loading, environment temperature, specific soil resistance and accumulation, and Πf is the sum of further possible reduction factors such as pipe-laying or covering.

Thus the current load ability lies above the maximum expected load current for the wind farm of 486 A.

The load at the connection point V for the worst-case point of view as weak load is set to zero. Take-off loads would reduce the current on the cable when taking the normal power factors of users into consideration.

10.4.2   Checking the Voltage Changes/Voltage Band

As the wind farm will eventually be used for voltage support and the grid operator has specified a defined reactive power range, the voltage changes Δu_a must achieve $\cos \phi = 0.95$ inductive and $\cos \phi = 0.95$ capacitive for the extreme values. The voltage changes at the connection point V can be calculated with regards to the reactive power through the wind farm by means of the following equation:

$$\Delta u_a = \frac{S_{Amax}\left(R_{kV} \cos|\phi| - X_{kV} \sin|\phi|\right)}{U^2} \tag{10.6}$$

where S_{Amax} is the sum of the maximum connected virtual power, R_{kV} is the short-circuit effective resistance at the supply point, and X_{kV} is the short-circuit effective reactive resistance at the supply point.

The voltage change is usually positive (voltage increase). If the second term in the numerator is larger than the first term, then it can become negative, which corresponds to a voltage reduction. However, this assumes a sufficiently small $\cos \phi$ and thus a correspondingly high delivery of reactive power of the WT. But since such a high reactive power substantially increases the power loss and reduces the transmission capacity of the lines, this case almost never occurs in practice in low- and medium-voltage grids.

With reactive power draw-off by the wind farm, the voltage increase at the connection point is strengthened. The following equation applies:

$$\Delta u_a = \frac{S_{Amax}\left(R_{kV} \cos|\phi| + X_{kV} \sin|\phi|\right)}{U^2} \tag{10.7}$$

Equations (10.6) and (10.7) are approximations in which the angle between the collector rail voltage and the voltage at the connection point V are assumed to be zero. In addition, the effects of the voltage changes on the current and the voltage of the wind farm at the connection point V are neglected (linearisation of the non-linear load flow). These approximations are sufficiently accurate for quick practical applications. Compared with results of complex load flow calculations the results are slightly higher values and offer the grid planner an estimate on the safe side.

Preferably, this type of calculation, especially for larger grids with more users and generating installations, should be carried out using complex load flow programs.

As, in practice, the high voltage grid changes in the range 96–123 kV, grid-coupling trans-formers are usually equipped with a step switch that keeps the voltage on the secondary side (medium voltage) sufficiently constant. Usually a control region of ±22% over a total of 27 steps is used (13 steps +, 1 step middle setting and 13 steps –).

The resulting voltage changes in the WF, according to the example in Figure 10.9, are cal-culated according to equations (10.6) and (10.7), simplified about the sum of the active and reactive power. The step switch for the transformer is assumed to be in the middle setting. A calculation of the voltage change for each branch is to be carried out for the connection of several reference plants and generating plants via various outputs in the transformer substa-tion. The summed effect of the take-off and feed power via the transformer is carried out using a phase-correct addition of the active and reactive power.

Impedance determination according to the equivalent circuit diagram Figure 10.8:

Calculating the impedance of the 110 kV grid:

$$Z_{Q20kV} = \frac{U_n^2}{S_{k110kV}} \tag{10.8}$$

where S_{k110kV} is the grid short-circuit power of the 110 kV grid, and Z_{Q20kV} is the impedance of the source (referenced to 20 kV).

$$S_{k110kV} = 2000\,\text{MVA}$$
$$Z_{Q20kV} = \frac{U_n^2}{S_{k110kV}} = \frac{(20\,\text{kV})^2}{2000\,\text{MVA}} = 0.2\,\Omega$$

In the high-voltage grid feed with rated voltages above 35 kV, fed via open lines, Z_Q can in most cases be seen as reactance (ind. reactive resistance $R \ll X$). If the resistance (active resistance) is not exactly known, then $R_Q = 0.1\,X_Q$ can be used. This then results in the follow-ing values:

$$X_Q = Z_Q = 0.2\,\Omega$$
$$R_Q = 0.1 X_Q = 0.1 \times 0.2\,\Omega = 0.02\,\Omega$$

Impedance of the transformer 110/20 kV:

Rated virtual power: S_{rT} 20 MVA
Rated voltage, overvoltage side: U_{rTOS} 110 kV
Rated voltage, undervoltage side: U_{rTUS} 20 kV
Rated value of the short-circuit voltage: u_{kr} 11.6%
Rated value of the active portion of the short-circuit voltage: $u_{Rr} = 0.82\%$

Calculation of the impedance Z_{T20kV} referenced to the undervoltage side of the transformer (20 kV):

$$Z_{T20kV} = \frac{u_{kr}}{100\%} \frac{U_{rTUS}^2}{S_{rT}} = \frac{11.6\%}{100\%} \frac{(20\,\text{kV})^2}{20\,\text{MVA}} = 2.3\,\Omega \tag{10.9}$$

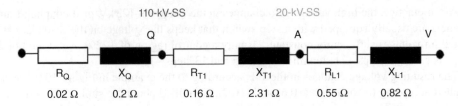

Figure 10.10 Equivalent circuit diagram referenced to the 20 kV level

The resistance R_{T20kV} of the grid transformer can normally be neglected but, if necessary, can be calculated from the active portion of the short-circuit voltage or from the coil losses of the transformer.

$$R_{T20kV} = \frac{u_{Rr}}{100\%} \frac{U_{rTUS}^2}{S_{rT}} = \frac{0.82\%}{100\%} \frac{(20\,kV)^2}{20\,MVA} = 0.16\,\Omega \qquad (10.10)$$

The reactance X_{T20kV} is determined as follows:

$$X_{T20kV} = \sqrt{Z_{T20kV}^2 - R_{T20kV}^2} = \sqrt{(2.32\,\Omega)^2 - (0.16\,\Omega)^2} = 2.31\,\Omega \qquad (10.11)$$

Active and reactive resistance of the 20 kV cable to the connection point C:

Cable type (aluminium conductor, VPE insulation, 20 kV): $3 \times 1 \times NA2XS2Y, 500^2$
[VPE = meshed nylon ethylene]
Length: $l = 8000\,m$
Resistance coating of the cable: $R_L' = 0.0681\,\Omega/km$
Reactance resistance coating of the cable: $X_L' = 0.102\,\Omega/km$

$$R_L = R_L'l = \frac{0.0681\,\Omega}{km} \times 8\,km = 0.55\,\Omega \quad X_L' = X_L'l\frac{0.102\,\Omega}{km} \times 8\,km = 0.82\,\Omega$$

Then the following values result for the equivalent circuit diagram (Figure 10.10 referenced to 20 kV).

As the transformer possesses a step switch and the medium voltage at the 20 kV collector rail is controlled to a fixed rated value, first the voltage increase from the 20 kV collector rail of the substation to the connection point V for the required operating points of the wind farm:

- $\cos \phi = 0.95$ inductive (reactive power draw-off)
- $\cos \phi = 1.0$ (no reactive power exchange)
- $\cos \phi = 0.95$ capacitive (reactive power supply)

is calculated according to Equations (10.6) and (10.7).

cos φ = 0.95 inductive:

$$\Delta u_L = \frac{S_{max}\left(R_L \cos|\phi| - X_L \sin|\phi|\right)}{U^2}$$

$$= \frac{16.84\,MVA(0.55\,\Omega \times 0.95 - 0.82\,\Omega \times 0.31)}{(20\,kV)^2} = 1.1\%$$

cos φ = 0.95 capacitative:

$$\Delta u_L = \frac{S_{max}\left(R_L\cos|\phi| + X_L\sin|\phi|\right)}{U^2}$$

$$= \frac{16.84\,\text{MVA}(0.55\,\Omega \times 0.95 + 0.82\,\Omega \times 0.31)}{(20\,\text{kV})^2} = 3.3\%$$

cos φ = 1.0 → sin φ = 0

$$\Delta u_L = \frac{S_{max}(R_L\cos|\phi|)}{U^2}$$

$$= \frac{16.84\,\text{MVA}(0.55\,\Omega \times 0.95)}{(20\,\text{kV})^2} = 2.2\%$$

Further, it is necessary to determine the voltage change at the 20 kV collector rail that is caused by the transformer. This occurs first without considering the step regulation. For the calculation according to equations (10.6) and (10.7) with the use of the transformer values $R_{T20kV} = 0.16\,\Omega$ and $X_{T20kV} = 2.31\,\Omega$, the results are analogous to the equation of the voltage change on the line:

cos φ = 0.95 inductive:

$$\Delta u_T = \frac{16.84\,\text{MVA}(0.16\,\Omega \times 0.95 - 2.31\,\Omega \times 0.31)}{(20\,\text{kV})^2} = 2.4\%$$

cos φ = 0.95 capacitative:

$$\Delta u_T = \frac{16.84\,\text{MVA}(0.16\,\Omega \times 0.95 + 2.31\,\Omega \times 0.31)}{(20\,\text{kV})^2} = 3.6\%$$

cos φ = 1.0 → sin φ = 0

$$\Delta u_T = \frac{16.84\,\text{MVA}(0.16\,\Omega \times 1.0)}{(20\,\text{kV})^2} = 0.64\%$$

The transformer possesses a step control with ±13 steps with a step size of 1.7%. A step change occurs first with a deviation of approximately 1% of the set rated value, thus in the upper half of the control range. The rated value of the collector rail voltage is controlled depending on the stepping of a control step and the control deviations. The following voltage values or voltage changes at the 20 kV collector rail and at the connection point in the fluctuating condition result from the interplay with the transformer control (simplified calculation without vectorial addition of the impedances and reactances).

cos φ = 0.95 inductive:

In underexcited operation of the wind farm, a negative voltage change or voltage decrease appears at the collector rail of the transformer substation. The voltage decrease is then smoothed over at the latest by the step switch when the stepping of the control coil falls below −1.7%. The translation relationship of the transformer is thus set down, i.e. the voltage of the 110 kV grid lies above the rated voltage of the primary coil. In this case, two control steps are carried out by the step switch (2×(+1.7%)). This results in a voltage of −2.4% + 3.4% = +1% at the 20 kV collector rail. At the connection point V the voltage increase, with a voltage increase on the line of 1.1%, is then approximately 2.1%.

cos φ = 0.95 capacitative:

For overexcited operation of the wind farm a voltage increase that is balanced because of the step switch control due to the choice of a larger translation relationship occurs at the collector rail. The rated voltage of the primary coil then lies above the grid voltage and the current fed into the grid becomes less. Two control steps are also carried out in this example. At the 20 kV collector rail this results in a voltage of 3.6% − (2 × (+1.7%)) = +0.2%. At the connection point V the voltage increase, with that on the line of 3.3%, is then approximately 3.5%.

cos φ = 1.0 (no reactive power exchange):

According to the calculation, here a voltage increase of 0.64% occurs at the 20 kV collector rail. A control step in this case is not carried out by the step switch. At the connection point V the voltage increase with that on the line of 2.2% is then approximately 2.8%. A summary of the results of all three working points is shown in Figure 10.11.

The influence of the prescribed reactive power working point can be clearly seen. The grid operator must now evaluate the calculation results. The following applies according to the BDEW guideline *Erzeugungsanlagen am Mittelspannungsnetz* [Generating plants at the medium-voltage grid] [12]:

In undisturbed operation of the grid, the amount of voltage changes caused by all generating plants with connections point at a medium-voltage grid may exceed a value of 2% at any connection point in this grid compared with the voltage without generating plants.

$$\Delta u_a \leq 2\% \qquad\qquad (10.12)$$

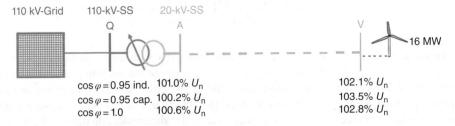

110 kV-Grid   110-kV-SS   20-kV-SS
Q   A   V   16 MW

cos φ = 0.95 ind.   101.0% U_n 102.1% U_n
cos φ = 0.95 cap.   100.2% U_n 103.5% U_n
cos φ = 1.0   100.6% U_n 102.8% U_n

Figure 10.11   Summary of the results of all three working points

Remarks:
- The generating plants with a connection point in the underlying low-voltage grids of this medium-voltage grid remain untouched by this. The limiting values of the guideline *Erzeugungsanlagen am Niederspannungsnetz* apply.
- A deviation of 2% from the value can occur in individual cases according to the needs of the grid operator and taking the possibility of the static voltage maintenance into account.
- Depending on the resulting displacement factor of all generating plants, the voltage change can be positive or negative, thus a voltage increase or reduction can follow.
- As the grid transformer usually possesses an automatic voltage control, the collector rail voltage can be seen as almost constant.

Preferably the voltage changes can be determined with the aid of complex load flow calculations.

Then only the underexcited operation would still be permitted (limiting range 2.1% to 2%). However, the grid operator can deviate from the proposed value. It must particularly be seen here whether customers are still connected to underlying low-voltage grids or are still to be connected. Even more decisive, however, is the question of whether further feeds are to be carried out into this low-voltage grid. It must then be checked whether this leads to exceeding the low-voltage tolerances at connection point V. If this is not the case, then possibly also the other two operating ranges of the wind farm are permitted. This decision is always the responsibility of the grid operator. Voltage stability is today one of the biggest problems in the operation of electrical grids. At present investigations are under way into whether in the future controllable transformers of medium voltage to low voltage may also be used. For this purpose no classical step switch is to be used, but electronic control on thyristor base is planned. First pilot projects are being tested.

10.4.3 Checking the Grid Reaction 'Fast Voltage Change'

Here the effects of the voltage are to be evaluated in the case that the the whole wind farm, for instance, is to be immediately separated from the grid because of a malfunction. The following then applies according to the BDEW guideline *Erzeugungsanlagen am Mittelspannungsnetz* [Generating plants at the medium voltage grid] [12]:

When switching off one, or for simultaneous switching off of several, generating plants at a grid connection point (here a wind farm), the voltage change at every point in the grid is limited to:

$$\Delta u_{max} \leq 5\% \tag{10.13}$$

In this, all generating plants, which can fail at the same time due to the switching-off operation as also for triggering the protection are to be taken into account.

Switch-off of the whole wind farm is calculated by taking the existing voltage change as the difference between the voltage with and without feed, without taking into account the voltage control of the grid transformers.

The largest voltage jump results in this case at the 20 kV collector rail of the transformer substation. According to the previous calculation this occurs due to the voltage change over the transformer that is only equalised again later by the step switch. Usually a time of a minute is required for the excitation and implementation of a step control.

In the example, the largest fast voltage change is 3.6% in the overexcited operation and is therefore permissible.

10.4.4 Checking the Short-Circuit Strength

The operation of a WT increases the short-circuit current of the grid, especially in the region of the grid connection point by the short-circuit current of the generating plant. Data for the expected short-circuit flows of the generating plant at the operation connection point is thus important information for the grid calculation.

If no exact values are known, then the values given in Section 10.3.1 can be used for approximately determining the short-circuit current contribution of a wind turbine. For an exact calculation, the impedances between generator and grid connection points (customer transformer, lines, and so on) must be taken into account.

If, due to the generating plant, the short-circuit current in the medium-voltage grid is increased above the rated value, then suitable measures must be taken in agreement between the grid operator and the power customer as to how the limitation of the short-circuit current from the generating plant is to be carried out (e.g. by the use of I_s limiters).

Figure 10.12 gives an example for the determination of the short-circuit power at the 20 kV collector rail of the transformer substation. The short-circuit power is made up of two parts:

- contribution from the 110 kV grid;
- contribution from the wind farm.

R_Q, X_Q and X_T are decisive for the short-circuit contribution from the 110 kV grid. For the active and reactive portions, when added together at point A:

$$R_A = 0.18\,\Omega \quad \text{and} \quad X_A = 2.51\,\Omega$$

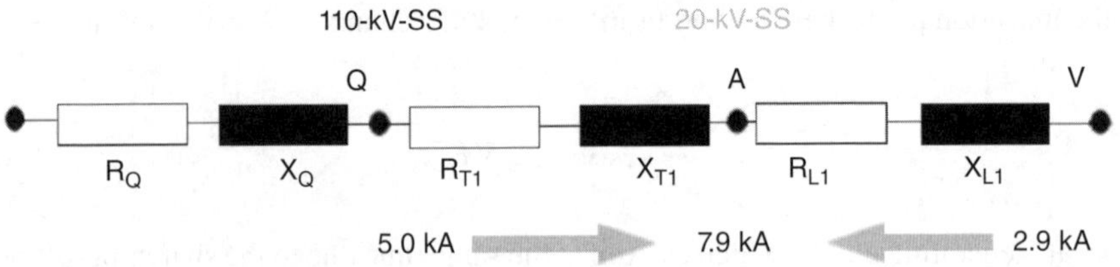

Figure 10.12 Short-circuit alternating current at the respective components

The software impedance is then derived as follows:

$$Z_A = \sqrt{R_A^2 + X_A^2} = \sqrt{(0.18\,\Omega)^2 + (2.51\,\Omega)^2} = 2.52\,\Omega \tag{10.14}$$

The maximum software power from the 110 kV grid via the transformer is calculated according to the following equation:

$$I_{kA}'' = \frac{cU_n}{\sqrt{3}Z_A} = \frac{1.1 \times 20\,kV}{\sqrt{3} \times 2.52\,\Omega} = 5\,kA \tag{10.15}$$

For a coarse and safe estimate, the short-circuit contribution of the wind farm is taken as six times the rated current (double-fed asynchronous machine) without consideration of the transformer and line impedance added linearly.

$$I_{kfarm}'' = 6 \times 486\,A = 2.9\,kA \tag{10.16}$$

At the 20 kV collector rail of the substation this then gives a value of:

$$I_{koverall}'' = I_{kA}'' + I_{kfarm}'' = 5\,kA + 2.9\,kA = 7.9\,kA$$

The switching installation is dimensioned for a starting short-circuit alternating current I_k'' of 16 kA and is thus sufficiently sized. The short-circuit strength for the impact short-circuit current i_p must be controlled similarly to the starting short-circuit alternating current.

10.5 Grid Connection of WTs

The basis for connection to the grid is first the determination of a suitable connection point, taking the legal requirements of the EEG into account. For the various possible variants there is especially the choice of the most economical connection overall (from a commercial point of view). For plant planners and grid operators the determination of the most suitable connection point is becoming more difficult due to the rapid growth of all generating plants based on regenerative energies. Wind turbines are to be erected and operated with consideration of the respective valid regulations in such a manner that they are suitable for parallel operation with the grid of the grid operator, and that impermissible reactions on the grid or other customer installations are excluded.

The following must be adhered to as a minimum for the erection and operation of electrical plants [12]:

- the legally valid and authoritative regulations;
- the valid DIN-EN and DIN-VDE Standards;
- the Operating Safety Ordinance;
- the working protection and accident prevention regulations of the responsible trade associations;
- the determinations and guidelines of the grid operator.

The connection to the grid is to be agreed in detail in the planning phase with the grid operator, before ordering the important components. Planning, erection and connection of the wind turbine(s) to the grid of the operator is to be carried out by specialist companies. The plant planner must provide informative documentation on the generating plant for the technical grid test by the grid operator [12].

On the basis of the documentation, the grid operator can determine a suitable grid connection point that ensures safe grid operation from the point of view of the wind turbine(s) and at which the required power can be taken up and transmitted. Decisive for the evaluation of the grid connection point is always the behaviour of the generating plant at the grid connection point as well as in the grid of the general supply. The evaluation of the connection possibilities from the point of view of the grid reaction occurs on the basis of the impedance of the grid at the point of connection (short-circuit power, resonances), the connection power as well as the type and method of operation of the wind turbine(s). Insofar as several generating plants are connected to the medium-voltage grid, their overall effect must be studied.

As a rule, wind turbine plants are connected to the grid by means of the customer's own medium-voltage stations or by means of separate cables in a transformer substation. For reasons of costs the connection mostly takes place as a so-called simple connection (spur terminal). Further uncoupling and protective functions are necessary. besides the short-circuit and overload protection for the wind turbine(s). Depending on the grid features, number and size of the wind turbine plants, the connection of the generating plants in the medium-voltage grid is carried out either by means of circuit breakers or a load interruptor switch/safety combination. The connection to the transformer substation of a grid operator occurs by means of a circuit breaker.

10.5.1 Switchgear

Besides the transmission capability of the switchgear for connection of wind turbines, the short-circuit strength of the switchgear is an important design criterion. The so-called transfer station provides the transfer from the customer installation to the grid of the grid operator. This station, with a medium voltage installation, is thus subject to a special consideration. The BDEW guideline *Technische Anschlussbedingungen – Mittelspannung* [Technical connection conditions of medium voltage] [16], the connection conditions of the grid operator and the generally valid regulations of medium-voltage plants (especially DIN VDE 0101, DIN VDE 0670 and DIN VDE 0671) are to be adhered to for the erection of the connection installation. In [16] this is described as follows:

Electrical plants must be designed, constructed and erected in such a manner that they are capable of safely withstanding the mechanical and thermal effects of a short-circuit current. The verification of the short-circuit strength for the whole of the transfer station is to be provided by the power customer. In the case that, due to the operation of the customer plant, the short-circuit current in the medium voltage grid is increased above its rated value, then suitable measures are to be agreed upon by the operation operator and power customer on the limitation of the short-circuit current from the customer plant (e.g. by the use of I_s- limiters).

Table 10.1 Typical values for the short-circuit strength of medium-voltage plants

Starting short circuit alternating current I_k''	Impact short-circuit current i_p
16 kA	40 kA
20 kA	50 kA
25 kA	63 kA
31.5 kA	80 kA

The grid operator provides the required characteristic values for the dimensioning of the connection installation at the grid connection point (e.g. rated voltage and rated short-circuit current). Further, at a request for dimensioning the power customer's own protective equipment and for grid reaction, the grid operator provides the power customer with the following data [12]:

- starting short-circuit alternating current from the grid of the grid operator at the grid connection point (without the contribution of the generating plant);
- malfunction repair time of the main protection from the grid of the grid operator at the grid connection point.

Typical values for the short-circuit strength of medium voltage plants are given in Table 10.1.

10.5.2   Protective Equipment

Protection is of great importance for safe and reliable operation of the grid as well as for the wind turbine(s), including the connection installation and the wind farm grid.

According to DIN VDE 0101, the electrical plants must be provided with independent apparatus for switching off short-circuits. The grid operator is responsible for the protection of his grid up to the connection/transfer point and the operator of the wind turbine plant for reliable protection of his plant (e.g. protection for short-circuits, earth fault, overload, protection against electrical impact). The protection concepts as well as the protection settings at the interface between grid operator and wind turbine plant operator are to be arranged so that a hazard to the neighbouring grids and plants is excluded.

For protection of wind turbine plant(s) and other customer installations, the wind turbine plants or their wind farm must be able to be separated from the grid, in circumstances such as grid malfunctions, islanding, or a slow build-up of the grid voltage after a malfunction in the transmission grid. This task is undertaken by the so-called uncoupling protection. The uncoupling protection can be integrated in an independent device as well as in the plant controls of the wind turbines as an independent logic. A failure of the supply voltage of the protection equipment or the plant controls must lead to the immediate triggering of the transfer switch.

Uncoupling protection devices are installed at the connection/transfer point of the wind farm and/or at the individual wind turbines. The uncoupling protection must be able to carry out the following functions:

- Voltage reduction protection $U <$ and $U \ll$
- Voltage increase protection $U >$ and $U \gg$

- Frequency reduction protection $f <$
- Frequency increase protection $f >$
- Reaction power low voltage protection $Q \to$ p & $U <$

The voltage protection and frequency protection equipment (see Fig. 10.03) has the task of protecting the customer's plant from impermissible voltage and frequency conditions in the case of isolation, as well as to ensure switching-off of the wind turbine(s) in the case of certain malfunctions in the grid. Besides the voltage and frequency monitoring, there is also reactive power undervoltage protection ($Q \to$ p & $U <$) as a special task. After a defined period of time, (usually 0.5 s) it separates the wind turbine from the grid when all three chained voltages at the grid connection point are less than 0.85 U_c (logically AND interlinked) and when the wind turbine at the same time takes up inductive reactive power from the grid of the grid operator. This protection monitors the proper system behaviour of the wind turbine(s) after a malfunction in the transmission grid. The rebuilding of the grid voltage after a malfunction in the transmission grid is strongly impaired due to the take-up of inductive reactive power from this grid. If a wind turbine does not behave correctly then it must be quickly separated from the grid. For better structuring of the grid voltage or for voltage support, it is necessary in the case of malfunctions in the transmission grid to additionally feed the grid with reactive current from the wind turbines.

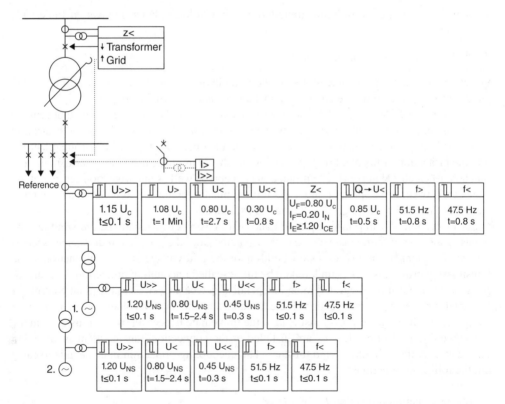

Figure 10.13   Example of a protection concept for the connection of wind turbines to the collector rail of a transformer substation

Short-circuit protection of the wind turbine(s) is necessary for switching short-circuits off in the connection installation and the wind farm. In the case of connections via a circuit breaker, at least an overcurrent protection is provided as short-circuit protection. The short-circuit protection with the connection over an interruptor protection combination occurs by means of a fuse.

10.5.3 Integration into the Grid System

Some well-known manufacturers of wind turbines offer a remote control over their own management systems. By this means, remote diagnostics can also be carried out and setting parameters can be changed. In addition, the operator of the plants also has the possibility of calling up the current status messages, operating conditions and actual online values.

Wind turbines with connection to the collector rail of a transformer substation are normally integrated into the grid management system of the respective grid operator. For this purpose there is also the remote control possibility in the case of emergency action, whether or not the required power reduction is carried out within the framework of the feed management (Eisman) or grid safety management (GSM) in order to protect the transmission grid from an overload. In this case, no controlled shut-down is carried out, but rather a so-called 'hard' switching-off by actuating the circuit breaker. For wind turbines or wind farms that are integrated into the medium-voltage supply grid, an integration into the grid management system of a grid operator can only occur when corresponding communication channels are available. The required protocols and technology are to be decided in individual cases.

10.6 Further Developments in Grid Integration and Outlook

At present, WTs feed energy independently of power requirements (usage) of the grid customers and without preparation of control energy into the public grid. However, in order to integrate further WTs into the grid, the power grid must adapt to new requirements. On the one hand this can be by means of grid expansion, and on the other hand, by means of intelligent storage and improved load flow control, the so-called 'smartgrid'.

The European Union defines a smartgrid in the *Strategic Implementation Concept for Smart Grids* as follows [17]:

A Smart Grid is an electrical network that intelligently coordinates the actions of all connected users – generators, consumers, storers – in order to ensure efficiency in the sustainable ecological, commercial and reliable current supply.

In order to coordinate the users intelligently, it is necessary to build up not only an electrical but also a bi-directional communication network between them (Figure 10.14).

In the definition, the three aims – environmental compatibility, supply reliability and commercial feasibility – or the target triangle of energy management law are included. For proper supply of energy, these aims must always be evenly balanced out.

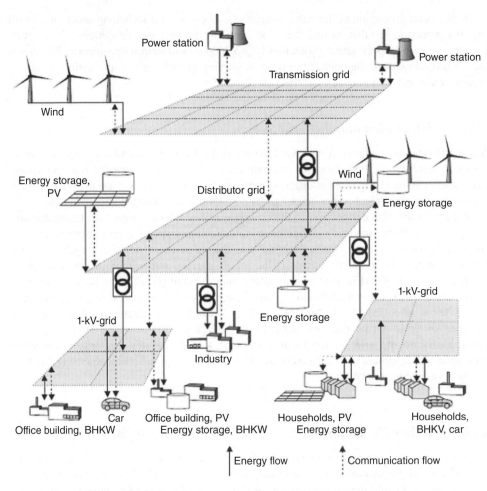

Figure 10.14 Development scenario of the electrical energy system

10.6.1 Grid Expansion

In recent years the generation from decentralised power plants (DPPs) has grown continuously and it is assumed that it will carry on doing so. In order that the grid can absorb this additional energy, it must be expanded not only in the medium- and high-voltage level, but also in the highest voltage level. The construction of such a system, however, can be delayed by up to 15 years due to difficult approval procedures. For this reason today for the 110 kV grids, and in future probably also grids in the medium-voltage range, will be subject to monitoring. With 110 kV exposed lines, for instance, transmission power calculations are made for the environmental conditions, temperature, wind velocity and wind direction, and the feed of the DPP reduced to this capacity according to the motto 'less copper, more intelligence'.

In order to generate a possibly user-oriented feed capacity from these fluctuating sources of power, they should in the future be combined over a wide area. Thus, for instance, the generation of power from solar power stations in North Africa is to be combined with those of wind

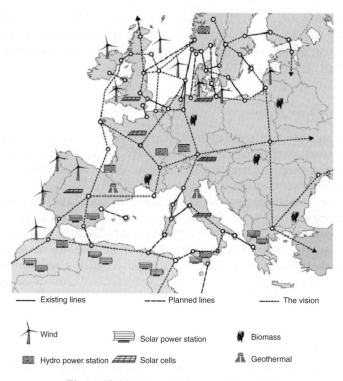

—— Existing lines	– – – Planned lines	------ The vision

Wind   Solar power station   Biomass

Hydro power station   Solar cells   Geothermal

Figure 10.15   Planned European Supergrid

turbines of the North and Baltic seas and hydropower stations in northern Europe. For this purpose grid expansions must not only occur in Germany but in the whole of Europe (Figure 10.15). In order to implement this 'Supergrid', transport lines must be built that can transmit power over long distances with the lowest possible losses. These transport lines can be operated either with three-phase current or direct current. A serious disadvantage of three-phase transmission is the frequency-dependent losses that are caused by remagnetising losses and the skin-effect due to the resistance increase of the conductor. These losses do not occur with direct current, so that for transmission of power over large distances, use is made of high-voltage direct current transmission (HDT). However, in order to convert the alternating current that is produced by the generators of the generating plants into direct current, the generated power must be rectified for transmission, and after transmission again converted into an alternating current.

This conversion is carried out by means of power converters that, however, incur losses. For this reason, and because of the high costs, it is worthwhile to make use of the HDT cables only for a length of more than 50 km and for an open line only from a length of approximately 600 km. The HDT technology is particularly suitable for equalising short-term feed fluctuations that, for instance, occur due to wind gusting, due to the fact that in the direct current circuit a power storage is available that compensates for the short fluctuations. For this reason this technology is especially used for the connection of offshore wind farms to the high and highest voltage grid. The disadvantage of the HDT, however, is the high investment costs for the current rectifier installations. Furthermore, the power converters are more prone to disturbances than the well-known operating media in three-phase transmission. But to transport significant amounts of power over great distances, HDT is still the best alternative at today's state of technology for three-phase current [18].

10.6.2   Load Displacement

Because of the unsteady feed power of fluctuating generators, it is difficult to cover the usage properly and if necessary to prepare positive or negative compensating power. Even before the boom in fluctuating energies, attempts were made to cope with short-term power bottlenecks (e.g. peak load) or power excesses.

For this reason, some large customers have closed contracts with the grid operators, for instance, to switch the operations of users such as large cold stores on or off for a certain period of time. In the future all generators and users will be linked by means of a communication network. Thus, for instance, the washing machine, the deep freezer or also the storage heater in a household should know when the regenerative feeders generate energy. In order to realise such a system, a complex data network must be installed between the operating media to sit alongside the power grid.

10.6.3   Energy Storage

In Germany it is primarily wind energy that has a main role in regenerative energy generation. However, a problem is that it is dependent on the weather. Therefore, in the case of strong winds, the surplus energy that is fed into the grid, despite load displacement, should be used to provide energy storages. The stored energy should then be made available in a lag in the energy supply. These energy storages can be of different types and are partly still in development.

The most mature and widespread storage technology is the pumped storage hydropower station. In the case of a surplus of energy it pumps water from a lower to a higher level. If there is a requirement for energy, electricity can be generated again by releasing the higher placed water through a turbine that is linked to a generator. This type of water storage is dependent on the landscape and for this reason the storage capacity in Germany is limited. However, a country that has many hydropower stations and still possesses much potential for expansion is Norway. Here 99% of the electrical power is generated by hydropower stations. For this reason it would be conceivable to export the surplus energy from German WTs via HDT sea cable to Norway and to import it again in the case of an energy requirement. The degree of efficiency of this process depends on the distance and the pumped storage hydropower station at approximately 70% [19].

A further possibility for storing electrical power is compressed air storage. A compressor is used to fill a compressed air storage that could, for instance, be situated in a washed-out salt cavity. If energy is required, the compressed air can be released and thus power can be generated. A disadvantage of this technology is that the degree of efficiency of the process is approximately 55% and suitable sites for storage must be available [20].

A chemical possibility for storing electricity involves generating natural gas from electrical power. In order to produce natural gas, hydrogen is generated by means of electrolysis. Methane can be produced by several chemical processes including the addition of carbon dioxide. This can be stored or fed into the natural gas network. If electrical power is required again, then the stored methane can be used to generate electrical power. At present plans for industrial conversions are just being prepared, so there is no experience available yet.

Batteries can be used to store electrical power. There are several concepts for such storage. On the one hand, large stores of some kW to MW can be erected centrally, which can then be used to equalise power peaks or bottlenecks. On the other hand, decentralised batteries of electric cars that are equipped with intelligent loading management can be used for storage. This allows the batteries to act as users in the case of surplus power, and as generators in the case of power requirement.

References

[1] IEC. 2002. *DIN IEC 60038*. IEC Normspannungen.
[2] IEC. 2010. *DIN EN 50160, Voltage characteristics of electricity supplied by public distribution.*
[3] Regelleistung (Energie). http://de.wikipedia.org/wiki/Regelleistung (Energie) [accessed 15 May 2011].
[4] amprion. *Marktplattform, Bilanzkreise.* Available at http://www.amprion.net/bilanzkreise [accessed 18 May 2011].
[5] VDN: Leistungsbilanz der allgemeinen Stromversorgung in Deutschland. Vorschau 2005–2015. Available at http://www.vdn-berlin.de/global/downloads/Publikationen/LB/VDNLB_VS_2005-2015.pdf.
[6] Institut für Solare Energieversorgungstechnik (ISET), *Summenganglinien für Energie 2.0.*
[7] Monitoring-Bericht des Bundesministeriums für Wirtschaft und Technologie nach § 51 EnWG zur Versorgungssicherheit im Bereich der leitungsgebun-denen Versorgung mit Elektrizität. Available at www.bmwi.de/BMWi/Redaktion/PDF/M-O/monitoring-versorgungssicherheit-elektrizitaetsversorgung,property=pdf,bereich=bmwi,sprache=de,rwb=true.pdf [accessed 30 August 2010].
[8] Spitzenlast. http://de.wikipedia.org/wiki/Spitzenlast [accessed 15 May 2011].
[9] amprion. http://www.amprion.net/grundlast-mittellast-spitzenlast [accessed 18 May 2011].
[10] amprion. http://www.amprion.net/netzfrequenz [accessed 18 May 2011].
[11] swissgrid: Versorgungssicherheit, Systemdienstleistungen. Available at http://www.swissgrid.ch/swissgrid/de/home/reliability/ancillary_services.html [accessed 7 May 2011].
[12] BDEW. 2008. *Technische Anschlussbedingungen für den Anschluss an das Mittelspannungsnetz.* BDEW Bundesverband der Energie- und Wasserwirtschaft e.V., Berlin.
[13] Verfügbarkeit. Available at http://de.wikipedia.org/wiki/Verfügbarkeit [accessed 4 June 2011].
[14] Technische Regeln zur Beurteilung von Netzrückwirkungen. 2. 2007. VDN Verband der Netzbetreiber, Berlin.
[15] VDE. 1995. *DIN VDE 0276 Teil 1000: Starkstromkabel: Strombelastbarkeit, Allgemeines, Umrechnungsfaktoren.* VDE Verlag GmbH, Berlin.
[16] BDEW. 2008. *Technische Richtlinie: Erzeugungsanlagen am Mittelspannungsnetz. Richtlinie für Anschluss und Parallelbetrieb von Erzeugungsanlagen am Mittelspannungsnetz.* Bundesverband der Energie- und Wasserwirtschaft e.V., Berlin.
[17] Fenchel, G. and Hellwig, M. 2010. *Smart Metering in Deutschland*. EW Medien und Kongresse, Frankfurt.
[18] ABB. Was ist HGÜ? Available at http://www.abb.de/cawp/db0003db002698/1969e8ef4e83cb62c125725f005 4bf10.aspx/
[19] Pumpspeicherkraftwerke. http://de.wikipedia.org/wiki/Pumpspeicherkraftwerk [accessed 23 June 2011].
[20] RWE: Innovationen, Stromerzeugung, Energiespeicherung, Druckluftspeicher. Available at http://www.rwe.com/web/cms/de/183732/rwe/innovationen/stromerzeugung/energiespeicherung/druckluftspeicher [accessed 30 June 2011].

11

Offshore Wind Energy

Lothar Dannenberg

11.1 Offshore Wind Turbines

11.1.1 Introduction

As fewer places become available for siting onshore wind turbines (e.g. Germany, Netherlands) or wind conditions prove to be disadvantageous, many countries are being forced to install their wind turbines at sea (offshore). Pioneers of this at present are Denmark (in relatively near-coastal regions with shallow water depth: Horns Rev, at approximately 15 m depth) and the UK (near the coast).

In Germany there are great doubts about placing such plants near the coast because, to a large extent, these areas are nature reserves (Wadden Sea Nature Reserve), or for tourism reasons because of the 'disturbing silhouettes'. For this reason, operators are forced to place offshore wind farms beyond the horizon, at distances of 30–120 km from the coast in water depths between 20 and 45 m. In the North Sea particularly, the conditions are almost as bad as deep-sea conditions, with wind and waves that are much stronger than in coastal regions.

Due to the environmental conditions and the great distances from the coast, the investments required for erection and servicing offshore wind turbines (OWTs) are very high. With the subsidies provided by the old German Renewable Energy Law (EEG) [Erneuerbare Energie Gesetz], this was not economical. This has changed with the updating of the EEG, but the grid operators have been additionally obliged to bear the costs of the connections to the grid from the wind farm.

However, the technical and economic risks are still very high, as up to now only little experience has been gained in the erection and operation of such plants over a longer period of time.

Understanding Wind Power Technology: Theory, Deployment and Optimisation, First Edition.
Edited by Alois Schaffarczyk. Translated by Gunther Roth.
© 2014 John Wiley & Sons, Ltd. Published 2014 by John Wiley & Sons, Ltd.

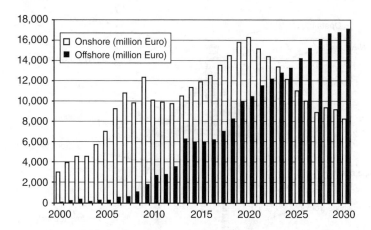

Figure 11.1 Prognosis of investments in Europe to 2030 in onshore and offshore WTs. *Source*: EWEA 2011

According to a prognosis of the EWEA (European Wind Energy Association) of 2011, by 2030 up to 5000 OWTs with a capacity of around 25 GW (1 gigawatt $= 10^{12}$ watts) will be erected in the German EEZ (Exclusive Economic Zone), and across Europe approximately 150 GW. The investment for this will be around 220 billion Euro (see Figure 11.1).

11.1.2   Differences between Offshore and Onshore WTs

The rotor blades, gondolas and towers of offshore wind turbine (OWTs) are subjected to similar loads as for onshore plants (wind, gusts and turbulences), but with medium-strong winds (less turbulence).

Whereas with onshore plants, and depending on the site, approximately 1700 to 2500 full power hours can be achieved per year, for offshore plants the figure is 4000 to 5000 hours. With a larger number of wind turbines (parks), the plants influence each other as they are positioned relatively close to each other due to the cost of cable connections. Thus, strong turbulences can occur that have an effect on the loading (size of the voltage amplitudes and the number of load changes).

In contrast to onshore plants, the offshore wind turbines are subject to further loads that concern mostly the foundations (see Figure 11.2). In this the loads are dealt with as static or quasi-static when their timed changes (fluctuation periods) are substantially smaller or larger than the eigen-periods of the observed structures, which are therefore not excited to fluctuations by the loads. The loads are:

Wave loads:

- Continuous waves in relatively great depths (relationship of wavelength to depth of water): the calculations are carried out according to the various wave theories such as Airy, Stokes or similar (see below), and loads caused by it according to the so-called Morrison equation (dynamic loads).
- Impact of breaking waves at relatively shallow water depths or for waves that are high in relation to the wavelengths in deep water (dynamic).

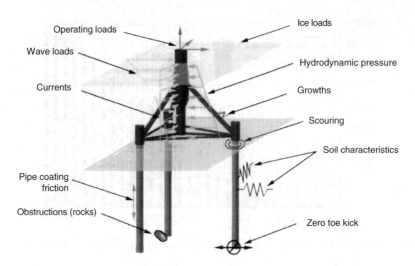

Figure 11.2   Loading of the foundations of wind turbines. *Source*: HDW, 2004

- Load changes: because of the high change-load numbers (medium wave period 3–10 s, approx. 3×10^8 load changes in 25 years) and the high density of the water, the wave loads are an important design criterion for the foundation structure.

Currents:

- Constant sea currents or currents caused by the wind (quasi-static loading).
- Tidal currents (quasi-static).
- Vortex shedding ($\rightarrow$ Kármán vortex street, dynamic).

Scouring:

- Reducing the stiffness of the foundation structures by means of scouring out the sea floor. This leads to higher stresses and lower eigen-frequencies.

Ice formation:

- Loading due to fixed ice, driving ice, ice formation due to spray water (quasi-static).
- In the North Sea the danger of ice formation due to the influence of the Gulf Stream except in the Wadden Sea is low, but ice formation must be expected in the Baltic Sea.

Marine growth:

- Because of the growth of organic materials (increase in cross-section and the masses), there is a constant increase of the load on the foundation due to waves and flows as well as a reduction of the eigen-frequencies (static and dynamic).

Corrosion:

- Increased due to the salt-rich sea water and moist air.

Ageing of fibre-reinforced materials:

- Strengths of the rotor blades are reduced due to UV-ray loading, moisture and abrasive effects of the salty air.

In the operation of offshore plants there is the added difficulty that the plants cannot be serviced as often and regularly as onshore plants. It is to be assumed that problems may not be able to be corrected immediately due to weather conditions. For this reasons suitable servicing and monitoring concepts are to be provided for offshore plants (CMS=condition monitoring system).

11.1.3   Environmental Conditions and Nature Protection

Offshore plants and their erection sites in the regions of the German EEZ must be approved by the BSH (Bundesamt für Seeschifffahrt und Hydrographie) [Federal Maritime and Hydrographic Agency]. The BSH has created comprehensive guidelines [1,2] for the approval of wind farms. Here the following is particularly checked:

- suitability of the erection site (impairment of shipping and fishing, protected areas, and so on);
- suitability of the sea floor;
- collision safety in case of collision with ships;
- influence on birds (bird migration);
- influence on sea fauna and flora (fish, marine mammals, seabed);
- sound emission into the water;
- turbine spacing within the OWT farm as well as farms themselves;
- characteristics of the turbines (optical, acoustic);
- marking on sea maps;
- paths of the cable terraces;
- possibilities for disassembly.

The proofs must be provided largely by the plant operators.

11.2   Currents and Loads

11.2.1   Currents

Ocean currents can cause appreciable static and dynamic loads on the foundations of OWTs. For this reason they must be taken into account in the dimensioning of the structures. The ocean currents are divided into:

- constant ocean currents (so-called thermosaline currents);
- tidal currents;
- wind-induced currents.

The loadings caused by these currents occur on a regular basis (quasi-static) as the change periods of the currents (e.g. tide periods of approximately 12.5 hours) are substantially greater than the fluctuation periods of OWT components.

A power equation is generally used for the tidal currents over the depth (u_{T0}=current velocity at the water surface ($z = d$), according to Germanischer Lloyd (GL) $n=7$ [3]):

$$u_T(z) = u_{T0} \left(\frac{z}{d} \right)^{1/n} \tag{11.1}$$

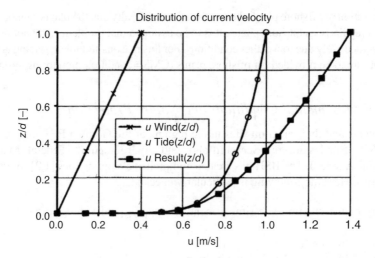

Figure 11.3   Velocity profile of wind-induced and tidal currents

For wind-induced currents, use is generally made of the depth z linearly in the equation:

$$u_{\text{Wi}}(z) = u_{\text{Wi}}(d)\left(\frac{z}{d}\right) \tag{11.2}$$

The current velocity on the surface of the water $u_{\text{Wi}}(d)$ is often used in the following approximation (U_{1h} is the wind velocity at 10 m height, averaged over 1 hour):

$$u_{\text{Wi}}(d) = 0.02U_{1h} \tag{11.3}$$

Comprehensive measurements are available for tidal currents but there are few data for other ocean currents. For u_{Wi} the equation generally uses 2 kn [3].[1] Then the result for the current velocity is u_c (c for current) over depth z in Equation (11.4). A typical distribution of both velocity types is shown in Figure 11.3. The direction of the two current portions can differ in this by the angle α.

$$u_c(z) = u_{T0}\left(\frac{z}{d}\right)^{1/n} + \cos\alpha\, U_{\text{Wi}}\frac{z}{d} \tag{11.4}$$

11.2.2   Current Loads

When a fluid is flowing around a body, and depending on the approach angle, drag and lift forces are created that can load the offshore foundations. The forces are differentiated in this between wave- and friction-induced drags. The wave-induced forces exist only at the boundary

---

[1] 1 kn (knot) = 1 sm/h (sea mile/hour) = 1.825 km/h = 0.5144 m/s.

surfaces between two differently dense media such as water and air (wave formation at the ship). For offshore foundations the wave drag can be ignored.

The friction drags for offshore structures are proportional to the square of the current velocity. Reynolds law is used for model investigations of the friction forces, such that the Reynolds numbers of the full-scale and the model must be the same. The Reynolds number is:

$$\text{Re} = Du/v \, [-] \tag{11.5}$$

where u is the inflow velocity, D is the diameter of the body subjected to the flow, and v is the kinetic viscosity of the medium.

Example of Reynolds number: $u = 2\,\text{kn}$; $D = 5\,\text{m}$; $v_{\text{seaw.15°C}} = 1.19 \times 10^{-6}\,\text{m}^2/\text{s}$; $\Rightarrow \text{Re} = 4.3 \times 10^6\,[-]$

For structures that have a slanted inflow, the current forces are divided into a lift portion and a drag portion vertical to it. For the friction drag the equation is:

$$F_\text{D} = -1/2\rho \int_{(L)} c_\text{D}\,|u|uA\,\mathrm{d}z \tag{11.6}$$

and for the lift portion:

$$F_\text{D} = -1/2\rho \int_{(L)} c_\text{L}\,|u|uA\,\mathrm{d}z \tag{11.7}$$

where c_D is the dimensionless drag factor of the section subjected to the flow, c_L is the dimensionless lift factor; ρ is the specific density of the flow medium $= 1025\,\text{kg/m}^3$ (seawater), A is the area subjected to inflow, and L is the length of the structure subjected to the inflow.

The factors c_D (D for drag) and c_L (L for lift) are dependent upon the form of the body subject to the flow, such as a cylinder or rectangle, and the flow velocity or the Reynolds number (laminar or turbulent currents). The term u^2 is to be split into $|u| \times u$ in order to maintain the prefix or direction of the forces of the current, as these act in the opposite direction to it. Numerical values for the drag and lift factors for different cross-sections can be found in, for instance, [4] and [5].

As the dimensions and shape of the cross-sections as well as the velocity of the current and thus also the factors of the current over the depth can change, the forces per unit of length must be determined and integrated over the length of the component:

$$f_\text{D}(z) = -1/2\rho_{\text{water}} c_\text{D}(z, A, \text{Re}) D(z)\,|u(z)|u(z) \tag{11.8}$$

In this the resulting drag becomes:

$$F_\text{D}(z) = -1/2\rho_{\text{water}} \int_{(L)} c_\text{D}(z, A, \text{Re})\, D(z)\,|u(z)|u(z)\,\mathrm{d}z \tag{11.9}$$

The calculation for the lift is analogous.

Example: the drag of a pile (cylinder) subjected to a wind-induced current in seawater is to be determined. Given are: $D = 5.0\,\text{m}$; $d = 30\,\text{m}$ (depth of water); $p_{\text{seaw}} = 1025\,\text{kg/m}^3$; wind velocity $u_{\text{1h}} = 15\,\text{m/s}$.

Results: current velocity at the surface $u_{Wi} = 0.3$ m/s; $Re = 1.25 \times 10^6$; $c_D \approx 0.35$; $F_D = 1210$ N.

Remarks: 1 kg = 1 Ns^2/m (mass) or 1 N = 1 kg $\cdot$ m/s^2 (force).

11.2.3   Vortex Shedding of Bodies Subject to Flows

When real bodies are subject to currents (even flows), the result can be the shedding of a vortex. Because of the changes of the pressure differences at the sides, forces lateral to the inflow are set up and these can excite the body subjected to the flow to oscillations.

These oscillation excitations can affect all components subject to the current in water as well as in the air. If the excitation frequencies lie in the neighbourhood of the component eigen-frequencies, it can lead to oscillation resonances that can mean an overloading of the component or that can lead to a reduction of the lifetime of the component. For this reasons resonances must be avoided at all costs.

The shedding frequency of the vortexes for cylindrical and similar bodies is:

$$f = \frac{Su}{D}[Hz] \tag{11.10}$$

where f is the vortex formation frequency (Hz), S is the Strouhal number, u is the undisturbed flow-in velocity, and D is the diameter of the body subject to the flow.

The Strouhal number is dependent on the Reynolds number of the currents (see Figure 11.4); a distinction is made between three regions for the turbulence (Re > 2320):

Subcritical region: $2320 < Re < 3.5 \times 10^5$ → $S \approx 0.65$
Supercritical region: $3.5 \times 10^5 < Re < 3.5 \times 10^6$ → $S \approx 0.65$ to 0.2
Transcritical region: $Re > 3.5 \times 10^6$ → $S \approx 0.2$

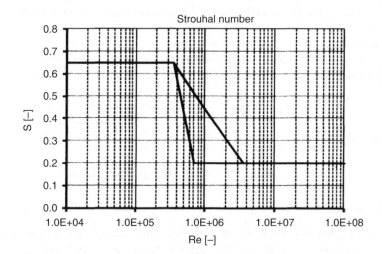

Figure 11.4   Strouhal number in relation to the Reynolds number

As the exact measurement of the Strouhal number is very difficult, the values given in the literature vary widely. Added to this is the fact that the vortex shedding frequencies at deviations of up to ±10–15% 'couple' themselves into the eigen-frequencies of the component, that is, they can adapt.

For instance, Strouhal numbers for various cross-sections have been summarised in the regulations of Det Norske Veritas (DNV) [6, 7]. The forces for vortex formation are negligibly small for components that are advantageously shaped for currents (current profiles).

Example: the vortex shedding frequencies and periods of a cylinder subject to currents in water or in air with a diameter of $D = 5.0$ m are to be determined. Given are:

- Flow-in velocity in water $= 2$ kn; kinetic viscosity of seawater (16 °C) $v = 1.135 \times 10^{-6}$ m²/s.

Results: Re $= 1.01 \times 10^7$ (transcritical) $\rightarrow S \approx 0.2$; $\rightarrow$ vortex shedding frequency $f = Su/d \approx 0.014$ s⁻¹; disengagement period $T \approx 24.3$ s.

- Flow-in velocity in air 10 m/s; kinetic viscosity of air (20 °C) $v = 1.59 \times 10^{-6}$ m²/s.

Results: Re $= 3.3 \times 10^6$ (transcritical) $\rightarrow S \approx 0.2$; $\rightarrow$ vortex shedding frequency $f = Su/D \approx 0.4$ s⁻¹; shedding period $T \approx 2.5$ s.

11.3 Waves, Wave Loads

11.3.1 Wave Theories

The ocean surface waves whipped up by the wind are so-called gravity waves. Due to the friction of the wind flows and the turbulences in the border region between air and water, the water particles are brought out of their rest condition and experience a wave motion as a result of the resettable gravity effect that, in the ideal case, is harmonically propagated. A vortex-free, incompressible and friction-free fluid is assumed in order to describe the dynamic of the waves.

There are various theories for describing waves according to Airy, Gerstner, Stokes, Fenton, the current function theory, as well as for individual waves, elliptical waves, and so on. Which theory is to be applied is determined according to the respective conditions of water depth, wave length, height and frequency (see below).

These theories describe the movement of a single wave in the x-z plane, where movements in the y direction, i.e. laterally to the direction of the wave, do not occur. Further information on the fundamentals and derivations of the different wave theories can be found in the literature (e.g. [4, 8, 9, 10]).

The most important terminology of a wave includes (see Figure 11.5): $h(x,t) =$ local wave height at point x at time t; $H =$ wave height; $L =$ wave length; $T =$ wave period; $c =$ wave advance velocity; $\omega =$ wave frequency $= 2\pi/T$; $k =$ wave count $= 2\omega/L$.

With the above preconditions and further assumptions (see below), the Bernoulli equation (one of the two fundamental equations in flow theories besides the continuity equation) can be linearised and the kinetic and dynamic limiting conditions can be fulfilled.

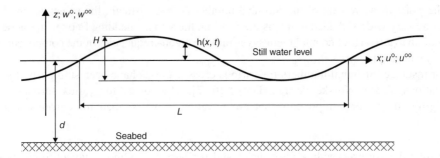

Figure 11.5   Terminology of the linear elementary or Airy wave

For incompressible, vortex- and friction-free media the two equations are:

Bernoulli equation
$$\frac{(u^\circ)^2}{2} + \frac{p}{\rho} + gz = \text{constant} \qquad (11.11)$$

Continuity equation
$$Q = Au^\circ = \text{constant} \qquad (11.12)$$

where u° is the current velocity $= du/dt$, p is the pressure in the current, ρ is the specific density of the medium, z is the height, Q is the volume flow, and A is the cross-section subject to the flow.

If a two-dimensional time-dependent velocity potential $\Phi(x,z,t)$ is introduced, then the potential equation is obtained from the continuity equation:

$$\Delta\Phi(x,z,t) = \frac{\partial^2\Phi}{\partial x^2} + \frac{\partial^2\Phi}{\partial z^2} = 0 \qquad (11.13)$$

If the velocity potential is known, then the velocities u° in the x direction and w° in the z direction are the result by derivations. The acceleration is obtained by means of renewed derivations (according to the time or the coordinates x and z):

$$u^\circ(x,z,t) = \frac{\mathrm{d}x}{\mathrm{d}t} = \frac{\partial\Phi}{\partial x}; \quad w^\circ(x,z,t) = \frac{\mathrm{d}z}{\mathrm{d}t} = \frac{\partial\Phi}{\partial z} \qquad (11.14a)$$

$$u^{\circ\circ}(x,z,t) = \frac{\mathrm{d}^2x}{\mathrm{d}t^2} = \frac{\partial^2\Phi}{\partial x^2}; \quad w^{\circ\circ}(x,z,t) = \frac{\mathrm{d}^2z}{\mathrm{d}t^2} = \frac{\partial^2\Phi}{\partial z^2} \qquad (11.14b)$$

If one inserts the velocities into Equation (11.11) then one obtains the special Bernoulli equation (Euler's motion equation) for vortex-free fluids:

$$\frac{\partial\Phi}{\partial t} + \frac{1}{2}(\deg\Phi)^2 + \frac{p}{\rho} + gz = 0 \qquad (11.15a)$$

$$\text{with } \deg\Phi(x,z,t) = \frac{\mathrm{d}^2\Phi}{\mathrm{d}x^2} + \frac{\partial^2\Phi}{\partial z^2} \text{ (two-dimensional)} \qquad (11.15b)$$

Equation (11.16) is a non-linear partial differential equation of the second order for whose clear solution the following limiting conditions are taken into account:

- On the seabed ($z = -d$) the velocity of the water particles vertical to the horizontal seabed is equal to zero.

$$w°(x,z=-d,t) = \frac{\partial \Phi}{\partial z} = 0 \tag{11.16}$$

- At the surface of the wave at $z=h(x,t)$ the pressure p is constant. The pressure can be set to zero if desired. In this case the Equation (11.15a) at this point is:

$$\frac{\partial \Phi}{\partial t} + \frac{1}{2}(\deg \Phi)^2 + gz = 0 \tag{11.17}$$

- For the velocity $w°$ at the surface $z=h(x,t)$ one obtains from the Equation (11.16):

$$w°(x,z=h(x,t),t) = \frac{\partial \Phi}{\partial z} = \frac{\partial z}{\partial t} = \frac{\partial h(x,t)}{\partial t}\frac{\partial x}{\partial t} = \frac{\partial h}{\partial z}u° + \frac{\partial h}{\partial t} \tag{11.18}$$

As the equations (11.17) and (11.18) are non-linear, a closed solution of the differential equation (11.15a) is not possible. Only approximations are possible for this.

11.3.1.1   Linear or Wave Theory According to Airy

The best known approximations solutions are those by Airy that assume a simple sine form of the wave and linearise the equation, and also the non-linearised theories by Stokes that use multi-member potential series.

In the linear water wave theory by Airy, a wave form in the shape of a sine or cosine form plus phase displacement (according to selected coordinate source) is assumed for the wave depiction under the following conditions:

- The water depth d is greater than the half wavelength L (deep water).
- The wave height H is small compared with the wavelength (low wave slope).
- The water surface is undisturbed.

Despite these limitations compared with actually occurring waves that are described more accurately with non-linear theories, it is possible with the Airy theories to depict the most important wave effects to a good approximation.

The Airy energy for the course of the wave height is:

$$h(x,t) = \frac{H}{2}\cos\left(2\pi\left(\frac{x}{L} - \frac{t}{T}\right)\right) = \frac{H}{2}\cos(kx - \omega t) \tag{11.19}$$

Because of the preconditions, the non-linear terms in equations (11.17) and (11.18) are smaller than the linear ones and can thus be ignored. Besides this, it is possible, in the border conditions equations (11.17) and (11.18), to replace the wave rise $h(x,t)$ by the constant $h=0$. With this simplification, the analytical solution of Equation (11.15a) with Equation (11.19) becomes possible, and one obtains for the velocity potential Φ:

$$\Phi(x,z,t) = \frac{H}{2}\frac{g}{\omega}\frac{\cosh[k(z+d)]}{\cosh(kd)}\sin(kx-\omega t) = \frac{H}{2}\frac{g}{\omega}\eta(z)\sin(kx-\omega t) \tag{11.20}$$

The term $\eta(z)$ represents the depth dependency of the velocity potential.

From the border conditions (Equation (11.18)) there is then obtained the so-called dispersion equation that describes the relationship between the wave frequency ω, wave count k and water depth d.

$$\omega^2 = gk\tanh(kd) \tag{11.21}$$

The wavelength is:

$$L = \frac{gT^2}{2\pi}\tanh\left(2\pi\frac{d}{L}\right) \tag{11.22}$$

for the wave period:

$$T = \sqrt{\frac{2\pi L}{g}\frac{1}{\tanh(2\pi d/L)}} \tag{11.23}$$

and for the wave propagation velocity:

$$c = \frac{L}{T} = \frac{\omega}{k} = \sqrt{\frac{gL}{2\pi}\tanh\left(2\pi\frac{d}{L}\right)} = \sqrt{\frac{g}{k}\tanh(kd)} \tag{11.24}$$

The velocity components are obtained from the partial derivations of the potential equation (11.20) according to the coordinates x or z:

$$u^\circ(x,z,t) = \frac{\partial\Phi}{\partial x} = \frac{gk}{\omega}\frac{\cosh[k(z+d)]}{\cosh(kd)}\frac{H}{2}\cos(kx-\omega t) \tag{11.25a}$$

$$w^\circ(x,z,t) = \frac{\partial\Phi}{\partial z} = \frac{gk}{\omega}\frac{\sinh[k(z+d)]}{\cosh(kd)}H\sin(kx-\omega t) \tag{11.25b}$$

and the acceleration from the second derivation according to the time.

The path curves of the individual water particles are obtained by integration of the velocities according to equations (11.25a) and (11.25b) over the time. They are:

$$u(x,z,t) = \frac{H}{2}\frac{gk}{\omega^2}\frac{\cosh[k(z+d)]}{\cosh(kd)}\sin(kx-\omega t) \tag{11.26a}$$

$$w(x,z,t) = \frac{H}{2}\frac{gk}{\omega^2}\frac{\sinh[k(z+d)]}{\cosh(kd)}\cos(kx-\omega t) \tag{11.26b}$$

The path curves in deep water $(d \to \infty)$ circles and in flat water $(d < \approx L/2)$ ellipses with the half axis:

$$\frac{H}{2}\frac{gk}{\omega^2}\frac{\cosh[k(z+d)]}{\cosh(kd)} \quad \text{(in } x \text{ direction)} \tag{11.27a}$$

$$\frac{H}{2}\frac{gk}{\omega^2}\frac{\sinh[k(z+d)]}{\cosh(kd)} \quad \text{(in } z \text{ direction)} \tag{11.27b}$$

With increasingly deep water z, the radii of the half axes correspond to hyperbolic functions, and the half axis in the z direction becomes zero at the sea bed, so that the water particles in a wave can only move horizontally backwards and forwards (boundary conditions).

Thus, there is obtained from equations (11.25) to (11.27) for the path curves, velocities and accelerations with $\cosh(\alpha x) = (e^{\alpha x} + e^{-\alpha x})/2$ and $\sinh(\alpha x) = (e^{\alpha x} - e^{-\alpha x})/2$:

$$u(x,z,t) = -\frac{H}{2}e^{kz}k\sin(kx - \omega t) \tag{11.28a}$$

$$w(x,z,t) = -\frac{H}{2}e^{kz}k\cos(kx - \omega t) \tag{11.28b}$$

$$u^{\circ}(x,z,t) = -\frac{H}{2}\omega e^{kz}k\cos(kx - \omega t) \tag{11.29a}$$

$$w^{\circ}(x,z,t) = \frac{H}{2}\omega e^{kz}k\sin(kx - \omega t) \tag{11.29b}$$

$$u^{\circ\circ}(x,z,t) = \frac{H}{2}\omega^2 e^{kz}k\sin(kx - \omega t) \tag{11.30a}$$

$$w^{\circ\circ}(x,z,t) = -\frac{H}{2}\omega^2 e^{kz}k\cos(kx - \omega t) \tag{11.30b}$$

The dependency of the waves on the depth of the water is contained in equations (11.21) to (11.35). For deep water with $d \to \infty$, Equation (11.21) becomes:

$$\omega^2 = gk \tag{11.31}$$

In practice, deep water is assumed when $d < L/2$. Thus approximations $\tanh(kd) \approx 1$ and $\sinh(kd) \approx \cosh(kd) \approx e^{kd/2}$ apply, and one obtains for deep water:
Wavelength:

$$L = \frac{gT^2}{2\pi} \tag{11.32}$$

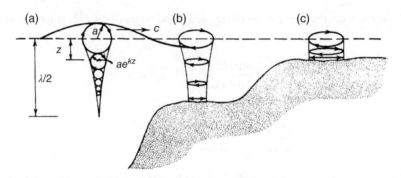

Figure 11.6 Movement of water particles for different depths of water: (a) for deep water; (c) for shallow water; (b) for the transition region (according to McCormick)

Wave frequency:

$$\omega^2 = \left(\frac{2\pi}{T}\right)^2 = gk \qquad (11.33)$$

Wave velocity:

$$c = \frac{L}{T} = \frac{\omega}{k} = \sqrt{\frac{gL}{2\pi}} = \sqrt{\frac{g}{k}} \approx 1.25\sqrt{L} \qquad (11.34)$$

Therefore, the parameters H and L or H and T are sufficient for a clear description of an Airy wave (harmonic wave) in deep water.

Shallow water is assumed when $d < L/20$ applies. Then because $\tanh(kd) \approx \sinh(kd) \approx kd$ and $\cosh(kd) \approx 1$ from Equation (11.24) for the velocity:

$$c = \frac{\omega}{d} = \sqrt{gd} \qquad (11.35)$$

meaning that the wave velocity is only dependent on the depth of the water.

For the transition region there then applies: $L/2 > d > L/20$. The radius r of the wave motion is quickly reduced with increasing depth of water:

$$r = r(z) = \frac{H}{2}e^{2\pi z/L} \qquad (11.36)$$

where z is the distance from the water surface (z is negative); at a water depth of $z = -L/2$, $r \approx 0.043H/2$ and at $z = -L$ it is $r \approx 0.0019H/2$, i.e. it is no longer perceptible. The movements of the water particles at different water depths are shown in Fig. 11.6

If waves flow into shallow water, then the wave heights change. On the condition that the waves have not yet broken, there is derived from the point of view of the energy (energy and

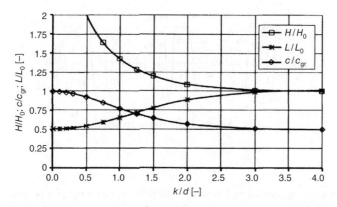

Figure 11.7   Changes of the wave speed, length and velocity when a wave enters shallow water

period are taken as constants) for the wave height H in shallow water (H_0=height in deep water):

$$\frac{H}{H_0} = \left[\frac{1}{\tan(kd)\left(1+\dfrac{2kd}{\sinh(2kd)}\right)} \right]^{1/2} \qquad (11.37a)$$

In the border region of 'shallow water' the relationship becomes (see Fig. 11.7):

$$\frac{H}{H_0} = \left[\frac{\sqrt{gT}}{4\pi\sqrt{d}} \right]^{1/2} = 0.5\sqrt{\frac{T}{\sqrt{d}}} \qquad (11.37b)$$

Wave group velocity
A swell is designated as 'regular' when it consists of many random superimposed harmonic waves of similar frequency, length and direction (wave group). The wave group moves forward with the velocity of the 'encapsulator' of the group velocity c_{gr} (see Figure 11.8). This is:

$$c_{gr} = \frac{c}{2}\left[1+\frac{2kd}{\sinh(2kd)} \right] \qquad (11.38a)$$

In deep water the group velocity, because of the rapidly growing term for sinh ($2kd$) of the depth d:

$$c_{gr} = \frac{c}{2} \qquad (11.38b)$$

For shallow water the group velocity becomes:

$$c_{gr} = c \qquad (11.38c)$$

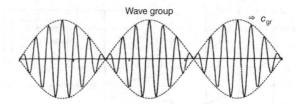

Figure 11.8 Wave group

Wave energy

The energy of a moving, undamped system with mass is made up of a potential and a kinetic portion. For calculating the wave energy, a fluid volume is considered that is limited in the x-z level by $x=0$ and $x=xL$ (over a wavelength) as well as by $z=h(x,t)$ and $z=-d$. A unit width of 1 is assumed in the y direction. The potential energy without form-changing parts (position energy) is:

$$V(t) = \iiint_{\text{vol}} m(x,y,z)\, g\, z(t)\, dx\, dy\, dz \quad \text{with } m = \text{mass/volume} \tag{11.39a}$$

By choosing a suitable reference system for z it can always be achieved that the potential energy is greater than zero. The following is obtained for the potential wave energy:

$$V = g \rho_{\text{seaw.}} \int_{(A)} z\, dA = \ldots - \frac{1}{2} \rho_{\text{seaw.}}\, g d^2 L + \frac{1}{4} \rho_{\text{seaw.}}\, g \left(\frac{H}{2}\right)^2 L \tag{11.39b}$$

The first term of Equation (11.39b) gives the potential energy of water at rest when this condition is selected as the reference level. As regards the wave energy, this is a random constant (wave parameters other than the length do not appear in this) and can therefore be left out. Thus, one obtains for the potential energy of a wave of height H, length L and width 1:

$$V = \frac{1}{4} \rho_{\text{seaw.}}\, g \left(\frac{H}{2}\right)^2 L \tag{11.39c}$$

The kinetic energy of motion is obtained according to:

$$T = \frac{1}{2} \rho_{\text{seaw.}} \iiint_{\text{vol}} m(x,y,z)\, [\underline{r}^\circ(x,y,z,t)]^2\, dx\, dy\, dz \geq 0 \tag{11.40a}$$

where $\underline{r}^\circ$ is the velocity vector. The kinetic energy, because of the square of the velocity, is $\underline{r}^\circ = dr/dt$ always ≥ 0. With this, the kinetic wave energy of all water particles in this region is:

$$T = \frac{1}{2} \rho_{\text{seaw.}} \iiint_{(V)} [\underline{r}^\circ(x,y,z,t)]^2\, dx\, dy\, dz = \frac{1}{2} \rho \int_{(A)} [u^{\circ 2} + w^{\circ 2}]\, dA$$

$$= \ldots = \frac{1}{2} \rho_{\text{seaw.}}\, g \int_{x=0}^{L} h(x)\, dx = \frac{1}{4} \rho_{\text{seaw.}}\, g \left(\frac{H}{2}\right)^2 L \geq 0 \tag{11.40b}$$

The sum of these energy portions is divided by the wavelength L and one obtains the average energy of a wave referenced to a horizontal unit surface with a length and width of 1:

$$E = \frac{T+V}{L} = \frac{1}{8} \rho_{seaw.} g H^2 \qquad (11.41)$$

The energy transport occurs with the same velocity of the wave group as according to Equation (11.38).

According to the Airy theory, the water particles carry out a pure orbital or circular movement, so there is no mass transport, only a transport of energy.

This is not the case according to the non-linear wave theory or with real waves. Whereas the individual wave particles have the same height $= z$ coordinates after a wave pass, an advance in the x direction occurs in the water particles as the velocity of the water particles in the wave are dependent upon, among others, the depth. In the wave crest the forward transport velocity is greater than the backwards velocity in the wave trough, as the water particles there have smaller z coordinates ($z = -H/2$), so the path curves are no longer closed.

The transport velocity of the water particles in the wave advance direction is:

$$u^{\circ}(z) = \left(\frac{\pi h(z)}{L} \right)^2 c \qquad (11.42)$$

It falls off rapidly with the depth due to $h(z) = 2r(z)$.

11.3.1.2   Non-Linear Wave Theories

Non-linear wave theories are required for more accurate calculation of the velocities and accelerations of water particles passing offshore structures in the form of waves, as the resulting velocities, accelerations and forces can be more accurately ascertained. Furthermore, the resulting loadings are generally higher than with linear theories. As these forces are one of the main loadings of such structures, a more accurate knowledge of them is important for the optimisation of the foundation structures.

In contrast to the Airy theories, the actual waveforms are not sine curves, but rather the wave crests and wave troughs have different lengths and forms. Furthermore, the Airy theory only applies to very flat waves. The waves that occur in a swell, however, have a much greater height to length relationship so that non-linear theories must be used for more accurate description of the real waves.

For description of a single wave, use is often made, for instance in ship building, of a trochoid (a special form of cycloid or roll-off curve) according to Gerstner [11] as a good approximation for surface waves in deep water.

In order to generate a trochoid one rolls a wheel with radius R on a straight (see Figure 11.9). The path curve is generated by the point P situated on the radius $r < R$ from the centre point of the wheel and which moves forward with the wheel. The circumference of the wheel with the radius R corresponds to the wave length L. Thus, a water particle at the surface carries out a circular movement with the radius r. If R and r have the same size, then a classic cycloid has been formed.

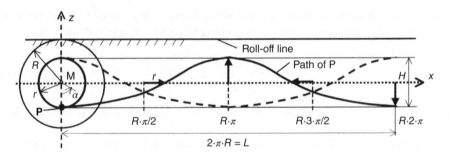

Figure 11.9 Trochoid-shaped wave according to Gerstner [11]

The form of the equations of the wave in parameter form (parameter is the angle α) depends on the position of the selected coordinate system; for the system used here (see Figure 11.9) they are:

Wave crest at $L/2$: $x = R\alpha + r \sin \alpha; z = -r \cos \alpha$ $(0 \le \alpha \le 2\pi)$
Wave trough at $L/2$: $x = R\alpha - r \sin \alpha; z = r \cos \alpha$

where $R = L/(2 \cdot \pi)$, L is the wavelength, $r = H/2$, and H is the wave height.

From the path of the wave profile it can be seen that the wave trough is longer than the wave crest ($l_{WT} = L/2 + 2H$; $l_{WB} = L/2 - 2H$). The larger the relationship H/L, the larger will be the wave trough. This is observed in reality.

Wave slope
The measure for the wave slope is the relationship H/L (wave height to wave length). At a slope greater than 1/7, the wave theoretically becomes unstable and breaks even in deep water. In real swells, the waves break at a slope of approximately 1/10. This causes them to lose a large part of their energy. In shallow waters the relationship of wave height to water depth is decisive; there the waves break at $H/d \ge 0.78$.

As the Airy theory is strictly used only for waves with a slope $H/L < 1/50$) and the actual waves have a substantially greater slope (H/L 1/30–1/10), it is necessary for the dimensioning of offshore structures to make use of wave theories that correspond to the relationships of water depth/wave length/frequency/wave height. In Figure 11.10 the wave profiles are presented according to the different theories, and Figure 11.11 shows their area of application. Computing programs are available for the calculation of non-linear waves and the resulting forces, and they make use of the above relationships depending on the suitable wave theory. Generally they are too complex to calculate manually.

If waves meet large obstructions or if the water depth is reduced, for instance in shelf seas such as the North Sea for long waves, then the propagation of the waves is hindered. The following effects can occur:

Shoaling: Due to the wave touching the bottom (at $d \lesssim L/2$) the wave velocity is reduced, the wavelength reduces, and the wave becomes higher (shoaling factor K_S).

Refraction: Waves that do not run vertically to the rising seabed, swing along the coast until their ridges are parallel with the depth contours. The wave heights may therefore be larger or smaller than the starting waves. They are described by the refraction coefficient K_R.

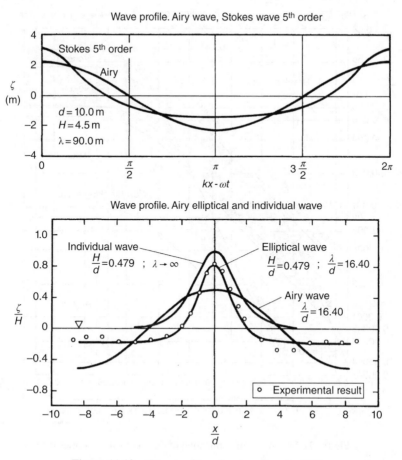

Figure 11.10   Wave profiles of various theories [12]

Diffraction: Waves that meet an obstruction, such as a large structure, a tongue of land or similar, swing around it and run into the shadow of the waves. The heights of the waves are generally reduced.

Reflection: If a massive obstruction prevents the propagation of the waves in deep water, then these are reflected (angle of entry = angle of exit). The heights of the waves remain the same.

11.3.2   Superposition of Waves and Currents

If waves and currents occur together, then the waves are distorted. The simplest case is the superposition of a deep water wave (wave propagation velocity c_0) with a current velocity $U°$ (e.g. tidal current) of constant depth exactly in the direction of the wave ($U° > 0$) or in the opposite direction ($U° < 0$).

Using a site-fixed coordinate system and the characteristic of the undisturbed wave size with the index = 0, one obtains the following assumption that the wave frequency or period remains constant in the superposition:

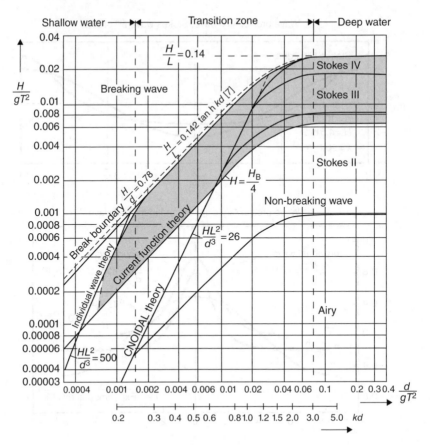

Figure 11.11   Areas of application of different wave theories [8]

$$\frac{L}{L_0} = \frac{1}{4}(1+\chi)^2 \tag{11.43a}$$

$$\frac{c}{c_0} = \left(\frac{L}{L_0}\right)^{1/2} = \frac{(1+\chi)}{2} \tag{11.43b}$$

$$\text{with } \chi = \left(1 + 4\frac{U^\circ}{c_0}\right)^{1/2} \tag{11.44}$$

At a velocity $U^\circ = 0$ or $\chi = 1$, Equation (11.43) passes into the undisturbed wave.

One can recognise from equations (11.43a) and (11.44) that for a current in the direction ($U^\circ > 0$), wavelength and speed become larger, while for ($U^\circ < 0$) they become smaller. Because of Equation (11.44), U° cannot become smaller than $-c_0/4 = c_{gr}/2$, ($\chi - 0$). At this current velocity, the velocity of the energy transport is equal to zero, so the wave height grows and the wave breaks.

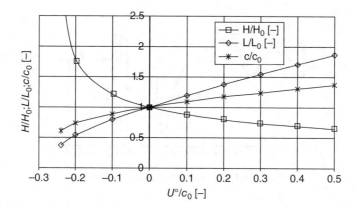

Figure 11.12 Changes to the wave heights, lengths and velocities for superposition with currents

The change of wave height H under the influence of the current in comparison to the undisturbed wave height H_0 can be determined by taking the energy stability into account. On the assumption that the wave energy in a current propagates with velocity $c_{gr} + U°$, one obtains:

$$E(c_{gr} + U°) = E_0 c_{gr0} \tag{11.45}$$

and from this the relationship of the wave heights becomes (see Fig. 11.12):

$$\frac{H}{H_0} = \sqrt{\frac{2}{\chi(\chi+1)}} \tag{11.46}$$

11.3.3 Loads Due to Waves (Morison Method)

The loads on an offshore structure caused by waves are influenced by the relationship of wave lengths to the size of the structure, as well as wave height or wavelength to the depth of the water. The size of the structure relative to the height of the wave is described by the Keulegan-Carpenter number, KC.

$$KC = \pi \frac{H}{2D} \tag{11.47}$$

where H is the wave height and D is the dimension of the structure lateral to the wave.

In the offshore technology a distinction is made between narrow and large-volume structures. Narrow structures with $D/L \lesssim 0.2$ or $KC \lesssim 0.4$ are termed 'hydrodynamically transparent' and it is assumed that the wave is not influenced by the structure when passing it by. Current drag (effects caused by toughness) and the acceleration force of the wave are about the same size.

Relatively large structures with $D/L \gtrsim 0.5$ are termed 'hydrodynamically compact'. Here the wave is deformed by the structure and the acceleration forces dominate with an increasing value of D/L.

The foundation structures of offshore wind turbines are taken as being hydrodynamically transparent or narrow structures, i.e. the waves pass around the structures almost uninfluenced.

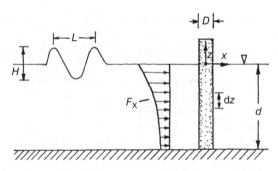

Figure 11.13   Loading of a pile by waves [4]

In contrast, offshore oil or gas transport platforms are generally so large (hydrodynamically compact) that they influence the waves. Then the wave forces must be calculated according to the diffraction theory (see, for instance, [8]).

The narrow foundation structures of OWTs can be calculated according to the superposition method (drag and acceleration forces) of Morison [13]. It is the method used most often for calculating the wave forces on narrow structures.

Morison method

The forces of a body subject to the flow of a wave according to the Morison method are divided into two components, a mass force F_M and a drag force F_D. The components are calculated separately and superimposed. The maxima of the two force parts occur according to the determining sizes (velocity $u°$ or acceleration $u°°$) at different times. They are displaced from each other by a quarter of the wave period T. The horizontal portion of the velocity of the water particles in a wave corresponds to the active current, and the horizontal component of the acceleration to the active acceleration (see Fig. 11.13).

The line load resulting from the current drag in a wave according to Morrison is:

$$q_D(z,t) = \frac{1}{2} C_D(z,\mathrm{Re}) \rho_{seaw.} D(z) \, u°(z,t) \left| u°(z,t) \right| \tag{11.48a}$$

and the line load of the acceleration is:

$$q_M(z,t) = C_M(z) \rho_{seaw.} A(z) \, u°°(z,t) \tag{11.48b}$$

where $C_D(z,\mathrm{Re})$ is the drag coefficient of the cross-section subject to the flow, ρ is the specific density of the water, $D(z)$ is the diameter of the cross-section at the considered depth, $A(z)$ is the cross-sectional area of the considered depth, $u°(z,t)$ is the horizontal velocity of the water particle, $u°°(z,t)$ is the horizontal acceleration of the water particle, $C_M = 1 + C_a = $ drag coefficient of the current acceleration, $C_a = m_a/m_0 = $ coefficient of the hydrodynamic mass of the cross-section, $m_a(z)$ is the hydrodynamic mass, and $m_0(z)$ is the water displacement of the body (volume × density of the water).

Thus, the line load acting on the structure resulting from the waves is then:

$$q_W(z,t) = q_D(z,t) + q_M(z,t) \tag{11.49}$$

Hydrodynamic mass

If a body is accelerated in a medium, then a part of the surrounding medium must be accelerated with it. This accompanying mass is known as the hydrodynamic mass or additional mass, as it apparently increases the mass of the accelerated body. This means that more force is required to accelerate a body in water than for the same acceleration of only the body mass. The size of the hydrodynamic mass is dependent to a large extent on the form of the body, but in, contrast, is mostly independent of the velocity or acceleration. It must also be taken into account in vibration processes in fluids. Only for very high frequency vibrations can it also be dependent on the frequency.

In the determination of the hydrodynamic masses it is assumed, for the sake of simplicity, that a certain portion of the fluid surrounding the body has the same acceleration as the body itself. In this determination, care must be taken whether the body is moistened by the medium on one or both sides.

Values for assumed hydrodynamic masses can be found, for instance, in [4]. In the use of the linear wave theory according to Airy and a vertical pile, one obtains the following according to Morison:

$$q_D(z,t) = \frac{1}{2}C_D\rho D\left(\frac{H}{2}\right)^2 \omega^2 \left(\frac{\cosh[k(z+d)]}{\cosh(kd)}\right)^2 |\cos\omega t|\cos\omega t$$

$$= \frac{1}{8}C_D\rho DH^2\omega^2\eta^2(z)|\cos\omega t|\cos\omega t \tag{11.50}$$

$$q_M(z,t) = C_M\rho\frac{\pi D^2}{4}\frac{H}{2}\omega^2\left(\frac{\cosh[k(z+d)]}{\cosh(kd)}\right)^2 \sin\omega t$$

$$= \frac{1}{8}C_M\rho\pi D^2 H\omega^2\eta^2(z)\sin\omega t \tag{11.51}$$

The resulting forces or lateral forces in the direction of the wave $\Rightarrow Q_x(z,t)$, which act on the offshore structure as per the integration of the overall line loads $q_w(z,t)$ over the water depth.

$$Q_x(z=z_0,t) = -\int_{z=0}^{z_0}[q_D(z,t)+q_M(z,t)]\,dz = -\int_{z=0}^{z_0}q_w(z,t)\,dz \tag{11.52}$$

In order to determine the bending moment about the y axis $\Rightarrow M_y(z,t)$, the line load in the integration must still be multiplied by the respective lever.

$$M_y(\bar{z}_0,t) = -\int_{\bar{z}=0}^{\bar{z}_0}q_w(\bar{z},t)\,\bar{z}\,dz \quad (\bar{z} \text{ is measured from the seabed}) \tag{11.53}$$

For a vertical pile with constant values for D and C_D over z, the integral of Equation (11.52) can be considered closed; the result for the lateral force is:

$$Q_x(z=d,t) = -\int_{z=0}^{d}[q_D(z,t)+q_M(z,t)]\,dz = -F_D(d,t)-F_M(d,t) \tag{11.54a}$$

$$\text{with } F_D(t) = \frac{1}{8}C_D\rho D\,H^2\,\omega^2 d\frac{\sinh(2kd)+2kd}{kd[\cosh(2kd)-1]}|\cos\omega t|\cos\omega t \tag{11.54b}$$

$$F_M(t) = \frac{1}{8} C_D \rho A H \omega^2 \frac{1}{k} \sin \omega t \qquad (11.54c)$$

The path of the bending moment according to Equation (11.53) must generally be numerically integrated, for instance according to Simpson's law.

Example: loading of a monopile by means of regular swells (Morison).
Water depth $d = 25\,\text{m}$; pile diameter $D = 6\,\text{m}$; wavelength $L = 50\,\text{m}$; wave height $H = 4\,\text{m}$; coefficient $C_D = 0.8$; coefficient $C_M = 2.0$; density of the water $\rho = 1025\,\text{kg/m}^3$:

$$\Rightarrow \text{conditions } d/L = 0.5 \Rightarrow \text{transfer region, shallow/deep water}$$

$$\Rightarrow k = \frac{2\pi}{L} = 0.1257\,\text{m}^{-1}; \quad T = \sqrt{\frac{2\pi L}{g \tanh (2\pi d/L)}} = 5.7\,\text{s}; \quad \omega = \frac{2\pi}{T} = 1.108\,\text{s}^{-1}$$

The coordinate z is to be measured from the ocean surface.

Velocity:	$u^{\circ}_{max} (z=0) = 2.233\,\text{m/s}$
Acceleration:	$u^{\circ\circ}_{max} (z=0) = 2.466\,\text{m/s}^2$
Current loading:	$q_{Dmax}(z=0) = 10.15\,\text{kN/m}$
Acceleration loading:	$q_{Mmax}(z=0) = 99.24\,\text{kN/m}$
Total force from current:	$F_{Dmax}(t=0) = 40.88\,\text{kN}$
Total force from acceleration:	$F_{Dmax}(t=T/4) = 880.6\,\text{kN}$
Bending moment in pile on seabed	$M_{ymax}(z=d) = 13,928\,\text{kNm}$

The actual bending amount of the 'cantilever monopile', however, is higher, as, due to the active insertion length of the pile in the seabed, the cantilever is 'lengthened' (by approximately 1 to 2 D) depending on the floor conditions.

Velocity, acceleration and line loads resulting from current and acceleration are depicted as a function of the depth z/d in Figure 11.14, and the time course of the maximum line loads on the seabed over a wave period t/T is shown in Figure 11.15.

Surface roughness
The roughness k of the structure's surfaces subject to the flow, due to corrosion, organic growth, and so on, has a substantial effect on the wave forces. The roughness is given by the relationship k = particle size to structure diameter (particla size for new steel ≈ 0.02–0.1; concrete ≈ 0.5–1). The coefficient C_D becomes larger with increasing roughness, and the coefficient C_M smaller.

Superposition of waves and constant currents
If, additionally to the waves, there is also a constant current, then the attacking current forces are increased in the wave crests (squared with increasing current velocity) and in the wave troughs they become correspondingly weaker. With a superimposed current the acceleration forces remain unchanged. It must be noted that waves and currents can move in different directions.

Loads from breaking waves
In the determination of the forces of breaking waves it is assumed, for the sake of simplicity, that the carrier forces are small compared with those of the current forces. The water particles, however, break on the structures with a relatively high velocity.

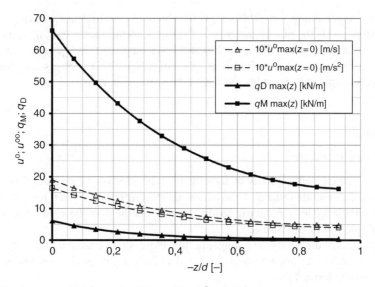

Figure 11.14 Current velocity, acceleration and load on a pile dependent on the depth, according to Morison

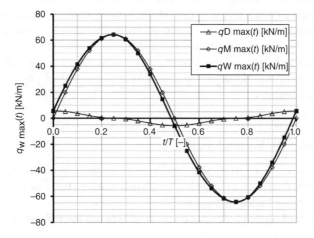

Figure 11.15 Time course of the maximum current and acceleration loads over a wave period

The breaking wave height H_B is to be taken as the wave height (dependent on the depth of the water). The horizontal velocity of the breaking wave approximately corresponds to the propagation velocity of the wave in shallow water.

$$u^\circ_{max} \approx \sqrt{g H_B} \tag{11.55}$$

The impact-type force, also called 'slamming', will then be:

$$F_S = C_S \frac{1}{2} \rho D \left| u^\circ_{max} \right| u^\circ_{max} \lambda h_B \tag{11.56}$$

where F_S is the slamming force (kN), C_S is the slamming coefficient $\approx 2\pi$, ρ is the density of the water (t/m³), D is the pile diameter (m), λ is the 'Curling' factor ≈ 0.5; h_B is the maximum height of the breaker above the smooth water level (approx. $0.7 \cdot H_B$) (m), and u°_{max} is the maximum horizontal wave velocity (m/s). (*Remarks*: Equation (11.56) is non-dimensionally neutral, i.e. the arguments must be given with the units given above).

11.4   Swell

11.4.1   Regular Swell

Ocean waves are created primarily by the effects of the wind on the surface of the water. If air impinges on the smooth surface of the ocean, then the friction between the air and water as well as air turbulence lead to the triggering of waves. In the first phase, low and short waves, so-called ripple waves, are formed.

With the effect of the wind over a longer period, these waves grow continuously and become faster. At the same time, small waves are created again that grow and superimpose themselves on the previously formed waves.

With 'regular' swells it is assumed that this exists only as sequentially following waves with similar wavelengths and heights. The cause of the wave here is only meant to be the wind.

If the wind blows for a longer period with a constant velocity and direction ('fetch time') over a longer distance ('fetch length'), then in deep water, waves of almost the same length, height and direction dominate (regular swells or 'mature wind sea'). This applies for greater wind strengths with correspondingly longer and higher waves, for example, in the North Atlantic, North Polar Ocean as well as Indian and Pacific Oceans south of the 40th parallel ('Roaring Forties'). In the North and Baltic Seas a mature swell can only be formed at relatively low wind strengths.

11.4.2   Irregular or Natural Swells

In addition to the waves caused by the wind with different lengths and heights, waves that come from different directions such as from a distant wind system can also be added. These superimposed waves are called 'natural swells'.

In the description of a swell a distinction is made between the following types:

- Long-crested swells: only harmonic waves of any frequency are superimposed, but with a dominating growth of neighbouring lengths and frequencies. They all only have one direction.
- Short-crested swells: here the waves can have different directions, lengths and frequencies, as other direction division functions are added. The frequencies generally have a Gaussian distribution and the wave heights a Rayleigh distribution (see below).

The swell can be described by means of a Fourier integral. Fourier integrals are analogous to the Fourier row development in which the development frequencies are no longer an integer multiple of a base frequency but are continuously distributed (precondition: the integral must be finite).

11.4.3 Statistics

The distribution of the parameters that describe the natural or irregular swell, such as wave height, frequencies or lengths, cannot be described by means of the above wave theories but only by statistical methods. The most important methods for this are:

Gaussian or normal distribution: This describes, for instance, the wave frequencies of the natural swell. The Gaussian distribution of a random process is (s = spread):

$$f(x) = \frac{1}{s\sqrt{2\pi}} \exp\left[-\frac{x^2}{2s^2}\right] \tag{11.57a}$$

The probability is then:

$$F(x_0) = \int_{-\infty}^{x_0} f(x)\,dx = \frac{1}{s\sqrt{2\pi}} \int_{-\infty}^{x_0} \exp\left[-\frac{x^2}{2s^2}\right] d\xi \tag{11.57b}$$

Weibull distribution: The Weibull distribution represents an extreme distribution referenced to an expected minimum of a random process. It is used especially for wind distribution, length of life of structures, or for materials investigations. By changing the parameters it can be adapted to many different distributions. The frequency of distribution of the Weibull distribution is:

$$f(x) = \begin{cases} \dfrac{m}{x_0}\left(\dfrac{x-x_u}{x_0}\right)^{m-1} \exp\left[-\left(\dfrac{x-x_u}{x_0}\right)^m\right] & \text{for } x \geq x_u \\ 0 & \text{for } x < x_u \end{cases} \tag{11.58a}$$

where x_0 is the nominal value of the size investigated, x_u is the lower expectation or failure limit of the investigated dimension, and m is the Weibull module (the larger the module, the 'steeper' is the distribution of the frequency or the steeper is the probability distribution: see Figure 11.16).

Then one obtains for the probability distribution:

$$F(x) = \int_0^x f(x)\,dx = \begin{cases} 1 - \exp\left[-\left(\dfrac{x-x_u}{x_0}\right)^m\right] & \text{for } x \geq x_u \\ 0 & \text{for } x < x_u \end{cases} \tag{11.58b}$$

Rayleigh distribution: The Rayleigh distribution describes certain random processes such as the distribution of the wave height of a natural swell. It is a special case of the Weibull module $m = 2$ and the lower expectation limit $x_u = 0$. The frequency of distribution of the Rayleigh distribution is:

$$f(x) = \frac{\pi x}{2x_M} \exp\left[-\frac{\pi}{4}\left(\frac{x}{x_M}\right)^2\right] \quad \text{for } x \geq 0 \tag{11.59a}$$

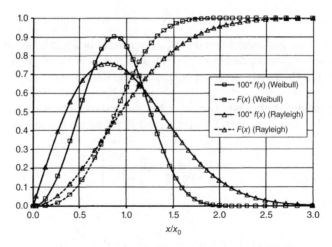

Figure 11.16 Sequence of the statistics distributions according to Weibull ($m=3$) and Rayleigh

where x_M is the reference value (e.g. average wind velocity). The probability distribution is then:

$$F(x) = 1 - \exp\left[-\frac{\pi}{4}\left(\frac{x}{x_\mathrm{M}} \right)^2 \right] \quad \text{for } x \geq 0 \tag{11.59b}$$

The moments from the distribution are then:

$$m_0 = \int_0^\infty f(x)\,\mathrm{d}x; \quad m_n = \int_0^\infty x^n f(x)\,\mathrm{d}x \tag{11.60}$$

where $f(x)$ is the frequency of distribution, and m_0 is the zeroth order moment, with the moments of higher (nth) orders are determined analogously.

11.4.4 Swell Spectra

The wave height of an irregular swell can be depicted by means of the following Fourier integral:

$$h^2(t) = \int_{\omega=0}^\infty S_\mathrm{h}(\omega)\,\mathrm{d}\omega < \infty \tag{11.61}$$

where $h(t)$ is the rise of the water surface at time t, ω is the swell frequency, and $S_\mathrm{h}(\omega)$ is the spectrum of the swell (energy spectra density).

By multiplying Equation (11.61) by the factor $\rho g/2$, one obtains the energy contained in the swell described. In this the left side represents the average energy of the swell and the right side the energy distribution over the swell frequency ω.

For long-crested swells (mature wind sea) it is assumed that all waves have the same direction. Then the swell can be dealt with as a two-dimensional problem.

In order to describe a short-crested swell, the spectrum of the long-crested swell must be multiplied by a distribution function of the wave direction. This leads to a three-dimensional problem.

$$S_h(\omega,\mu) = S_h(\omega)\, G_h(\omega,\mu_e) \tag{11.62}$$

where $G(\omega,\mu)$ is the direction function of the wave propagation, μ is the angle of the main direction compared with a fixed coordinate system, and μ_e is the angle of an elementary wave measured from the main wave direction.

The following applies for the wave direction function:

$$\int_{\mu=\pi/2}^{\pi/2} G(\omega,\mu_e)\, d\mu_e = 1 \tag{11.63}$$

An equation in the following form is used on the assumption that the distribution of the wave energy over the direction of run is independent of the frequency:

$$G(\mu_e) = \begin{cases} k_n \cos^n(\mu_e) & \text{for } -\dfrac{\pi}{2} \le \mu_e \le +\dfrac{\pi}{2} \\ 0 & \text{otherwise} \end{cases} \tag{11.64}$$

with the most often-used values being $n=2$ or $n=4$.

The consideration of the direction distribution function generally leads to a lesser loading (load change count and stress amplitudes) of OWT foundations as the maximum stresses, depending on the wave direction, occur at different places in the foundations. For this reason they can be constructed with low material strengths. This also applies for distribution and frequency of the wind direction and wind velocities and the resulting load.

General swell spectra according to Bretschneider are:

$$S_h(\omega) = \alpha\omega^{-5}\exp(-\beta\omega^{-4})\ [\text{m}^2\text{s}] \tag{11.65}$$

where $S_h(\omega)$ is the spectral density of the swell, ω is the swell frequency, α is a coefficient (m^2/s^4) and β is a coefficient (s^{-4}) (see below).

For a mature wind sea (so-called long-crested swell, wind velocity u), one obtains the derived Pierson-Moskowitz spectrum (P-M spectrum) with the coefficients α and β found by observations and measurement of swell in the North Atlantic:

$$S_h(\omega) = 0.0081\, g^2\, \omega^{-5}\exp\left[-0.74\left(\frac{g}{u\omega}\right)^4\right] \tag{11.66}$$

Here $\alpha=0.0081\times g^2$, and $\beta=0.74\times(g/u)^4$.

In practice, the two-parameter distorted P-M spectrum, the 'modified' P-M spectrum or ISSC spectrum (ISSC=International Ship Structure Conference) is also often used. This spectrum can also be used for 'immature' swells, when the wind-impinging length and/or duration are insufficient to generate a mature wind sea (example: storm in the North or Baltic seas). For this purpose, further dimensions must be introduced that characterise such wind seas.

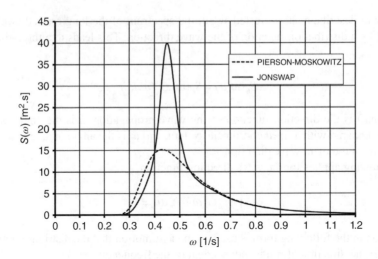

Figure 11.17 Pierson-Moskowitz and JONSWAP swell spectra (wind velocity = 20 m/s, α and ω_p are equal)

These are: $H_{1/3}$ = average height of the waves that are higher than two-thirds of all waves (also designated as 'significant' wave height H_s); T_0 (period of the zero passes ($h(t) = 0$) or T_m (average wave period) or TP (period of the maximum of the spectrum).

The parameters $H_{1/3}$ and T must be determined from wind sea statistics. The ISSC spectrum is:

$$s_h(\omega) = 173 H_{1/3}^2 \, T_m^2 \, \omega^{-5} \exp\left[-692\left(\frac{1}{T_m \omega}\right)^4\right]$$

$$= 124 H_{1/3}^2 \, T_0^2 \, \omega^{-5} \exp\left[-496\left(\frac{1}{T_0 \omega}\right)^4\right]$$

(11.67)

Between the different periods, the following relationships apply for the P-M and the ISSC spectra:

$$T_m = 0.7716 \, T_P = 1.0864 \, T_0$$ (11.68)

From the spectrum according to Figure 11.17, the wind velocity u = 20 m/s for ω_p or T_p can be read off:

$$\Rightarrow \omega_p = 0.43 \, \text{s}^{-1} \Rightarrow T_p = 2\pi/\omega_p = 14.6 \, \text{s}; \quad T_0 = 0.7102 \, T_P = 10.4 \, \text{s}; \; T_m = 11.3 \, \text{s}$$

The significant wave height $H_{1/3}$ is obtained from the value m_0 according to Equation (11.60) by means of integration of the swell spectrum.

$$H_{1/3} \approx 4\sqrt{m_0} = 4\sqrt{\int_0^\infty S_h(\omega) \, d\omega}$$ (11.69)

The result for $u = 20$ m/s: $H_{1/3} = 9.89$ m. However, this is a theoretical value as, for instance, with a wind velocity of 20 m/s $\approx$ 39 kn for a mature wind sea, a wind-impinging length of more than 4000 km would be required.

For seas with restricted wind-impinging length, a depth influence (transition region between deep and shallow water) such as, in the North, the so-called JONSWAP spectrum (Joint North Sea Wave Project) can be used.

$$S_{\rm h}(\omega) = \alpha\, g^2\, \omega^{-5} \exp\left[-1.25\left(\frac{\omega}{\omega_{\rm p}}\right)^{-4}\right]\gamma^p \tag{11.70a}$$

where α is the Phillips constant (see Equation (11.71a), in the P-M spectrum it has the fixed value 0.0081), and γ is the enlargement factor for the P-M spectrum; the factor lies between 1 and 7, and the value $\gamma = 3.3$ is often used (average JONSWAP spectrum).

$$p = \exp\left[-\frac{(\omega - \omega_{\rm p})^2}{2\sigma^2 \omega_{\rm p}^2}\right]$$

$$\sigma = \begin{cases} \sigma_{\rm a} & \text{for } \omega \leq \omega_{\rm p} \text{ (size for the width of the spectrum lef of } \omega_{\rm p}) \\ \sigma_{\rm b} & \text{for } \omega > \omega_{\rm p} \text{ (size for the width of the spectrum right of } \omega_{\rm p}) \end{cases} \tag{11.70b}$$

$$\sigma_{\rm a} = 0.07; \quad \sigma_{\rm b} = 0.09 \text{ (often-used values)}$$

The parameters σ and $\omega_{\rm p}$ are dependent on the active wind length x, and with the evaluation of measurement data there is often used:

$$\alpha = 0.076\, \overline{x}^{(-0.22)} \tag{11.71a}$$

$$\omega_{\rm p} = 0.557\frac{g}{\overline{u}_{10\rm m}}\, \overline{x}^{(-0.33)} \tag{11.71b}$$

where $\overline{x} = g/\overline{u}_{10}\, x^2$ and $\overline{u}_{10}$ is the wind velocity at 10 m height.

With the values for $\alpha_{\rm JONSWAP} = \alpha_{\rm P\text{-}M} = 0.0081$, and $\omega_{\rm p,P\text{-}M} = \omega_{\rm p,JONSWAP}$ (values see above), the two spectra in Figure 11.17 are presented for comparison.

If one compares the JONSWAP and a Pierson-Moskowitz spectrum with the above equal values for α and $\omega_{\rm p}$, the result for the relationship of the maxima of both spectra is the enlargement factor:

$$\gamma^p = \frac{S_{\rm hmax}^{\rm JONSWAP}}{S_{\rm hmax}^{\rm P\text{-}M}} \approx 2.7 \text{ (for the above example)} \tag{11.72}$$

From the respective energy spectra the result is the excitation loads on an offshore structure. If one knows the vibration behaviour (eigen-frequencies), then the structural answer can be calculated taking the damping behaviour into account. From the structural answer one can then determine the occurring stresses and, with the load alteration, the length of life from the swell loads (see further literature, e.g. [4,8]).

11.4.5   Influence of Currents

The superposition of a swell with a current $U°$ has a modifying result on the swell spectra. By taking into account the non-linear energy transport of the interactions between swell and current, one arrives at the following relationship:

$$S_h(\omega,U°) = \frac{4}{\chi(\chi+1)^2} S_h(\omega) \quad \text{with } \chi \text{ from } Eq.(11.44) \tag{11.73}$$

11.4.6   Long-Term Statistics of the Swell

The implementations of the previous section apply only for constant values of the significant wave height $H_{1/3}$, the wave period T_0 or T_p, and the main wave direction μ_H. These parameters are statistically different for longer periods of time and must be taken into account when setting up a long-term concept for the swell.

As $H_{1/3},\ T_0$ and μ_H appear to be stochastic over a longer time period, the four-dimensional distribution $f(H, H_{1/3}, T_0, \mu_H)$ is decisive for the long-term concept and the following can be assumed:

$$f(H,H_{1/3},T_0,\mu_H) = f(H/H_{1/3},T_0,\mu_H) \times f_L(H_{1/3},T_0,\mu_H) \tag{11.74}$$

This leads to the long-term excess probability of the wave height $H > H^*$:

$$P_L(H > H^*) = \int_0^{2\pi} \int_0^{\infty} \int_0^{\infty} P_K(H > H^*) f_L(H_{1/3},T_0,\mu_H)\, dH_{1/3}\, dT_0\, d\mu_H \tag{11.75}$$

This will not be detailed further within the framework of this chapter, but see, for instance, [4,12].

11.4.7   Extreme Waves

The determination of the maximum expected wave height (50- or 100-year wave) during the lifetime of an offshore plant is required for the following reasons:

- A maximum wave height is an extreme case that represents the maximum loading by waves; it occurs only once.
- If there is a platform that is not to be flooded, then this must be higher than the maximum occurring water level. The highest water level is the result of the water level at 'normal zero' (referenced to NZ, the average low-water level) + average tidal lift + wind backup + half the maximum wave height.

The maximum expected wave height during an hours-long or days-long storm can be estimated from the presented swell spectrum with the corresponding wind velocity:

$$H_{max} \approx 1.86\, H_{1/3} \quad \text{(hours-long blowing wind)} \tag{11.76a}$$

$$H_{max} \approx 2.23\, H_{1/3} \quad \text{(days-long blowing wind)} \tag{11.76b}$$

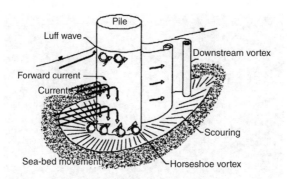

Figure 11.18 Scouring action [14]

11.5 Scouring Formation, Growth, Corrosion and Ice

11.5.1 Scouring

Scouring occurs when, because of the eroding effect of the flowing water, the sea bed is loosened and carried away. When structures are subjected to currents, they produce an additional increase in the current velocity and cause vortexes. The resultant scouring partly exposes the foundations (see Figure 11.18). The causes of the currents can be ocean, tidal, wind-induced and wave currents. Depending on the properties of the sea bed and the existence of current velocities, scouring can occur in so-called non-binding soils (see below), but for binding soils this process can take longer.

The question of scour formation or scour protection plays a large role in the planning of the foundations of offshore structures. The carrying and operating behaviour of the whole OWT is influenced by the scouring of the sea bed:

- The stresses on the foundation structure are increased (the 'cantilever' foundation and tower become longer).
- The loading of the bed (the penetration length of the foundation structure below the sea bed becomes shorter and thus the forces to be absorbed by the bed become greater).
- Eigen-frequencies of the carriers (foundation plus tower) become smaller, which changes the operating behaviour of the OWT (possibly displacement of the critical frequency regions).
- These effects can reduce the life, and can even lead to the early failure, of the OWT.
- With gravity foundations the contact area due to scouring can be reduced so that the stability of the plant is reduced.

For this reason it is necessary when designing the foundations to take a maximum expected scouring depth into account, or to take scouring protection measures (e.g. rock fills with several layers of differently sized rocks, synthetic sea-grass mats, and so on). Scouring protection is carried out before or after installing the foundations. If scouring protection measures are foreseen, they must also be controlled during the operation of the OWT and possibly repaired.

The cable connections for transferring the generated electricity must be especially protected in such a manner that they are on no account exposed, as they are generally not designed for the wave and current loads.

For the determination of the scouring depth the only comprehensive experience and calculations are for small pile diameters, such as bridge piles, that have also partly been proven by

Figure 11.19   Mussels growth on an offshore foundation (photo: M. Klaustrup)

experiments. For OWTs the above influences on the scouring depth have been little investigated. The calculation equation used up to now provides very different results. According to these one obtains scouring depths from 0.3–3 × D [15]. The equation [3] in its guidelines assumes a scouring depth of 2.5 D. Because of the wide range of results, the calculation equations developed up to now are hardly usable in practice.

11.5.2   Marine Growth

Structures that are submerged in the water for a long time are occupied by numerous animals and plants. The growth especially influences the dynamic and also the static behaviour of the structures. A distinction is made between:

- 'Soft' growth such as algae, seaweed, sea anemones, and similar: this causes an increase in the current drag and wave forces due to the increase in the cross-section and roughness. An increase in the static and hydrodynamic masses occurs only in small amounts, as the soft growth follows the movement of the structures only in part. The density of the growth is approximately $1 \, t/m^3$.
- 'Hard' growth such as mussels, barnacles, feather duster worms, and similar (see Figure 11.19). They increase the current drag and wave forces due to the increase in the cross-section and roughness. In addition, the static masses and, for vibration processes, the hydrodynamic masses, are increased as this type of growth has the same movement as the structures. The density of the growth is approximately $1.3–1.4 \, t/m^3$.

The strength of the growth is influenced by the following factors:

- Water depth: the greater the water depth the lesser the growth. At a water depth of 0–10 m the growth mass can be approximately $250–300 \, kg/m^2$, and at a depth of >50 m up to approximately $20 \, kg/m^2$.
- Distance from the coast: the greater the distance from the coast, the less the growth and the later the structures are occupied.

- Current velocity: the higher the current velocity, the more difficult it is for algae, barnacles and mussel larvae to attach to the structures.
- Temperature: higher temperatures promote faster growth.
- Clarity of the water: the clearer the water, the deeper the sunlight penetrates and the stronger the growth is promoted.
- Nutrition of the water: the higher the nutritional content of the water, the better are the growth conditions for algae, mussels, and so on.

11.5.3   Ice Loads

Water freezes at $0\,°C$ (sweet water) and at $-1.8\,°C$ with a salt content of approximately 3.5%. With continuing frost the salt content in the ice is reduced and the strength of the ice increases. It is dependent on the ice composition, the ice temperature and the speed of the icing process.

Ice that does not melt with the warm times of the year is known as multiyear ice; ice that reforms in winter is known as first-year ice. The strength of multiyear ice is generally higher than first-year ice, as the salt content of ice is slowly reduced.

The following forms of ice are distinguished in the design of offshore structures: closed ice cover, ice floes, pack ice fields and icebergs. The different forms of ice are created by local and regional environmental conditions, wind, currents and waves. The strength values are dependent on ice temperatures, salt content and air entrapments. The following values can be assumed for the compressive strength of ice at $0\,°C$.

North Sea:	$1.5\,MPa$ ($1\,Megapascal = 10^6\,N/m^2 = 1\,N/mm^2$)
Baltic Sea:	$1.8\,MPa$
Sweet water:	$2.5\,MPa$

The compressive strength rises by approximately $0.25\,MPa$ per degree of lower temperature. When fracturing, the ice behaves like a brittle material.

A dimension equation for the horizontal acting line load on the structure (according to Khorzavin) at the waterline is:

$$F_H = k_K\, k_F\, Dh\sigma_0 \tag{11.77}$$

where F_H is the horizontal force on the structure, k_K is the contact coefficient (ice/structure), k_F is the form coefficient (form of the structure), D is the width of the structure, h is the thickness of the ice, and σ_0 is the single-axis compressive strength.

Example: ice load on a vertical structure in the Baltic Sea with the following data:
Ice temperature $= -10°C$; $k_K = 0.33$; $k_F = 1.0$; $D = 5.0\,m$; $h = 0.5\,m \rightarrow \sigma_0 \approx 1.8 + 10 \times 0.25 = 4.3\,MPa \rightarrow F_H = 0.33 \times 1.0 \times 5.0 \times 0.5 \times 4.3 = 3.5\,MN$.

11.5.4   Corrosion

Seawater is an aggressive medium, especially combined with intensive UV rays, high moisture and frequent falling below the dew point (condensate formation with salt concentration). For this reason, special attention must be paid to sufficient corrosion protection for offshore plants with a life of approximately 20 years and limited access. Defects have a short- or long-term effect on the availability of plants, often quickly become visible on the outer surface and can optically damage the reputation of offshore plants. The corrosive effect of the salty and moist air can reach up to over the gondola of the OWT.

Corrosion is the primary term for material changes of components. These can be material losses (reduction of thickness) or changes to the material properties (chemical or metallurgical). The most important types of corrosion are:

- Oxidation: with steel this leads to the formation of rust (transformation of the iron into iron oxide with completely different material properties) and thus to the reduction in the material thickness capable of carrying the load. For this, access of oxygen to the surface is necessary. With aluminium, a layer of oxide forms that is impermeable to oxygen and thus protects the material. If this layer is destroyed, such as by mechanical handling, then another layer forms immediately. In seawater this protective layer is generally not stable.
- Galvanic or contact corrosion: if two conducting materials with different electrical potentials (e.g. steel and aluminium) are wetted by an electrically conducting fluid such as salt water, then a material erosion occurs in the material that has the lower electrical potential, here the aluminium. The greater the difference in the potential, the stronger the galvanic corrosion.
- Further types of corrosion are: pitting, splitting, stress, vibration cracks and biochemical corrosion.

With welding, the corrosion resistance of normally corrosion-proof steels can be reduced in the region of the weld, sometimes substantially (increased tendency to tensile welding eigenstress crack corrosion in the welding seam region, and separation of the alloy parts).

In the dimensioning rules of the certifying institutes (GL, DNV, etc.), corrosion allowances are provided for endangered components. They are meant to ensure that a component, after longer operation, in which rusting and reduction of the material thickness is unavoidable, still possesses sufficient material strength.

Depending on the corrosive load (seawater) and servicing, it can be assumed that with paint-protected components there is a reduction in the material thickness due to rust of approximately 0.1 mm per year, for unprotected components of 0.3 mm per year, and in the border region of air/water (water line) up to 0.5 mm per year.

There are numerous standards and guidelines for corrosion protection. The two most important for offshore plants are:

- DIN EN ISO 12944, *Corrosion protection of steel structures*, 1998.
- NORSOK M-501, *Surface preparation and protective coating*, 2004 [16].

The DIN standard differentiates several categories of corrosion loading. For OWTs in the atmospheric region the categories are 'C3 light' to 'C5 very strong' depending on place of erection (near or far from the coast with salty atmosphere). For foundations in soil or water the categories are Im1 to Im3. For the offshore region in the atmospheric region the category is 'C5 very strong' and in the region of continuous water loading (underwater and spray water zone) the categories Im2 and Im3 are provided.

The Norwegian NORSOK standard M-501 [16] has been created for offshore plants of the oil and natural gas industry. It describes comprehensively and in great detail the corrosion protection to be undertaken for this type of plant and can also be used for offshore wind turbines.

11.6 Foundations for OWTs

11.6.1 Introduction

The types of foundations for OWTs depend very strongly on the size of the WT, the depth of the water, the properties of the sea bed (carrying capacity) and the environmental conditions such as the currents, tides, waves, ice loads, and so on.

The following can be used for depths of water up to approximately 60 m:

- Monopiles: up to approximately 30–35 m depth of water.
- Tripods: up to approximately 40 m depth.
- Jackets/framed: up to approximately 60 m depth.
- Suction buckets: up to depths of approximately 25 m.
- Weighted or gravity foundations: up to depths of approximately 20–30 m.
- Variations and combinations of the above foundations.

Because of the high costs of foundations for WTs, only large wind turbines in wind farms with 30 or more WTs from the 3.5 MW class are used. The deeper the water, the larger the plants should be. In Denmark, the UK and Ireland, numerous wind farms have already been installed, each with a 3.5 MW capacity, in water depths up to approximately 25 m at relatively short distances from the coast. For the German wind farms that are to be installed at greater water depths (up to approximately 40 m) and much further from the coast (up to 120 km), use will be made of mainly the 5 MW class and larger. In addition, the number of turbines per farm must also be larger due to the high grid connection costs. Most of them are planned for 80 turbines and more.

For water depths deeper than approximately 60 m, so-called floating foundations that are anchored to the seabed must be selected. In the oil and gas industry, such foundations are the state of the art today. There, plants for water depths greater than 1500 m have been installed, with a correspondingly large use of material. These technologies can also be used for wind turbines, but they differ in the design requirements. The following can be used for floating offshore wind turbines:

- 'Tension legs': large floating bodies that are kept underwater by means of anchoring on the sea bed and, due to pre-stressing in the anchors, equalise the forces and moments acting on the OWEC (offshore wave energy converter).
- 'Spar buoys': vertical floating towers that are anchored to the sea bed. By means of corresponding weight distribution and pre-stressing of the anchoring, the tower is kept vertical even under loading of the wind turbine.
- Variations of the above floating foundations.

Also large WTs are used because of the amount of material used for floating foundations. The economics of such plants are not clear at present.

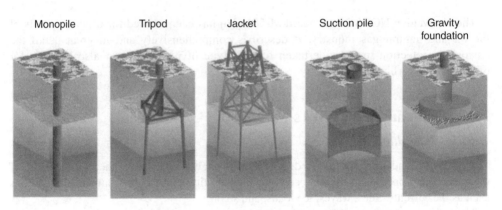

Figure 11.20 Fixed types of foundations

11.6.2 Fixed Foundations

Fixed foundations are for OWTs that stand on the sea bed and which transfer the loads they are subjected to down to the sea bed (Figure 11.20). Such foundations have long been used by the offshore oil and gas industry, and they have been installed there up to ocean depths of approximately 400 m.

The limiting criteria for fixed foundations are:

- height h of the foundation because of the required bending strength;
- stresses increase with h or h^2;
- deformation or slant of the gondola increases with approximately h^4 or h^3; the slant of the gondola in operation must not be greater than 1–2°;
- eigen-frequencies decrease with approximately $1/h^2$;
- sea area (wave heights, swell spectra, currents, wind conditions);
- sea-bed carrying capacity;
- ramming load (required ramming energy);
- weight of the foundation structure (production, transport, crane capacity);
- environmental loading (e.g. noise when ramming).

11.6.2.1 Monopiles

The foundation of a monopile consists of the lower part of a cylindrical pipe that is rammed approximately 15–35 m deep into the seabed, depending on the size of the OWT, the loading and bed carrying capacity. Other techniques such as illuviation or boring (in the rocky subsurface) as well as combinations are also possible but are more expensive. For large pipe diameters, the part of the pipe above the seabed is either formed conically or cylindrically with or without reducing wall thickness as the bending moment is reduced higher up. The actual tower of the WT is then placed on the pipe above the water.

Advantages of the monopile compared with other foundation methods:

- Simple and economical production (pipe up to 8 m in diameter and wall thicknesses up to 90 mm).

- Simple and quick installation, as only one pile needs to be rammed.
- No preparation of the sea bed.
- Good mooring possibilities for service ships.
- Relatively safe collision with ships.
- Simple removal, as the pile is flushed free approximately 3 m under the sea bed and capped there.

Disadvantages:

- Relatively low bending strength and thus limitation of the above-mentioned water depths.
- Larger pipe dimensions (diameter, wall thickness) are difficult to ram (limitation due to size of the ramming hammer and ram loading of the pile; alternatives are illuviation, boring or a combination).
- Large amount of material usage.
- Substantial loading of the monopile head when ramming (influence on length of life).
- Very high noise level when ramming.

The monopiles can be produced as pipes with constant or changing wall thickness corresponding to the momentary load. The advantage of the latter is the lesser use of material; the disadvantage is the poorer ramming characteristics as the ram energy is partly reflected at the thickness jumps and no longer acts completely at the foot of the pipe.

Diameter, wall thickness and penetration length are oriented towards the depth of the water, the loading by the wind turbine, the waves, the current and the bed properties. The diameters vary between 4 and 8 m, the wall thicknesses between 30 mm and 90 mm, the penetration depth up to approximately 35 m. Included in the overall length is the depth of the water plus the part that projects beyond the water surface, so that the overall length of the pile can be up to 80 m.

Ramming

The ramming energy or size of the ram hammer is adapted to the respective pipe dimensions and sea-bed conditions. As the bed characteristics are mostly known from sea-bed investigations, a ram analysis is carried out before ramming, i.e. it is determined how many blows and what ram energy are required during the pile insertion and what the load on the pile head will be during ramming. During the ram procedure the penetration depth per ram blow is monitored. It should be 25 blows per 1 cm penetration as a maximum. If more blows are needed, then there is the danger that the pile head will be stressed too highly. In total, the number of ramming blows is between 3000 and 6000. After successful ramming, approximately 25% of the operating strength of the pile is used up.

There is a high level of noise during ramming; it can rise to 260 dB. Such levels are fatal for fish and sea mammals. For this reason, ramming procedures usually begin with a so-called soft start, that is, starting the ramming with low ram energy and allowing time for sea life to flee.

The BSH permits a noise level of 160 dB at a distance of 750 m. As these limitations cannot generally be adhered to for large monopiles, noise protection measures such as bubble curtains, pile covers and similar are necessary for the ramming procedure. The effectiveness and costs of such measures are being intensively investigated at present.

Grouted joint

The connection of the monopile with the actual tower of the OWT is usually carried out by means of a so-called grouted joint. In this, a pipe (transition piece) with a larger diameter of 200 mm to 400 mm and a length of approximately 5–8 m is pushed over the monopile, arranged vertically, sealed at the bottom, and the space between the monopile and pipe filled with special seawater-resistant and quick-hardening concrete. The grouting material should have strong adhesion to the steel parts. In this way, possible alignment errors of the monopile that occurred during ramming can be equalised. The tower is connected to the pipe by means of a flange.

This type of connection has already been successfully used in the offshore oil industry, but the OWT is subjected to other loads (high load changes, alternating stresses). So-called shear keys are arranged at the connection pipe and the monopile in order to secure the grout connection (recommendation of the DNV [7]). The shear keys are meant to stop the tower sliding off in the case that the connection between the grout and monopile has loosened.

Bolted flanged connections or similar between the foundation pile and tower for rammed piles are not used as the pile flange may be damaged during the ramming and alignment errors cannot be compensated.

Materials

Ferritic steels with high ductile values (high impact strength at temperatures of −40°C according to GL regulations) and good welding properties are used in the production of monopiles. The steels mostly have flow limits ($\sigma_{0.2}$ limit) of 355–450 MPa (1 megapascal = 1 N/mm^2). Higher flow limits are generally not necessary as, besides the sustainable stresses, also safety against buckling (compressive stresses) of the piles is taken into account. For this it is not the strength of the material that is decisive, but the thickness and the modulus of elasticity, which is the same for all ferritic steels.

Examples of steels used:
S355EM, S450EM, S355G7 + M, S355G7 + N to EN 10225; the letter S stands for steel, the following number for the $\sigma_{0.2}$ limit, and the following marks for the quality (ductility, welding ability, temperature in use, etc.).

The implementation of ramming, the grouting connection and the materials also apply for the foundations described below with tripods, jackets and suction buckets.

11.6.2.2   Tripods

Tripod foundations are steel structures that are erected on the seabed with three legs. The legs are joined at a central node below or above the surface of the water and connected to the OWT by means of a flange or grouted joints. The tripods are anchored into the bed of the ocean either by means of rammed piles or suction buckets (see below). The tripod is then erected on them. The erection of the tripods is carried out in one piece complete on land and then transported to the site by pontoons and erected by large cranes.

The forces generated by swells, currents and winds are distributed over the individual foundation elements of the supporting structure. This leads to a reduction of the load on the individual elements. Because of the legs of the tripod construction, the bending moments are converted mainly to tensile and compressive forces. Alternatively the rammed piles can also be inserted into the sea bed in a slanted direction, and thus the horizontal forces are introduced into the sea bed in the form of tangential forces.

Advantages of the tripod foundation:

- Lower overall weight (use of material).
- Greater stiffness for deeper water depths than is suitable for monopiles.

- Small ramming piles or suction buckets.
- Lower ramming energy and noise emissions during ramming.
- Less ground load.

Disadvantages:

- Higher production costs (manufacture and welding of thick-walled nodes).
- Sea-bed preparation required (levelling).
- Collision danger when mooring service ships due to the slanted supports.
- Less collision safety.
- Larger scouring protection surface required.
- Higher ramming risk, if large rocks are present in the floor (the last of the rammed pile encounters a large rock).

Besides the classic tripod, other foundation forms have also been developed, even the tripile (see below), that at present have not been used.

- Tripile of the BARD Company: the three legs of the foundation are joined above the highest water level on a supporting cross and the tower is placed on this. The legs rest on rammed piles or suction buckets. The advantages are simple construction as no nodes need to be welded together; critical welded seams and the largest part of the surfaces subject to corrosion are above the water. A disadvantage is the greater weight compared to the classic tripod (greater material thickness in order to achieve the necessary stiffness).
- Asymmetric tripod: in this construction the main pipe rests vertically on a foundation pile, the other two smaller pipes are connected to the main pipe. The main pipe reaches to the seabed. The foundation can be carried out as ramming piles or suction buckets.

11.6.2.3 Jackets

The foundation structure of a jacket consists of a spatial framework of steel tubes with sleeves arranged at its lower end-points through which the piles (one or more piles per end point) are rammed. It is also possible to use suction buckets in place of the piles. Jackets have been used successfully for decades as foundation structures for offshore platforms of the oil and gas industries for water depths up to 400 m.

The jacket framework is produced completely on land, towed to the erection site using a pontoon, and erected there by large cranes.

Advantages of jacket foundations:

- Low weight (material usage approximately 50–60% of a comparable monopile).
- Small dimensions of the individual parts during production.
- High degree of stiffness, thus suitable for greater water depths.
- Smaller ramming piles or suction buckets.
- Lower ramming energy and noise emissions.
- Smaller cross-sections at the water level when the main node is above the surface, thus lower loading due to currents, waves and ice.
- Lower floor loading.

Disadvantages

- Higher production costs (manufacture and welding of many nodes).
- Sea-bed preparation required (levelling).
- Higher corrosion protection effort because of many small surfaces.
- Low collision safety for ships.
- Larger scouring protection surface required.
- Higher ramming risk when there are large rocks in the bed (see Section 11.6.2.2).

11.6.2.4 Gravity Foundations

The principle of the gravity foundation is that of placing a heavy body on the seabed, and because of its great weight and large erection area, to thus ensure the stability of the OWT. The material used for this is seawater-resistant concrete. The concrete body is manufactured on land and transported to the erection site on a pontoon, or if produced in a floating manner, towed to site and then sunk. After sinking, the chambers provided for this are filled with ballast (e.g. sand) in order to absorb overturning moments.

Advantages of the gravity foundation:

- Economical manufacture when made in series and with the use of concrete, which is very cheap compared with steel.
- Economical foundation for water depths of 10–20 m.
- Experience with the principle and the material from bridge-building and the offshore oil industry.
- No ramming required except possibly a guide pile for controlled sinking.
- No noise emissions.
- Low corrosion protection measures.
- High degree of loading for ice formation (simple arrangement of an ice-breaking cone at the level of the water line).
- Simple removal.

Disadvantages:

- High start-up costs for production.
- Great weight, and when transporting by pontoon large cranes are required (weight up to 3000 t).
- Large erection surfaces required.
- High degree of floor loading especially at the edges.
- Danger of saddle formation (compaction of the floor and the edges of the circumference) with changing directions of loading.
- Substantial floor preparation is necessary (levelling a large surface, possibly additional increase of the carrying capacity due to compaction of the floor, filling up with sand, or similar).
- Difficult scouring protection.

Besides the classic gravity foundation, there are further developments such as the 'ocean brick system', which is structured modularly and uses separated structures as well as systems that instead of a circular shape have a cross-shaped arrangement of larger concrete blocks. The aim is to save weight for the same stability.

11.6.2.5 Suction Buckets

In the suction bucket type of foundation, a large body, closed on top and open at the bottom (overturned bucket), is pressed with an internally-generated underpressure into the sea bed. In order to obtain sufficient stability, the diameter and length of the cylinder must be large. Stability is to be ensured by the coating friction of the suction bucket in the soil. The forces caused by the underpressure must not be taken into account in the stability, as it is slowly nullified by the entering water.

Advantages:

- No ramming work required.
- No noise emission.

Disadvantages:

- Increased buckling danger due to external pressure; stiffeners for increasing the buckling stability are necessary.
- Large dimensions, thus high material usage and heavy weight.
- Floor preparation required (levelling).
- Careful floor investigation to see whether the underpressure is sufficient, to press the pipe against the coating friction sufficiently deep into the floor.
- Disadvantageous relationship against stress failure and against buckling failure.

An estimate of the circumferential stress of the cylinder under external or internal pressure is:

$$\sigma = \frac{pR}{t} \text{ (boiler equation)} \tag{11.78}$$

An estimate of the buckling pressure of a cylinder under external pressure is [17]:

$$p_{crit} = \frac{\pi E R}{9l}\left(\frac{t}{R}\right)^{5/2}\sqrt[4]{\frac{36}{(1-v^2)^3}} \tag{11.79}$$

where p_{crit} is the critical buckling pressure = maximum permissible pressure difference outside/inside, E is the modulus of elasticity of the material, v is Poisson's ratio, R is the radius of the cylinder, l is the length of the cylinder, and t is the wall thickness.

11.6.3 Floating Foundations

11.6.3.1 Tension Legs

In a tension leg foundation, floating bodies (often triangular in shape) of steel or concrete are provided that possess a high degree of buoyancy in a fully submerged condition (Figure 11.21). This is compensated by anchors at the corners of the body with which chains or cables of the floating body are kept completely under water in the correct position. The tower is placed over pipe uprights in the middle of the floating body.

In the case of heeling of the immersed body by the wind and wave forces, one or two of the pre-stressed anchor cables or chains become loose due to the excess of lift (depending on the direction of the plant to the wind). The cable forces that fall away cause a lifting moment of

Figure 11.21   Floating types of foundations

the floating body. As long as the lifting force is greater than the heeling force, the plant remains vertical. The tensile forces in the anchor cables caused by the uplift must be so large that the heeling moment from the overall load of the WT can be equalised in all directions.

Because of the large amount of material used for immersed bodies and anchoring as well as the difficult erection, these foundations are only worthwhile for large WTs at great water depths (> approximately 60 m).

11.6.3.2   Spar Buoys

'Spar buoys' can be used for great water depths (> approximately 200 m). They are also mainly used in the offshore oil and gas industries. They are vertical floating hollow bodies whose centre of gravity is substantially below the centre of buoyancy and thus generate a stabilising moment. The excess of buoyancy is compensated by the tension of the anchoring with cables and additional sliding weights. These add additional stability to the vertical floating position. The possibilities of use for OWTs at great depths are being intensively investigated at present.

Here, too, because of the high levels of material use for floating bodies and anchoring, as well as the difficulties in erection, such foundations are worthwhile only for large WTs (5 MW and more) at great water depths (approximately 200 m).

11.6.4   Operating Strength

Besides the maximum loading of an OWT from the 50- and 100-year gusts and the 50- or 100-year wave, the high load cycles from the swell and wind loads of up to 3×10^8 load cycles in 25 years are a further decision criterion for the dimensioning of the OWT.

There are several methods for calculating the lifetime, and the most commonly used is the linear damage accumulation method according to Palmgren-Miner (see, for instance, [3]).

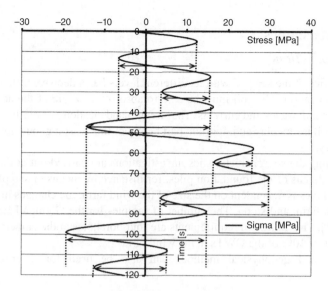

Figure 11.22   Rainflow diagram

According to this, the timely loading of components of the foundation is divided into k so-called load collectives. Each collective is defined by means of occurring load cycle counts, stress amplitude and average stresses. Corresponding part amounts d_i are collated into a damage sum D for the load collective that has been defined:

$$D = \sum_{i=1}^{k} d_i \leq \eta \leq 1 \qquad (11.80)$$

The IIW (International Institute of Welding) has determined so-called detail categories (reference stresses) for components of steel or aluminium [3] for the constructive and production of components (type of component, type of welding seam, edge processing, and so on). The reference stresses are corrected with the aid of influencing characteristics such as the type of welding method, thickness of the component, and importance of the component. The detail categories do not take the strength of the material into account; this is only considered with the use of an influence factor in the correction of the reference stresses.

Mostly $\eta = 1$ is assumed. This also means that the factor of safety against failure is equal to 1. However, the Palmgren-Miner rule displays certain weaknesses, for instance, the loading history is not taken into account. Thus, initial higher loads have a stronger reducing influence on the lifetime than lower loads than would be the case in a reverse order. Therefore the damage sum should remain $\eta < 1$. DNV [7], in its regulations for η values, gives 0.1 to 1 depending on the importance of the component.

The limiting value $D \leq 1$ should also only be used for components of steel. For components of aluminium or FRP (fibre-reinforced plastic) the limiting value to be used is rather $D \lesssim 0.5$, as such components fail much earlier than is shown by Equation (11.80).

The rainflow method (see Figure 11.22) has shown itself to be particularly well suited for determining the load collective (see above). It is shown in more detail in [18].

11.7   Soil Mechanics

11.7.1   Introduction

The characteristics of the sea bed with its different layers has a decisive influence on the stability of the foundation. The soil must be able to absorb the whole of the active forces and moments from wind, waves and currents. The strength and elasticity or plasticity properties of the individual soil layers play an important part here. For this reason good knowledge of them is very necessary.

For determining the soil characteristics, investigations are carried out at the sites down to the penetration depths of the foundation piles, for instance by means of sample drilling. The number of boreholes is dependent on the size of the wind farm, the changes in soil properties within the wind farm, and so on. The number of samples for the German EEZ is defined by the BSH, see [1]. The minimum requirements are four boreholes at the comers and one in the middle of a WF, or 10% of the OWTs in the wind farm.

Geotechnical and geophysical methods can also be used for determining the soil properties.

11.7.2   Soil Properties

As already mentioned, the soil properties are decisive for the carrying capacities and thus the stability of the OWT. The layers and the properties can be very different in both in the North Sea and also in the Baltic Sea. Generally, one differentiates soils as being binding (cohesive) or non-binding (non-cohesive). Whilst both types of soils can transmit large compressive and shear stresses, the transferable tensile stresses in the non-binding soil types in dry conditions are equal to zero, and in the wet condition, very small (example: dry and wet sand). Binding soils can transmit low tensile stresses. The property is described by the cohesion factor c'.

Binding types of soil are loam, clay, silt, till and peat. They generally have a platelet type of structure and a small particle size of approximately 0.001–0.06 mm. Their pouring weights in water are between 1 (peat) and 12 kN/m^3 (marl) according to Schmidt [19].

Non-binding types are sand and gravel with a rounded structure and particle sizes of 0.06 (sand) to 65 mm (gravel). Their pouring weights in water are between 10 (gravel) and 11 kN/m^3 (dense sand) [19]. Further types of soil are rock and organic slime as well as mixtures of the above types.

Depending on the density of the soils, one differentiates between loose, medium dense, dense and very dense soils. The spaces between the particles are filled with water.

In order to determine the carrying capacity of a foundation, the following mechanical particle characteristics of the soils are required (numerical values according to [19]):

- compressive strength σ (flow limit);
- shear strength τ;
- internal friction angle ϕ' (similar to the angle of repose): in water approximately 30–40° for non-binding soils, approximately 15–25° for binding soils;
- cohesion factor c': approximately 2–8 kN/m^2 for non-binding and 5–25 kN/m^2 for binding soils;

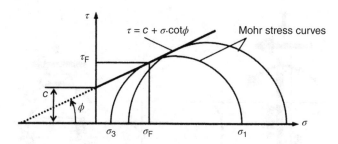

Figure 11.23 Determination of the sea-bed strength (Mohr-Coulumb boundary conditions)

- secant modulus E_s: approximately 20–300 MN/m² for non-binding soils, approximately 0.5–100 MN/m² for non-binding soils;
- permeability k: approximately 5×10^{-5} m/s for loose sand and approximately 10^{-3} for dense clay.

The permeability (penetrability of water) indicates how quickly water flows through the type of soil. For dynamic loading, for instance vibrations of passing waves, the water pressure changes in the sea bed with time. If it does not adapt correspondingly quickly to the surrounding static pressure with too little permeability, then an over- or under-pressure will exist (the so-called porewater pressure). The change with time of the porewater pressure is described by the gradients $p(t)$. If this value exceeds the flow limit of the soil, then the shear strength of the soil becomes zero, i.e. the soil loses its carrying capacity completely, and it 'liquefies'. The calculation of this effect is at present insufficiently accurate, but it must be fundamentally taken into account in the design of the foundations.

The determination of sea-bed characteristics are mostly experimental, through seismic methods or laboratory experiments on samples from local boreholes. A proven method is investigation with a triaxial device. With this the most important sea-bed characteristics can be determined by variations of the parameters (stresses in the three directions, porewater pressure, wet or dry experimental conditions see Fig. 11.23). More information on the laboratory experiments can be found, for instance, in [19].

Continuously alternating loading of the sea bed leads to a reduction of its carrying capacity and stiffness. If a sea-bed layer is overloaded, then it deforms permanently (plastic). Methods for determining the limiting carrying capacity of a soil type (plastic failure, surface fracture) can also be found in [19].

11.7.3   Calculation of Load-Bearing Behaviour of the Sea Bed

The loads acting on the foundation of the OWT resulting from the weight of the plant and the wind, wave, current and ice forces must be able to be absorbed by the sea bed. The weight of the plant causes vertical forces, while the rotor movement and the forces from the wave, currents and ice are horizontal forces; these, multiplied by the respective lever arm, cause additional moments. Basically, the equilibrium principle also applies here, such that the sum of the forces from the loadings must be equal to the forces, and the sums of the moments from the loadings must be equal to the moments, that act as reactions in the sea,bed.

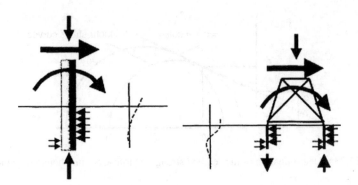

Figure 11.24 Load absorption and deformations for monopile and jacket foundations

Various loadings of the sea bed occur depending on the type of foundation:

Monopile foundation (see Figure 11.24): The bending moments and shear forces are taken up by horizontally directed pressures, the weight of the plant by vertical pressures (peak pressure at the foot of the pile and wall friction). The horizontal displacement on the sea bed is greatest and causes a high soil loading on the lee side (away from the wind) of the pile. On the luff side (towards the wind) no forces are transferred as the transferrable tensile stress between pile and soil is almost zero. A monopile 'only leans on one side against the soil'. The maximum bending moment in the pile, due to its 'elastic clamping', is below the seabed. Depending on the stiffness of the material in the uppermost soil layers, the maximum is 2 to 3 times the lower pile diameter.

  If, in a monopile, a wind or wave direction dominates or if the carrying capacity of the sea bed, for instance, due to a 100-year wave, is overloaded (plastifies), then a luff-side gap can form between the sea bed and pile that does not close after the load has been relieved. This leads to a slanting of the plant that will increase over the time of operation and which can reduce the life of the OWT.

Tripod and jacket foundations: the loads to be absorbed are distributed to the various reaction forces. The bending moments and vertical loads caused by peak pressure and wall friction forces on the piles (on the luff side directed upwards, and on the lee side downwards) and the shear forces are introduced by horizontal pressures into the sea bed. In this way the sea bed is loaded less and more evenly that for the monopile.

Gravity foundations: Weights and bending moment loads of the OWTs are distributed by means of vertically directed normal loads, spread over the surface area and absorbed, and the horizontal shear forces by means of horizontally directed shear stresses. The bending moments are absorbed by the distribution of the normal pressures in the direction of the load; on the lee side they increase, and decrease on the luff side.

Suction buckets: The loads are mainly absorbed by the wall friction on the inside and outside of the cylinder. The tangential stresses are relatively small due to the large surface area of the suction bucket in comparison to the tripod or jacket.

A possibility for the calculation of the sea-bed reaction under the external loading is to model the sea-bed behaviour by several linear or non-linear springs (see Figures 11.25 and 11.26). The spring force characteristics are described by stress displacements or force

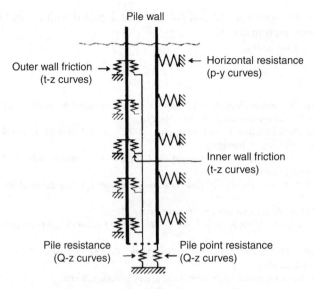

Figure 11.25 Spring model for calculating a pipe foundation

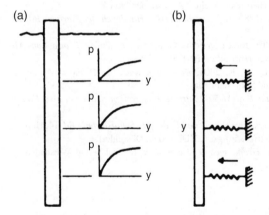

Figure 11.26 (a) p-y curves; (b) modelling by means of non-linear springs

displacement curves. Thus, for instance, the *p-y* curves represent the sea-bed deformation in the horizontal direction under the acting normal stresses (pressure) *p*, the *t-z* curves the seabed deformation in the *z* direction under the tangential stress *t*, and the *Q-z* curves under the vertically acting force *Q*. Tensile stresses cannot be transferred by the springs.

For sea-bed layers with different mechanical behaviours, each layer must be represented by the corresponding curves or spring characteristic curve. Thus the stresses and deformations of a pile can be calculated as a beam with linear (elastic) or non-linear bedding (i.e. with FE methods).

A different and difficult possibility for calculating is FE modelling of the sea-bed layers by means of volume elements with their respective mechanical properties (E-modulus, Poisson's ratio, stress deformation characteristics, etc.). The pile is modelled by volumes or areas. The boundary surfaces between sea bed and pile must be depicted with contact elements so that only pressure forces are transferred.

References

[1] BSH (Bundesamt für Seeschifffahrt und Hydrographie). 2003. *Standard-Nr. 7003*, Mindestanforderungen für Gründungen von Offshore-Windenergieanlagen, Hamburg.

[2] BSH (Bundesamt für Seeschifffahrt undHydrographie). 2007. *Standard-Nr. 7007*, Konstruktive Ausführung von Offshore-Windenergieanlagen, Hamburg.

[3] Germanischer Lloyd (GL). 2005. *Guidelines for the Certification of Offshore Wind Turbines*. GL, Hamburg. Available at www.gl-group.com

[4] Hapel, K.-H. 1990. *Festigkeitsanalyse dynamisch beanspruchter Offshore-Konstruk-tionen*. Vieweg Verlag, Stuttgart.

[5] Eck, B. 1966. *Technische Strömungslehre*. Springer-Verlag, Berlin.

[6] Det Norske Veritas (DNV). 2000. *Environmental Conditions and Environmental Loads*. DNV, Oslo. Available at www.dnc.com

[7] Det Norske Veritas (DNV). 2004. *Design of Offshore Wind Turbine Structures (DNV OS-J101)*. DNV, Oslo. Available at www.dnc.com

[8] Clauss, G. 1988. *Meerestechnische Konstruktionen*. Springer-Verlag, Berlin.

[9] Sarpkaya, T. 1981. *Mechanics of Wave Forces on Offshore Structures*. van Nostrand Reinhold, New York.

[10] Wehausen, J.V. 1960. *Surface Waves, Handbuch der Physik Bd. 9*, Springer-Verlag, Berlin.

[11] Gerstner, F.J. 1809. Theorie der Wellen. *Annalen der Physik*.

[12] Kokkinowrachos, K. 1980. *Offshore-Bauwerke, Handbuch der Werften, Bd. XV*. Schifffahrts-verlag Hansa, Hamburg.

[13] Morison, I.R. 1950. The force exerted by surface waves on piles. *Transactions AIME*, 189.

[14] Hamil, L. 1999. *Bridge Hydraulics*. E&F Spon, London.

[15] Richwien, W. and Lesny, K. 2004. *Kann man Kolke an Offshore-Windenergieanlagen berechnen*, BAW-Workshop Boden- und Sohl-Stabilität.

[16] NORSOK. 2004. *Standard M-501, Surface Preparation and Protective Coating*. Available at www.nts.no/norsok

[17] Ebner, H. 1967. *Festigkeitsprobleme von U-Booten, Schiffstechnik Bd. 14*, Heft 74.

[18] Haibach, E. 2002. *Betriebsfestigkeit*. VDI-Verlag, Düsseldorf.

[19] Schmidt, H.-H. 2006. *Grundlagen der Geotechnik*. Teubner Verlag, Wiesbaden.

Index

Absorptions, 189, 374, 452
Acceleration, 164, 348, 414, 416, 417, 421, 425–9
Acceleration equation, 348
Acceptance, 53, 54, 55, 57, 120, 121, 123, 124, 371
α-component, 321, 348
Acoustic level, 102
Acoustic power level, 101, 102, 104, 121
Acoustic wave propagation, 105
Acoustic weighting, 101, 102
Active component of the current, 290
Active space vector, 315
Active stall blade, 140–141
Actual value, 173, 182, 185, 320, 321, 327, 332, 343, 355, 364
Actuating activity, 321
Actuating level, 342, 343
Additives, 188–92, 329
Aerodynamic damping, 180, 183
Aerodynamic optimising, 139
Aerodynamic profile, 8, 13, 127, 139, 147, 151
Aerodynamic torque, 217
Aeroelastic, 29, 34, 133, 149–51
Aeroman, 58
Aeronautic obstructions, 120
Ageing of fibre-reinforced materials, 408–9
Aims of extension, 41
Air
    air heat exchanger, 226
    gap, 227, 229–31, 234, 249, 284, 293, 308
    gap stability, 227
    water heat exchanger, 227
Airy theory, 421, 422
Allgaier, Erwin, 20
Alternative circuit diagram
    induction machine, 275, 284–300, 302–4, 318, 323–6
    single-pole, 380
    slip ring rotor, 295–303
    synchronous machine, 273, 275, 284, 297, 303–10, 318, 326, 330, 331
    transformer, 92, 273, 275, 278–83, 286, 287, 298, 304, 332, 334, 340, 342–4, 367, 368, 370–372, 379, 381–3, 385, 386, 388, 389, 391–8, 400, 401
Amin, Adnan, 62
Amortisation, 54
Amount optimum, 321, 324, 329
Amplitude and frequency conversion, 310
Anchor cables, 447, 448
Anchoring, 57, 115, 254, 255, 266, 267, 272, 441, 444, 447, 448
Andreau, 22, 23, 26
Angle of attack, 9, 18, 127, 144, 148, 151, 153, 332
Annual energy yield, 97
Apparent output, 427
Application program, 81
Approval process, 49, 113–21
Aramid fibres, 189, 190
Area designation, 106
Argentina, 20, 47, 68

Understanding Wind Power Technology: Theory, Deployment and Optimisation, First Edition.
Edited by Alois Schaffarczyk. Translated by Gunther Roth.
© 2014 John Wiley & Sons, Ltd. Published 2014 by John Wiley & Sons, Ltd.